Werner Kunz

Species Conservation in Managed Habitats

Werner Kunz

Species Conservation in Managed Habitats

The Myth of a Pristine Nature with a Preamble by
Josef H. Reichholf

WILEY-VCH
Verlag GmbH & Co. KGaA

Author

Werner Kunz
Heinrich-Heine-University
Institute for Genetics
Universitätsstraße 1
40225 Düsseldorf
Germany

Translator
Translated from original German text into English by Ron Shankland/Mörlenbach

Illustration credit
Color plates contain photos of the author; graphically designed with the collaboration of Monika Dörkes/Nettetal

Cover credit
Colored background by "fotolia_© pia-pictures"; sepia background by "fotolia_© l-pics"

Library of Congress Card No.: applied for

British Library Cataloguing-in-Publication Data
A catalogue record for this book is available from the British Library.

Bibliographic information published by the Deutsche Nationalbibliothek
The Deutsche Nationalbibliothek lists this publication in the Deutsche Nationalbibliografie; detailed bibliographic data are available on the Internet at <http://dnb.d-nb.de>.

Print ISBN: 978-3-527-33845-0
ePDF ISBN: 978-3-527-68886-9
ePub ISBN: 978-3-527-68885-2
Mobi ISBN: 978-3-527-68887-6
oBook ISBN: 978-3-527-68884-5

Cover Design Adam-Design, Weinheim, Germany
Typesetting SPi Global, Chennai, India
Printing and Binding Markono Print Media Pte Ltd, Singapore

Printed on acid-free paper

Contents

Opening Remarks: Preservation of Rare Species – Breaking Grounds for a New Approach

Many species are declining in Central Europe right now in our time. Most of them are threatened despite the fact that they are protected by law. A substantial number have withdrawn completely from the countryside in recent decades. Some have found an alternative to make their living in cities, on air fields and industrial areas, however. These non-natural habitats became more attractive even than protected areas. What is going on?

Why is our modern and seemingly highly enforced nature conservation legislature so insufficient in general, though some larger species of mammals and birds are expanding their ranges and increase in abundance? For globally threatened birds such as the white-tailed sea eagle, Germany now ranges among the most important countries with a subpopulation approaching a thousand breeding pairs. There is also a soaring population of wild boar numbering in many thousands. Grey wolfs are thriving in a small but a well-doing population in Eastern Germany which has been the German Democratic Republic a quarter of a century ago. And in the winter months, nearly everybody can make the experience of seeing a snowy white heron standing in the snow out in the countryside. The numbers of wintering great white egrets now vastly exceed the former breeding population of this rare European bird species which managed to survive half a century ago mainly just at the inaccessible border areas of the time of the Cold War, for example, on the Lake of Neusiedl southeast of Vienna. Ornithologists have been very happy to catch a glimpse of the then shy and elusive white herons by approaching the Austrian–Hungarian border as close as security allowed. What a difference just a handful of decades later with great white egrets outnumbering the 'normally' much more common grey heron in most regions of Central Europe in winter. In winter! This matter of fact challenges the long standing view of the 'niche' of this 'tropical bird' and many other bird and mammal species according to conventional ecological wisdom of former times. Even in textbooks of ecology a lot of species has been misplaced according to the limited knowledge of where they can or could live, which more often than not is different from where we did find them under the local or regional conditions at a certain time.

So again the question, what is going on 'out in nature' as we tend to say, because we, the human people of the modern world, usually separate our immediate realm of living from the so-called free nature. Why became so many species rare or vanished at all from our neat and clean and progressively healthier environment despite such an extraordinary amount of conservation efforts and so many million Euros, which have been spent by our nature conservation organisations?

The answers given point at a still too much spoiled and poisoned environment, claim unbearable high levels of human disturbance and the excessive construction of roads, spread of settlements and other forms of 'development' which generally means destruction of nature. The birds and bees, the butterflies and bats, they cannot find enough suitable habitats for sufficiently large, extinction-resistant populations. Or so the saying goes. This may be true in a number of special cases, but certainly not in all, however. The common view, as expressed by most nature conservationists and widespread in the public opinion as a result of their campaigning, is highly biased in fact. It is this bias of a much too human-centred view of nature and how it should look like which resulted in such a low success or failure of the species conservation efforts. A closer look onto the demise of the vast number of smaller species, best exemplified by butterflies, moths and songbirds due to extensive field studies available from recent decades, reveals the shortcomings and simultaneously makes understandable the success of the few larger ones, which are expanding.

The spread of grey wolfs and white-tailed sea eagles, of great white herons and black storks, of peregrines and beavers and also the finally successful reintroduction of the lammergeyer into the Alps beyond any doubt is the result of a very much reduced hunting pressure posed on that and other increasing species compared with the situation of the pre-Second World War times. They could recover due to reduced losses by shooting and poisoning and also due to a much greater tolerance of human presence they achieved in recent times. The great winners in fact have been such species which found their way into the relatively high safety of the cities. However, a striking geographical pattern is visible in central Europe in the distribution and abundance of many species of larger mammals, birds and invertebrates. It shows the separation of a species rich Eastern and an impoverished Western part by what has been the Iron Curtain a quarter of a century ago. There is no natural ecological condition or constraint existing for this once political separation, but quite a massive one with respect to the intensity of the agricultural use of the countryside. In the former East major tracts of land, many fine scale structures of unused sites and still a lower expectation on how high the yields of the fields should become characterise the East and contrast to the Western attitude of maximum exploitation. Moreover, abandoned villages, cottages, once used sites for industrial production and highly contaminated areas from socialist times are in existence and probably will last for some decades more, before they become subjected to the cleansing according to the Western style. There even small gravel pits have been viewed as 'wounds in nature' which have to be healed as soon as possible by re-cultivation, which invariably meant planting trees to cover the 'offence to the eyes' without taking the needs of so many species of animals and plants into consideration. This is one of the major points, which is made in this book, and which will make it hard to digest without putting aside all the preconceived views how (clean) nature should be.

The other central point deals with the invisible stuff called nitrogen compounds which floods the whole of Central Europe since decades due to over-fertilisation of the arable fields and flushing the meadows with liquid manure, the latter in a total amount which greatly exceeds all the peoples' sewage in Germany. This general eutrophication promotes the growth and ecological displacement strengths of a very limited number of plant species and has led to a much wetter and cooler microclimate down at the vegetation level of the countryside than is indicated by the official meteorological measurements and 'expected' as a result of climate warming. Most of the warmth-loving (thermophilic) species are in severe decline even in protected areas and they form the greatest bulk of species in the 'Red Lists'. In fact, most habitats became cooler in the last half of a century, not warmer as expected. Therefore, and this is the great message of the book, we are in bad need of open, dry and 'unproductive' areas, unfertilised and managed in a proper way to act against the accumulation of plant nutrients due to the influx of nitrogen compounds and nutritive dusts from the sky by wind and weather. The small scientific community which has the privilege to look into the military training areas at some times knows that these are the most diverse sites with the highest numbers of rare and endangered species – and not those protected by law officially. Second to the military training areas in species richness are major cities such as Berlin or the vast Rhine-Ruhr-Agglomeration where a lot of rarities found their place of living amidst the densest human population concentrations.

Such are the 'good news', though not from a conventional nature conservationist's point of view, because it means we have better possibilities to preserve species on up to now largely neglected and under-rated areas such as open mining sites, abandoned industrial areas and highly disturbed places. They are the last heavens for quite a lot of species which a great many of our nature lovers would not like to miss such as the singing skylark, which is best seen, under uncomfortable acoustic conditions however, on the major airports. Summing up, we as conservationists should apply much closer looks onto where the rare and endangered species still exist rather than hanging on concepts how we would like to see the so-called free nature with a low intensity of agriculture close to the hunger level of the peasants. Such concepts which are based on the condition of the nineteenth century are out of date in modern times not only due to the fact that everything has changed over the last one and a half century, but also because they provide no realistic outlook for the future. The best guides certainly still are the species themselves. Conservationists should follow their guidance rather than their outdated, more or less romantic ideals. This is the combined message of all the threads and new approaches which Werner Kunz explores in his outstanding and demanding book. If read carefully and open minded, it could become a similar benchmark publication like the 'Silent Spring' by Rachel Carson, because of the 'new song' it offers to nature conservation.

Prof. Dr Josef H. Reichholf
München, Germany

Preface

Nature conservation is a good thing. Who would dare to raise objections against it? However, it is the moral–ideological and exaggerated approach, closely associated with the conservation movement since the second half of the last century that leads to the drawing of wrong conclusions. Nature and wildlife conservation were originally oriented on specific objectives, namely the protection of nature and species. In the 1970s and 1980s, however, a new concept was propagated, which embedded nature conservation and species protection into the vision of a clean environment and into the health of the population. This development expanded an area (that previously had reflected the interests of only some specific individuals) into a generally binding moral postulate. People had to be made aware that the protection of species presupposed an intact nature and that rubbish and pollution in the environment really endangered species. Since it was a moral duty to avoid environmental pollution, everyone had to automatically combine environmental cleanness with nature conservation and the protection of species.

However, the protection of some species has nothing at all to do with nature conservation and even less with a clean environment. Hygiene, cleanliness and orderliness are things that man needs; but they are not what many species necessarily need, at least not in the form in which human beings would like to see them. Exactly the opposite is often true. In recent centuries, hygiene and orderliness in our homes have put many formerly common animals into the endangered species categories. For example, the house rat (*Rattus rattus*), not to be confused with the brown rat (*Rattus norvegicus*) has become extremely rare. It had to be classified in Germany's Red List of Threatened Animals as 'critically endangered' (Category 1). The populations of the common bed bug have declined sharply in Central Europe in comparison to those of previous centuries. The same goes for fleas and lice. All these species used to be very common in Central Europe and they still are today in other countries around the world. They became rare in Central Europe thanks to hygiene, cleanliness and measures that serve the health of human beings. Most of us have of course no love for these creatures in our homes, but it is exactly this perception that reflects the anthropocentric standpoint. It is what human beings want, not what the animals want.

Just as in earlier times, civilisation saw homes being thoroughly cleaned of debris and dirt, and today's modern agriculture is 'cleaning' the landscape to an ever-increasing extent. We have recently begun to transfer a process, which formerly served domestic culture and health, to the countryside. During the last half century in Central Europe, farmland and pastures have been optimised for maximum yield and cleaned for machine processing. Agricultural land was cleared of stones and weeds and sandy or muddy surfaces and uneven ground were eliminated. The last square yards of unused gaps, corners and edge areas were incorporated into the production areas, and waste and crop residues were no longer left lying. Modern farmland is absolutely clean, homogeneous and as flat as a table.

However, animals have also been cleared from the fields, to the same extent as the fields have been cleared of 'refuse'. The modern field is almost species-free. Being clean and hygienic, it makes a 'proper' impression of course, but it is actually a desert that is hostile for species. Just as the cleaning of human domiciles removed the possibilities of survival for rats, bedbugs and fleas, the cleaning of the fields has left no room for the species. The brow hare, grey partridge, skylark and corn bunting can no longer find areas for food and nesting. If you take a walk over fields today, you will not hear the singing of the skylark at most locations; and it is for similar reasons that no more crickets chirp in our homes and no more rats frequent our cellars. It is not just the toxins that have made a desert out of our agricultural areas. A bit more dirt and grubbiness would have been good for the animals, an insight that is expressed by the provocative saying: 'A lazy farmer promotes biodiversity to a greater extent than ten hardworking nature conservationists'.

Orderliness and cleanliness have contributed considerably to the extermination of many species on agricultural land; but the popular belief that a clean environment also benefits species cannot be eradicated. Environmental cleanliness does indeed promote many aquatic species that breathe in the water, but a clean environment simply does not work for many terrestrial species. Of course, it is unhygienic and not at all aesthetic when rubbish is dumped in parking spots, or when food scraps are simply thrown out of cars to lie on village streets. This should be prevented in the interest of the vast majority of people; but this would only be in the interest of our own human needs for order and hygiene. In the same breath, we should not then regret that the sparrow is disappearing from the villages and the yellowhammer from the fields (Meyer, Eilers and Schnapper, 2003). The occurrence of the purple emperor butterfly on the streets in the midst of some Romanian villages is only thanks to the fact that there are no sewer systems in these locations.

Unpaved roads cause dust and dirt. If the roads are paved, we no longer make our shoes dirty, and there is not so much dust in the air; but then the swallows cannot find any more mud to build their nests. There are no more puddles on clean, paved roads; so there are also no more mosquitoes; but only we humans like this state of affairs, swallows do not.

Crumbling walls are a testament to the neglect of buildings and they are perceived as being 'not orderly'. 'Orderly' and hardworking homeowners ensure that their walls are well-plastered. However, a large number of mason bee species and other hymenoptera can no longer live on properly plastered walls; and they are on the Red List in Germany today. Cavities under house roofs and on their outer facades are the breeding grounds of the bat, the swift and the owl, but this presupposes that such houses are old and in need of renovation, so the cavities and holes in these buildings have mostly been closed and filled to save energy and avoid losing heat. Every measure taken for the purpose of environmental protection is directly opposed to species protection, and the list goes on.

Besides cleanliness and orderliness, there is another ideal, which in the minds of many people equates with the protection of species, but in many ways is the opposite of species protection – and that is undisturbed, unspoiled nature. Its importance is much too highly ranked by nature conservationists.

The decision to write this book was made many years ago, when I first realised something the consequences of which I had not hitherto been aware of. While searching for rare bird and butterfly species as a student in the decades after the Second World War, I very soon noticed that to find any remarkable species I had to visit the destroyed military airfields of that period. In these areas, I encountered birds and butterflies that could only be rarely seen elsewhere, if at all. It was only much later that the full meaning of this came home to me. In a nutshell, this meant that many endangered species are found in habitats that are definitely not wildlife sanctuaries. These ex-airfields were habitats that had nothing to do with nature and only owed their existence to the fact that nature had been destroyed in them.

These were areas that were characterised by several features. First of all, the former military airfields were tree and shrub-free almost as far as you could see. Second, they were levels with very heterogeneous surfaces (exacerbated by the many bomb craters), and third, the ground was only sparsely covered with vegetation, and the grass layers were interspersed with bare earth and rocky and sandy areas (intensified by the former concrete runways, now destroyed). These areas had no similarities with the current landscape of Central Europe. They had a greater resemblance with the landscapes of earlier centuries that had been devastated by human overexploitation at a time when there was no afforestation and no mineral fertilisation.

These ruined airfield areas were home to many birds and butterflies which had to make way for afforestation and eutrophication elsewhere. On the dry areas of these airfields, I found skylarks brooding in large numbers; common snipe, redshank and crakes were nesting in the rushes of the wetter areas and little ringed plovers had found suitable places to lay their eggs and rear their young on the destroyed runways. The wheatear populated the ruins of the former airport buildings and tawny pipits were breeding in considerable numbers on the sandy earth walls of the former airport boundaries (Kunz, 1959). On this landscape that was only sparsely covered with vegetation, silver-studded blue butterflies could be seen everywhere, alcon blues found suitable living conditions on the wetter surfaces and silver-bordered fritillaries were common.

In later years, I again came across an area to which many Red List species had retreated, because they could no longer live on the modern-day, over-fertilised agricultural fields and overgrown areas of non-agricultural land. Here again, this area was not a wildlife sanctuary, nor did it meet the criteria, which would have classed it as 'natural'. It was the large-scale, brown coal open-cast mining excavations west of Cologne in Germany. The nature and environmental conservation associations paraded these areas to the public as a terrible example of the destruction of nature, an example that should discourage us all from allowing such landscapes to exist. Again, these areas were characterised by the three features mentioned earlier, which are the conditions for the occurrence of many rare species: wide and treeless open spaces, a heterogeneous surface structure and bare earth.

Meadow pipits, corn buntings, wheatears and woodlarks were breeding on the untreated, nitrogen-poor soils that had been brought up from the depths of the mine and on the subsequent stages of these soils (provided that they were not already covered with dense vegetation), and all of these birds are now on the Red List in Germany (Südbeck *et al.*, 2007) (Plate 1). Swallowtail and small heath butterflies were very common there, and more than 15 species of orchids, the remaining populations of which have shrunk in Germany (and the locations of which are top secret) were also growing throughout the area (Albrecht *et al.*, 2005) (Plates 2 and 3).

Many nature observers got together on this landscape, having quickly realised that here they could find the rare species that have disappeared elsewhere. However, in the minds of a large part of the population, the open-cast excavations are a devastation of nature that arouses disgust, and they should really never have been allowed to exist. Some people fail to realise that the preservation of many endangered species in Central Europe has nothing to do with the aesthetics of a landscape, nor can species protection be achieved by leaving nature untouched.

One objective of this book is to make it clear that the protection of rare and endangered species in Central Europe in many cases has nothing to do with the conservation of nature. Indeed, it is correct that protecting Mother Nature first and foremost means protecting her from the encroachments of man, because human interventions lead to disruptions, since they interfere with natural development processes. However, this in itself is not always species protection. If a rare species still lives in a particular habitat, the nature conservation associations strive to place that habitat under protection, avoiding any attempts man may make at changing it. It is indeed true that altering that habitat would eventually drive out the species which is to be protected. However, what many people do not understand is that in many cases, it is not man but nature itself that changes the habitat over time, making it uninhabitable for the endangered species.

Almost all the habitats in Central Europe would become overgrown with trees if they were left to themselves without human intervention. Huge areas of Central Europe would become woodland. You could say that this is a good thing, a desirable natural condition. Indeed, this is an argument that cannot be contradicted from the viewpoint of nature conservation; but species protection pursues a different objective. In Central Europe, there is hardly an endangered bird or butterfly species that lives in forests, apart from some specialist species that need very specific forest structures (now missing from our forests) (Südbeck *et al.*, 2007). Today, nuthatches and almost all species of owls and woodpeckers are more common than they have been for a long time. It is mostly the species of the open country that are endangered in Central Europe today. These include many species of birds and almost all the butterfly species. Open landscapes are the habitats that are lacking today. We have enough forests. Anyone who really wanted to save the greatest possible number of threatened species would never advocate the creation of new forests.

In this respect, Central Europe differs from the rainforest areas of the world where the loss of the forests is jeopardising species that live there. In Central Europe, however, many species do not benefit from too many forests, they are more likely endangered by the current abundance of forests. This is because Central Europe was deforested by mankind for thousands of years and as a result is now mainly inhabited by species that have adapted to open habitats. Species protection is facing a tough dilemma: the open land species were not threatened for centuries in Central Europe, because they could colonise agricultural land with no problems. Fields, meadows and pastures provided enough suitable habitats in which open land species could live and propagate. However, the situation has no longer existed for half a century, because it was then that agricultural land became yield-optimised and offered neither a home nor food for the species as a result. This is why the open land species have retreated from the agricultural and meadowland areas to non-farmed open areas; but these refuges, which include mountain slopes and valley bottoms, are starting to become overgrown, simply because they are no longer used and are also over-fertilised by nitrogen raining down from the atmosphere. Almost the only sparsely vegetated landscapes that remain for the threatened open land species today are wastelands in cities, industrial areas, port facilities, traffic areas (such as motorway embankments), gravel quarrying sites, brown coal open-cast mining sites and military areas (Plate 4).

The problem of the current biodiversity loss in Central Europe cannot be adequately solved by taking away the economical utilisation of habitats in which rare species still live, declaring these habitats to be 'nature reserves' and essentially leaving them well alone. Protected areas like this must be protected from nature itself. To achieve this, permanent habitat management with technical equipment is required. This approach must be adapted to the habitat needs of particularly endangered species and provide them with the food and reproductive opportunities they need; and this sometimes requires intensive technical engineering of the landscape.

But therein lies the problem. Public awareness has to be fully behind this type of species protection action in order to ensure its implementation, and it is this awareness that is woefully lacking. The emotional desire for an undisturbed nature which must be left alone is deeply ingrained in the consciousness of the population – in excessively ideological manner. This is why there is a considerable resistance against accepting the fact that leaving Mother Nature to her own devices does not encourage the survival of many species in Central Europe. As soon as efforts were made to save endangered species by cutting down trees or partially burning certain areas, or forestry machinery were used to remove the topsoil over large areas in order to recover lost heaths and dry grassland landscapes, storms of protest from the population are more or less preprogrammed. These measures are perceived as being a destruction of nature (which they indeed are) and would arouse incomprehension and indignation. The sense that nature conservation and species protection are one single entity is so deeply entrenched in the public's consciousness that any attempts to dislodge it are doomed to failure, because it would be an attack against human emotions. This is why any large-scale technical habitat management projects are very difficult to implement politically in this day and age. In Germany in particular, the political conditions for intensive technical biotope management seem to be non-existent, due to insufficient information of the public. One can only tiptoe carefully forwards in very small steps (as is being done by some local nature conservation associations today); but these few steps may be too late to rescue some species.

References

Albrecht, C., Dworschak, U.-R., Esser, T., Klein, H., and Weglau, J. (2005) Tiere und Pflanzen in der Rekultivierung – 40 Jahre Freilandforschung im Rheinischen Braunkohlenrevier. *Acta Biologica Benrodis*, (Suppl. 10), 1–238.

Kunz, W. (1959) Die Vogelwelt des Kreises Bersenbrück. *Schriftenreihe Kreisheimatbund Bersenbrück*, **6**, 1–159.

Meyer, W., Eilers, G., and Schnapper, A. (2003) *Müll als Nahrungsquelle für Vögel und Säugetiere*, Westarp Wissenschaften, Hohenwarsleben.

Südbeck, P., Bauer, H.-G., Boschert, M., Boye, P., and Knief, W. (2007) The Red List of breeding birds of Germany, 4th edition. *Berichte zum Vogelschutz*, **44**, 23–81.

1
Introduction: Rare Species and Near-Natural Habitats in Central Europe

The progress in the environmental protection measures of recent decades has hardly stopped the decline of many endangered species.

The highly lauded, showpiece successes in saving some flagship species, such as the white-tailed eagles, cranes, black storks and peregrine falcons actually create a misleading picture about the real recent threat to species in general.

The preservation of many wildlife sanctuaries is not being threatened by man, but by nature itself.

Many red-listed species inhabit extreme habitats, which in Central Europe mostly bear no resemblance to pristine and unspoiled nature.

The loss of species in Central Europe is due to very different causes than the loss of species in the rainforest areas of the world. Consequently, the species protective measures being taken for Central Europe must be different from species protection in the tropics.

It is an illusion to believe that past agricultural practices could once again be revived for the purpose of species conservation. Organic farming methods would be of little help.

In certain cases, nature conservation associations must enforce species protection, even if it goes against their own nature conservation ideals.

1.1
Preliminary Remarks

This book deals with Central European species decline, mainly shown by the examples of selected bird and butterfly species. In the last few decades, Central Europe has lost more than half of its birds and a much larger number of butterflies (Thomas *et al.*, 2004). The book gets to grips with a phenomenon, namely that although the media-effective activities of the conservation organisations did change the public's ecological awareness in the 1970s and 1980s, the disappearance of many species is apparently progressing inexorably, even despite the

Species Conservation in Managed Habitats: The Myth of a Pristine Nature, First Edition. Werner Kunz.

visible results in environmental protection that have been achieved by many political measures.

However, the progress in the environmental protection measures of recent decades has hardly stopped the decline of many endangered species. The disappearance of many species is continuing unabated, and it is unfortunately the majority of the red-listed species that are benefiting less than expected from the upward trend in environmental protection. The widespread belief that a clean environment almost automatically benefits species is being put to the test. Environmental conservation (and even nature conservation) is not the same as species protection.

Butterflies are less well known and not as popular as birds, so they are less effective tools in the promotion of nature and environmental conservation goals. This is why the nature conservation associations devote less attention to them. Who knows the different species of fritillaries, and regrets their dramatic disappearance in recent decades? Upon hearing the term *species protection*, most people associate it with the decline of many species of birds; but the focus on bird protection efforts and their partial success gives us a false impression of the real and recent threat to species in general. The highly lauded, showpiece successes in saving some flagship species, such as the white-tailed eagles, cranes, black storks and peregrine falcons actually create a misleading picture.

When birders or insect experts in Central Europe want to see rare (and therefore desirable) species, they often prefer not to go to national forest parks or even to the nature reserves. In many cases, those who are looking for uncommon unusual breeding birds or migrants, or those who would like to see rare butterflies do not go to the near-natural habitats. They travel instead to the man-made open spaces such as sandpits, sewage farms and the areas where open-cast mining excavations have been carried out; because it is here that they will find the rare red-listed species, and not primarily in the areas which approximate pristine nature in Central Europe.

This book is a plea for the protection of species. Many of my conclusions are focussed on the fact that the species which must be saved in Central Europe are those that live in habitats which have little to do with untouched nature. These habitats are primarily endangered by ecological succession; they can only be preserved or optimised by technical interventions, some of which must be carried out thoroughly. This will prevent unspoiled nature (and especially the forest) from being given top priority; but that does not equate with nature conservation in the eyes of many people, so they do not want it; instead they want a primordial nature (and especially the forest). That is understandable, and it is not dealt with negatively in this book. However, this book makes it clear that the desire for nature and the desire for species richness in Central Europe are two different issues, which often cannot be achieved by the same measures. People who desire unspoiled nature and as much forest as possible will have to admit (albeit rather unwillingly) that they do not want species richness.

1.2
A Plea for Open Landscapes

An ecological movement was established in the 1970s and 1980s. Its objective was to tackle the lack of environmental awareness in our industrial society (Engels, 2006). In Germany, the effective media appearance of personalities such as Heinz Sielmann, Bernhard Grzimek, Horst Stern and others founded an ideology of alleged intact ecosystems being destroyed by human intervention. Ecological importance was attached to species and a sense of awareness arose that the extinction of certain species would cause entire ecosystems to collapse. The threat to many species was associated with a hazard to human health and poisoned food. The eco-classic 'Silent Spring' by Carson (1962) became a 'nature bible' for many. The question of what an 'ecological balance' and an 'intact ecosystem' actually are (if these two concepts even exist) was not asked by many people, or was repressed, because a non-contentious and valuable conviction lay behind the ideology from the outset. The extinction of several species, however, has nothing to do with natural balance. The value of a species equals the value of a historical monument (Reichholf, 2010). The loss of many species is of an ideational nature; in most cases, no balance is endangered.

Man was seen as the main enemy of many endangered species. Targeted human intervention for habitat regulation was frowned upon, and the popular definition of nature conservation areas was seen in the protection of the remaining 'residual nature' from further human intervention. For many nature reserves, however, it emerged that their preservation was not being threatened by man, but by nature itself. Gravel and sand surfaces become overgrown, dry grasslands become covered with bushes and shallow waters silt up. Nature is untamed ecological succession. It turns many currently existing protected areas into bush-covered, overgrown areas, which makes them worthless for many species, and it is exactly such habitats which could be described (with much more justification) as being real nature. Nature itself reduces the value of many currently existing protected areas by turning them into habitats which are no longer refuges for many endangered species. Many wildlife sanctuaries in Central Europe are not virgin nature, they are man-made habitats; and these man-made qualities are exactly the properties of the nature reserves that must be protected – not from human intervention but from nature itself, because nature would reclaim the land by ecological succession if it was not prevented from doing so.

Recent decades have seen many nature reserves (seen as refuges for endangered species) facing considerable competition from areas which earlier would never have been expected to become important for the conservation of threatened species. These are areas that were created for the military, the economy and transport; in other words, the planning and creation of these landscapes did not pursue the goal of setting up a refuge for endangered species. For several decades now, wastelands in towns and industrial sites, major road embankments, gravel pits, sewage plants, open-cast mining sites and military training areas have been the sites on which many rare birds such as the grouse, the grey partridge, the quail, the lapwing, the tree and meadow pipits, the woodlark, the wheatear and the corn bunting can most likely be seen, together with butterflies such as the swallowtail, the pale clouded yellow, the clouded yellow, the wall brown and other rare species (Plates 1, 2, 5 and 6). None of these species owe their preservation to any active species protection measures; their preservation evolved passively as a by-product of man's landscaping, which was intended for completely different purposes. Habitats like this have nothing to do with nature and would not exist in Central Europe if man himself had not created them.

What these areas have in common is that humans keep them free of vegetation for their own purposes; but the fact is that specific plant and animal species benefit from these areas, the purpose of which was certainly not species protection when they were created. Earlier, sparsely vegetated open habitats like this existed almost everywhere; but nowadays they have become scarce as habitats in Central Europe and are limited almost exclusively to industrial, residential, traffic and military areas (Plate 4). While it is correct that the current expansion of industry, housing and traffic pushes nature back more and more, it would be wrong, however, to associate the decline of the species in all cases with these factors and to complain about them. Species do not always need nature.

Many red-listed species inhabit extreme habitats, which mostly bear no resemblance to pristine and unspoiled nature in Central Europe. This seems to be a paradox, because these habitats are usually in a condition that makes the environmentalists want to avoid them as much as possible (and usually to eliminate them) (Anonymous, 2008). The lapwing and the grey partridge (both Red List species) breed on brownfields and on the terrains flattened by machines on the terrain of inland ports on the Rhine and Elbe rivers, where heavy cranes and trucks dominate the landscape. Some rare plant species have retreated to brittle asphalt surfaces in the parking areas between department stores because these spaces are more or less prevented from becoming overgrown by lush vegetation. Endangered plants which have a strong affinity for salt (halophytes) have again found favourable living conditions on the edges of major roads and have been able to propagate there, because they benefit from the use of road salt (Feder, 2014). These are probably the best examples that show why species protection, nature conservation and environmental protection are not the same, but are often at opposite poles of purpose.

National park and nature reserve policies aim to preserve or to create an ecosystem that is as near-natural as possible. However, it is hard to justify what a near-natural ecosystem really is in Central Europe; and after its millennia-long exposure to man's activities, it is doubtful if pristine nature can be restored at all in Central Europe. One thing, however, would appear to be certain: the habitats inhabited by a major proportion of the endangered Red List species are not near-natural.

In Central Europe, many Red List species live in open, sparsely vegetated areas where few trees and shrubs grow. These species need open ground or stone surfaces, escarpments with crumbling earth or gravel banks, that is, surfaces that warm up quickly when exposed to sunlight. Thick grass growth, which looks so beautifully green and healthy to the human eye (and gives the impression of intact nature), offers no possibilities for life for many Red List species, because the ground beneath the dense grass growth is too moist and cool. This is why sparsely vegetated grasslands are more species-rich than green meadows (Plate 4).

In past centuries, a great variety of flowers, butterflies and other insects flourished on the overexploited and nutrient-poor soils. It has almost become an ecological basic rule in Central Europe that nutrient-poor areas produce a great variety of species, but nutrient-rich areas produce a low number of species. This basic rule alone makes it easy to understand why so many species in Central Europe today have become rare. We are losing the nutrient-poor areas. Intensive fertilisation in modern agriculture and the nitrogen raining down from above (even far from the agricultural land) have taken away the chances of existence for many species during the last half century. The loss of species in Central Europe is due to very different causes than the loss of species in the rainforest areas of the world. Consequently, the species protective measures being taken for Central Europe must differ from species protection in the tropics. It is wholly remarkable that this receives minimal emphasis in the public statements on species protection.

Even in the early Middle Ages, the forest in Central Europe could no longer be called *wilderness*. Wilderness was the heathland. In north-western Central Europe and in many mountain ranges, dwarf shrub growth spread relatively quickly, and in the warm regions with calcareous soils it was the steppe-like juniper heath that propagated profusely. Areas that had steppe to semi-desert-like characteristics, gravel plains and sand dunes were the dominant landscapes in northern Germany and also on the plateaus of the Central German Uplands (Mittelgebirge), in Rheinhessen and on similar terrains elsewhere in Central Europe. Paintings and watercolours from earlier centuries show the land, almost bereft of shrubs and trees, reaching to far horizons in flat land, or barren, stony dry areas on the slopes of the mountains (Tauch, 1974). The landscape was more reminiscent of the Mediterranean area, the eastern steppes or the timberline of the north than the heart of Central Europe today.

It is not surprising that the Central Europe of past centuries and millennia was settled by many open land species of the north, the east and the Mediterranean area. In Central Europe today, many species such as brow hares, skylarks, buntings and most butterflies are not forest species; they were only able to spread when the sedentary, crop-cultivating humans started to partially re-establish the glacial primeval landscape in the midst of the expanding forests during the post-glacial Holocene period. Today, the forests are regaining the upper hand and that is why it is the open land species that are on the Red Lists, from the black grouse to the hoopoe. The re-forestation of Central Europe since the nineteenth century must be mentioned at the forefront if an explanation for the current loss of species is sought. Central Europe's modern landscape is completely different from what it was many centuries ago. Forestation, bush encroachment and ground-covering grasses and herbs have taken up a considerable proportion of land in the last half century and they bring about a frightening species decline and of insects in particular.

The transformation of Central Europe from an open land to forest and scrubland has taken place almost unnoticed for many people and it is not a risk potential for many species that the nature and environmental protection organisations wish to denunciate with top priority. If a man could go back in time for only 100 years, he would believe that he was in an entirely different country. In many locations, he would be able to look to the far horizon. He would see the neighbouring village from the edge of his own village. He would see castle ruins that still exist today standing on bare rock and not on wooded hills. He would see bare riverbanks and gravel banks, and above all he would see sandy and earthy areas with hardly any vegetation in the middle of meadows and fields. These macro- and mini-habitats have disappeared across the board in many places, and this has happened to a dramatic extent, particularly in the last 60 years. The Central European landscape has not looked as it does today for many centuries, neither in its detail nor in its overall picture.

Since the availability of fossil fuels made it unnecessary to cut the firewood from the forests, cattle are no longer driven into the woods to graze and the eutrophication of the landscape has increased thanks to excessive nitrogen, a process is happening which is very threatening for many species: Central Europe is becoming overgrown. Nature is reclaiming the areas that were kept open for centuries and millennia because of food deficiency; open rock, stone, gravel and sand areas are disappearing, because the vegetation is no longer being removed and the landscape is being fertilised by an excess of nitrogen. Large parts of the Central European landscape resemble a neglected garden that is no longer tended and in which wild growth has returned. Heaths and moors were once open areas with almost no bushes and no trees. In today's terminology, sandy pine forests are called *heaths and marshy birch forests are designated as moors*. There are people who no longer know what a moor or heath is, even though certain areas and even protected areas are still given these names.

Some believe that the species richness of earlier centuries had something to do with the more caring, more resource-conserving and, therefore, more sustainable relationship with Mother Nature we had in the past, while our present-day affluent and throwaway society simply exploits her. This is simply not true. The landscape was ruthlessly plundered in the past centuries. In the prevailing hardships of those days, the soils were drained, the humus layers of the heaths and forests were cut out and carried off, the emerging vegetation was removed, the slopes of the hills were overgrazed, and the bushes and forests were cleared or at least thinned out in almost all the accessible places (Paffen, 1940; Plachter, 1997; Küster, 2008). The fields were only used for a few years in succession, until nitrogen and the other minerals were exhausted; only sparsely vegetated areas remained.

However, it was exactly this exploitation of nature that suited numerous species in many ways. The centuries of agricultural and forest exploitation of nature and landscape in Central Europe was accompanied by an abundance of species, the pinnacle of which was reached about 1850 (Blab *et al.*, 1984; Häpke, 1990). The plundering of the countryside did not harm the majority of species – on the contrary, it even caused the explosion of species richness seen in the past. Nowadays, we can only dream about many of the bird and butterfly species that existed in Germany at that time. The hen harrier used to breed in numbers on the high plains of the Eifel; the marginal zones of the rocky crater lakes in the south-eastern Eifel and the embankments along the Rhine were the breeding places for many rock thrushes; the Rhine and Mosel valleys played host to the short-toad eagle, which found plenty of reptiles there and the arid heaths of north-western Germany – with an unobstructed view stretching to the horizon – harboured great numbers of black grouse. The great bustard also bred in Germany in its thousands (Schulze-Hagen, 2005).

The sparsely vegetated soil surfaces of the past were able to bask in hot, direct sunlight and were, therefore, more suitable for thermophilic species than today's overgrown, bushy and forested landscapes in this era of 'global warming' (Plates 1–4, 7 and 8). In the 'Little Ice Age' that lasted until the early nineteenth century, there was a greater abundance of heat-loving butterfly species in Central Europe than we find today in the midst of climate warming.

Anyone who searches for rare species (e.g. rare birds or butterflies) today will soon understand that the nature and environmental conservation organisations are leading a fight against the destruction of nature through gravel quarrying and open-cast mining, so they do not want to encourage important habitats which are actually refugia for rare species. A discrepancy is revealed here, one which is based on the fact that the fight for a pristine and unspoilt nature is not the same as the struggle to preserve the Red List species.

It is difficult for most people today to understand that the deforested, exploited landscape of Central Europe became an ideal habitat for a great variety of plants and animals that are threatened species today. The reason why it is so difficult to associate former species richness with a destroyed landscape can be found in the fact that no distinction is being made between nature conservation and species protection. It is regarded as a paradox that a destroyed nature was the prerequisite for the occurrence of many species. If, however, we can understand that nature conservation and species protection are two different entities, then species richness in the earlier ruined landscape of Central Europe will no longer be perceived as a contradiction. In the landscapes of Central Europe, which have been anthropogenously overformed for millennia, the nature that has been destroyed cannot be equated with a corresponding destruction of biodiversity.

Since most people do not know the difference between nature conservation and species protection, they find it paradoxical that military training areas have become perfect habitats for many Red List species. They are amazed that rare species occur in areas where tank treads destroyed shrubs and bushes and explosions ripped open the earth. Public opinion must become fully aware that the rare species do not flourish on military areas '*despite* the fact that tanks drive around there', but '*because* of the fact that tanks drive around there'. The tanks flatten and uproot the vegetation and explosions clear vegetation. This provides open ground, creating the habitats which have become very scarce indeed.

1.3 Central Europe Is Not Brazil: A Plea for Technical Biotope Engineering and Management

There are very different reasons behind the loss of species in Central Europe and the loss of species in the rainforest areas of the world. In the rainforests, human intervention is occurring in (an almost) pristine nature; here in Central Europe, the opposite is the case for many Red List species: we see insufficient human encroachment upon a landscape that was formerly open, but is now becoming more and more overgrown with vegetation.

Due to millennia of persistent anthropogenic landscape design, many Central European species are today threatened by quite different factors than the species in many other parts of the world, especially in the rainforest areas, and nature conservation in Central Europe is faced by tasks other than those faced by nature conservation in Latin America, Africa and South East Asia. Even in prehistoric times, Europe was a country which was intensively landscaped by humans. Hardly any other large region in the world has modern-day landscapes which differ so greatly from its landscapes of earlier centuries.

Protecting primary nature is important for many disappearing species in the tropical areas of the world; but the situation is quite different in Central Europe: there has been almost no unspoiled nature here for thousands of years and there are consequently only a few species that really need virgin nature. Central Europe is not natural country, although many people advocate this view (Bibelriether, 1997). It is a region that has been continually formed by man since the post-glacial reforestation era and the land has almost no residual nature that could be classified as primary or original (i.e. not man-made). Not only the open landscapes such as heaths bear the traces of prehistoric destruction, even the forests of Central Europe cannot be described as 'primary' in terms of the composition of their tree species (Küster, 2008). This is why many species, especially those on the Red List, cannot be helped by the conservation of 'pristine' nature. Quite the opposite, in fact: the most endangered species need constant human intervention against too much natural growth. In order to fight the current loss of species, human encroachments upon nature must be continued. Promoting a 'nature untouched by humans' is a basic approach that is wrong.

The dilemma in which species protection finds itself is due in part to the fact that many Central European species owe their existence to former agricultural practices. Such practices, however, are no longer feasible today. Today's agriculture is made for and by machinery, the fields are heavily fertilised, every square yard is used and hardly any food is left in the fields after harvesting, so not much is left for the animals.

The skylark, the lapwing and the grey partridge are farmland birds and their numbers are dropping dramatically in Central Europe; their populations have been more than halved since 1980 and the trend is continuing unabated. Since 1990 in Germany alone, the song of the skylark can now hardly be heard, since more than 1 million have disappeared; the overall numbers of the lapwing have been reduced during the past two decades to one-third of its earlier population and the populations of the grey partridge have declined by 90%. But this is not just about the birds; biodiversity as a whole is threatened to an even greater extent.

There are two opposing strategies to protect agricultural landscape species in Central Europe: either (i) by extensive farming methods that are much more ecologically focussed or (ii) by establishing special areas of sufficient size which would be made available parallel to the economically utilised agricultural areas and which would primarily serve as habitats for species. These areas would have to be designed by technical means and artificially tailored to the needs of the species.

The first strategy would imbed species protection into agriculture, which, however, would have to be subjected to significant reforms. This strategy is described as the integration of species protection into the agricultural production sector, or 'land sharing'. The same piece of land is made available for two different objectives at the same time: for agricultural production and for species protection. The second option, 'land sparing', aims at the segregation of agricultural land and species protection areas. Special areas would be created which would be intended specifically for the conservation of certain species. However, these would have to be areas on which the early destruction of nature must be simulated. The advantages and disadvantages of both strategies are being hotly debated. The strategy of 'land sharing' has its proponents (Tscharntke *et al.*, 2012) and its opponents (Phalan *et al.*, 2011). The proponents of 'land sharing' include the German nature conservation associations.

The first strategy, 'land sharing' seems to promise little success, because to persuade species such as the ortolan or corn bunting to return, the yield of the agricultural land would have to be considerably scaled back and poor fields with sparse stem growth would have to prevail once again (Plate 9). It is an illusion to believe that past agricultural practices could once again be revived for the purpose of species conservation. The old practices produced low yields and were often accompanied by famine. Organic farming methods would be of little help. What we do need are barren areas in which crops only grow sparsely, and where the weeds grow in the fields and on unused marginal land. The entire agricultural lands of today, the fields, meadows and pastures, seem to be lost to most of the species.

The second strategy, the separation of land use and species protection ('land sparing') has already proven to be effective (Schäffer and Flade, 2013). However, the production of artificially managed species protection areas would need significant financial means, consisting mostly of taxes; but this strategy will probably be cast aside from the outset, because the population sees the necessary technical engineering involved as a destruction of nature (which indeed it is, if based on the logic of the term *nature*). Such interventions must prevail against the popular ideology in Germany that equates species protection with nature conservation and warns against 'making a zoo out of nature'. Countries such as the United Kingdom or Holland have a much more pragmatic philosophy regarding nature

(Schäffer and Flade, 2013) (see Sections 6.4 and 8.2.1). The preservation of several vanishing species in Central Europe can only be achieved through measures which have almost nothing to do with nature conservation.

The awareness that artificial habitats have their own intrinsic value and that a decoupling of nature conservation and species protection is necessary to rescue especially endangered species will be very difficult to 'sell' in Germany, thanks to the nature conservation ideology of the last decades which has been guided in an opposite direction. This problem will probably never be solved as long as the big 'nature conservation' associations representing the goals of species protection and those of environmental and nature conservation are under one common roof. Conflicting goals are inevitable and they can hardly be solved as a compromise.

This dilemma can only be ended if the nature conservation associations can finally bring themselves to enforce species protection, and in certain cases even against their own nature conservation ideals. This also means tackling a common public awareness that species protection is only possible in a nature that is as untouched as possible. It is damaging to the cause if the impression is given that this is one and the same task. Species protection should be more strongly separated from environmental conservation and a 'healthy ecology'. The ultimate aim here should be that the different objectives of groups with opposite interests must be fought out against one another. To use an example here, we must choose between the wind turbine as a renewable form of clean energy and the red kite as a species that requires a special responsibility on the part of Germany. The two together simply cannot work.

References

Anonymous (2008) Ausverkauf des Niederrheins stoppen – NABU kündigt Widerstand gegen Kiesabbaupläne an. *Naturschutz in NRW – Mitteilungen des Naturschutzbund Deutschland*, **19**, 7.

Bibelriether, H. (1997) *Naturland Deutschland*, Franckh-Kosmos Verlag, Stuttgart.

Blab, J., Nowak, E., Trautmann, W., and Sukopp, H. (1984) *Rote Liste der gefährdeten Tiere und Pflanzen in der Bundesrepublik Deutschland*, 4th edn, Kilda-Verlag, Greven.

Carson, R. (1962) *Silent Spring*, Houghton Mifflin Company, Boston, MA, New York.

Engels, J.I. (2006) *Naturpolitik in der Bundesrepublik: Ideenwelt und politische Verhaltensstile in Naturschutz und Umweltbewegung 1950–1980*, Ferdinand Schöningh Verlag, Paderborn.

Feder, J. (2014) *Feders fabelhafte Pflanzenwelt: Auf Entdeckungstour mit einem Extrembotaniker*, Rowohlt, Berlin.

Häpke, U. (1990) Die Unwirtlichkeit des Naturschutzes – Böse Thesen. *Kommune*, **2**, 48–53.

Küster, H. (2008) *Geschichte des Waldes. Von der Urzeit bis zur Gegenwart*, 2nd edn, C.H. Beck, München.

Paffen, K. (1940) *Heidevegetation und Ödlandwirtschaft in der Eifel*, Ludwig Röhrscheid, Bonn.

Phalan, B., Onial, M., Balmford, A., and Green, R.E. (2011) Reconciling food production and biodiversity conservation: land sharing and land sparing compared. *Science*, **333**, 1289–1291.

Plachter, H. (1997) Naturschutz im Abseits? *Biologie in unserer Zeit*, **27**, 306–316.

Reichholf, J. (2010) *Naturschutz. Krise und Zukunft*, Suhrkamp Verlag, Berlin.

Schäffer, N. and Flade, M. (2013) Elektrozaun, Prädatorenbekämpfung, Ablenkfütterung: Welchen Vogelschutz wollen wir? *Der Falke*, **60**, 396–403.

Schulze-Hagen, K. (2005) Allmenden und ihr Vogelreichtum – Wandel von Landschaft, Landwirtschaft und Avifauna in den letzten 250 Jahren. *Charadrius*, **40**, 97–121.

Tauch, M. (1974) *Rheinische Landschaften – Gemälde und Aquarelle aus dem 19. und 20. Jahrhundert*, Gesellschaft für Buchdruckerei A.G, Neuss.

Thomas, J.A., Telfer, M.G., Roy, D.B., Preston, C.D., Greenwood, J.J., Asher, J., Fox, R., Clarke, R.T., and Lawton, J.H. (2004) Comparative losses of British butterflies, birds, and plants and the global extinction crisis. *Science*, **303**, 1879–1881.

Tscharntke, T., Clough, Y., Wanger, T.C., Jackson, L., Motzke, I., Perfecto, I., Vandermeer, J., and Whitbread, A. (2012) Global food security, biodiversity conservation and the future of agricultural intensification. *Biological Conservation*, **151**, 53–59.

2
Looking Back to the Lost Landscape Structures of the Past

2.1
The Species Richness of Former Habitats

The numbers of meadow waders that were prevalent in the past in Central Europe will probably only remain as a memory.

Older lepidopterologists who can look back over the past 50 years have observed a massive decrease in the number of individuals and species in the case of the blues and fritillaries.

In Central Europe, the 'total biomass' of birds and insects is steadily decreasing.

Many European Red List species live in today's habitats, which nature lovers regard as being contrary to nature.

The ornithologist Ernst Hesse (1874–1945) provides us with an image of Central European species richness of earlier centuries by describing a spring day in 1914 in the Havelländisches Luch, which is a large, waterlogged, open lowland area close to the River Havel in Brandenburg/Germany (Hesse, 1914). Here is a freely translated excerpt from his publication:

> What an "enjoyment of nature" and a delight it was to see different harriers and short-eared owls carrying out their courtship flights, and to hear the winnowing display of dozens of common snipe, interspersed with the wonderful fluting, yodelling calls of curlews, redshanks and godwits. The rasping calls and distant warbling of the sedge warbler, the courting behaviour of the meadow pipit, the singing of the whinchat and the reeling of the grasshopper warbler seemed never-ending, while the lekking of the black grouse could be heard from the mating grounds all around. When twilight had fallen, the "whit-whit" whipcrack calls of the astonishing numbers of spotted crakes assailed the ears. Even in the night, the sound of a many-voiced bird concert carried from the nearby marshes across to the inn.

In Germany today, an impression like this is now (and will probably remain) a thing of the past. Many birds have returned to some meadow landscapes, thanks to the partial flooding of these areas. However, the rich prevalence of meadow waders that existed here 50 years ago will never return. Older ornithologists are

Species Conservation in Managed Habitats: The Myth of a Pristine Nature, First Edition. Werner Kunz.

finding that the earlier numbers of godwits, redshanks and common snipes are no longer to be seen in wetlands that have been rewetted (Zöckler, 2007). Experienced observers of nature remark on today's meadow structures: 'I can't help but see that meadows today look totally different' (Manfred Kipp, verbal communication). He is of course referring to the high, dense grass cover, which today covers wetlands which have been 'restored to their natural states' and which has driven out the former patchy vegetation everywhere one looks.

In looking back to the past with nostalgia, it must, however, be borne in mind that the rich bird life in the Central Europe of 100–200 years ago in no way reflected unspoiled nature; it was the wealth resulting from man-made changes in nature, the outcome of man's destruction of the truly pristine nature. Before man intervened, the vast wetlands in north-eastern Germany's Elbe and Oder lowlands were overgrown with shrubbery and swamp forest areas. It was the extensive land improvements and drainage projects of Frederick the Great 250 years ago that transformed the forest and scrub areas of the wetlands of Brandenburg into an open landscape with dry meadows and large wet sedge meadows (Blackbourn, 2007), making the land into a species-rich habitat.

However, many bird species also bred in the more westerly northern German lowland plain, and ornithologists can hardly believe this today, since the species there today are almost or totally extinct (http://www.nw-ornithologen.de/downloads/projects/project_2_RL_gefaehrdete_brutvogelarten_nrw.pdf).

The hen harrier, black grouse and golden plover were common breeding birds in these wide open spaces of northern Germany, while the ruff, the dunlin and the great snipe bred in the damp fens. The hoopoe, the woodchat shrike and the European roller could be found on the outskirts of villages near orchards and pastures throughout almost all of Central Europe. Shortly before the turn of the nineteenth and twentieth centuries, the breeding population of the great snipe in the state of Schleswig-Holstein in Germany numbered more than the breeding population in Sweden today, and the total number of aquatic warblers in Brandenburg at the beginning of the last century was probably greater than its world population today (Schulze-Hagen, 2005).

All these species have almost disappeared from Central Europe today. Today, the hen harrier breeds almost exclusively in the sand dunes in the North Sea islands of Germany and Holland. In earlier centuries, such vast sand dunes were normal for the large inland heaths, and the fact that the hen harrier was a common breeding bird there is a clear indication of what the heath landscape looked like in earlier times. What we call heaths in northern Germany today are tiny parcels of land in the oak-birch forests – and they hardly deserve the name of 'heath'.

Nowadays, in order to experience similar impressions of a species-rich wet meadow and fen landscape as Hesse did, you must travel up to the Biebrza fen country in the outermost corner of north-eastern Poland. What we can see and hear today in these wetlands on both sides of the Biebrza River in north-eastern Poland could be seen and heard in many places in the northern part of Central Europe 100 and 200 years ago, more or less right on our doorsteps. However, this species richness in the Biebrza region will also come to an end soon; with Poland's entry into the European Union, the meadow birds will suddenly disappear and the development that took place in Germany 50 years ago will occur again in Poland. The EU subsidy policy is not compatible with a rich meadow structure.

Entomologists also experience a similar disillusionment in a nostalgic look back at the insect-rich past of 50 – 100 years ago. Weitzel notes: 'The massive decline in species and individuals in the case of the Blues and Fritillaries is very obvious for any lepidopterologist who knew the "Mainzer Sand" in earlier times [a sand dune area designated as a conservation area in the middle of Germany]. During my first entomological outing in July 1966, clouds of Blues were flying around on the Mainzer Sand and we recorded fourteen Lycaenidae species in just two hours' (Weitzel, 2012).

Retzlaff cites the century-old literature on the marsh fritillary (*Euphydryas aurinia*) in the district of Lippe in North Rhine-Westphalia, Germany: 'In 1893 and 1894 the species could be found from the 9th of May on, and the number of individuals increased gradually until one could collect many hundreds of specimens in the marshy meadows around the city' (Retzlaff and Robrecht, 2011). Today, the marsh fritillary is extinct there. In the same area, Retzlaff writes about the small pearl-bordered fritillary (*Boloria selene*), remarking: 'In the Lichtenau bog area on the second of July, 1973, I counted approximately 5,000 butterflies [and that was just one of more than 30 habitats in the district of Lippe] ... The species now seems to have largely disappeared ... '

To experience similar numbers of butterfly individuals and species today, we must travel to the Bükk Mountains on the edge of the Carpathians in a corner of north-eastern Hungary. The sun-drenched, flower-rich forest paths and dry slopes are what we used to see there 100 – 200 years ago in many locations in Germany, but the species richness in Hungary will also end soon. With Hungary's entry into the European Union, the grazing of dry grass has been abruptly reduced and the nitrogen input is soaring. Many butterfly species will decline or disappear, and the development that has taken place in Germany during the last 100 years will be repeated in Hungary. Again, the EU subsidy policy is not compatible with a rich meadow structure.

2.2
The Decline of the 'Biomass' in Central Europe

One important aspect is the question of whether we have less of a 'bird mass' today in Central Europe than we had in the past. Could there be statistical facts behind the subjective impression that we have fewer birds today than in the past? Is the decline in the number of starlings, skylarks, swallows, sparrows and other species being quantitatively offset by the increase in the number of crows, tits, warblers and other species? Since the monitoring programme developed for common breeding bird species in Germany by the umbrella organisation of German Avifaunists Dachverband Deutscher Avifaunisten (DDA) has only existed for a few years and consequently contains no long-term information, this question can only be answered unsatisfactorily at present. There is no information about this in the fourth version of the Red List of breeding birds in Germany published in 2007 (Südbeck *et al.*, 2007).

The first information about the 'total biomass' of breeding birds is provided by some long-term measurements that have quantified the 'mass' of certain bird species every year. A comparative quantitative assessment of the numbers of individuals of many species of birds has shown that the total bird mass in Central Europe has declined considerably in the last three to four decades (Lemoine *et al.*, 2007). For many years now, birds have been caught for ringing in the bird traps of the Radolfzell Ornithological Station in Mettnau at Lake Constance in Germany and these trapping results were compared with one another over a period of 20 years. The quantitative comparison of the trapping figures with nets and traps in controlled and comparable conditions shows a sharp decline in the number of captive birds from year to year (Berthold *et al.*, 1993; Bauer *et al.*, 2002).

Very accurate quantification procedures were performed in the United Kingdom and Holland (Thomas *et al.*, 2004) and the results showed that the United Kingdom has lost 54% of its total domestic bird mass, not counting migratory birds and winter visitors, during the past 20–30 years. Measurements of local bird densities in the Netherlands resulted in similar numbers. The surveys in the United Kingdom were based on more than 15 million separate counts performed by more than 20 000 volunteers. The frightening thing here is that these enormous loss rates are prevalent in wide geographical areas of England, meaning that they affect different ecosystems. So this is neither a local phenomenon, nor is it simply a decline of the bird populations in very specific habitats.

However, it is not only the birds that are losing their habitats. In contrast to popular opinion, insects are disappearing to an even greater extent than birds and mammals. The United Kingdom has shown that about 50% of all bird species there are declining in their individual numbers; but in the case of all the butterfly species, the percentage decline is more than 70% (Pennisi, 2004; Thomas *et al.*, 2004). Other groups of insects such as bees, wasps, bumblebees and dipterans have declined to a similar extent. It is becoming increasingly obvious that birds and mammals are more stable than invertebrates, so they are the wrong yardstick to use for assessing the alarming decline of 'biomass'.

2.3
Lost Landscape Structures – Where Can They Be Found Today?

Hermann Hesse (1877–1962), German poet and winner of the Nobel Prize for Literature in 1946, describes in inimitable fashion the feeling of happiness that the sight of butterflies (as incomparable aesthetic experiences) can trigger in humans (Hesse, 1979):

> (free translation): The butterflies of which we speak are like flowers for many people, a very preferred piece of creation, an especially valued and

> effective object that engenders astonishment, a particularly lovely occasion to experience or to sense the great miracle and to worship life itself. Like the flowers, it seems to have been devised by exceedingly friendly, graceful and witty genii as a specific ornament, as a jewelled gem, as a small, sparkling work of art and a song of praise, and created with tender and creative sensuousness. One would have to be blind or very hardened indeed not to feel pleasure, a remnant of childish delight, or a fleeting trace of Goethe's astonishment at the sight of butterflies.
> […] He dreamt of hunting butterflies, running in breathless joy over neighbouring meadows and slopes, and from out of the now unveiled depths of long-forgotten memories, the sleeper again found a nostalgic desire from his childhood days – to see an Apollo butterfly. The image of the beautiful butterfly king, snow white with red spots and the goal of his boyish eagerness, hung before him in the blue yonder.[…]
> Suddenly a white apparition glided past him. Pausing, he looked up, listening. Inaudibly and calmly, a bright butterfly came down from on high in an elegant curve, flew about the ground, fluttering and searching, finally landing gracefully on the steep, sun-drenched surface of a rock. It seemed to be listening, moving its delicate antennae, and then it spread all four wings widely and peacefully in the warm light. Apollo! On the silken, white wings, darker veins were apparent, drawn in delicate lines with a metallic lustre, and in the middle of the white, silken background, the magnificent eyes gleamed bright scarlet. The Apollo beat its wings gently together, clearly showing its beautiful, elongated form with the impeccable rounding of the upper wings, unhurriedly spread these glorious objects once again as if drawing a breath, turned completely around and ascended gracefully into the air. It flew from the rocks to the tip of a high, purple thistle, and thence down in the direction of the lake, shrouded in the dark and shadowy depths. It then ascended again, hovered undecidedly for a moment, performed a series of jubilant wing beats and disappeared up into the deep, glowing sky.

The Apollo is a spectacular butterfly in Europe. It inhabits stony, eroded areas on hills and mountains. A little more than half a century ago, there were easily 100 areas (apart from the Alps) in Germany where the Apollo could be seen. This magnificent butterfly has now disappeared from all these areas, apart from four remaining pockets of land where it can still be seen (Ebert and Rennwald, 1991). These four remaining habitats are located in the Swabian Alb near Ulm, in the Altmühl valley of the Franconian Mountains, in the Jura Mountains of the Upper Palatinate near the Czech border and on the hill slopes surrounding the Mosel River. Otherwise, the Apollo only lives in the Alps. Afforestation and eutrophication have stolen almost all of its habitats.

The rapid disappearance of the Apollo outside of the alpine regions is one of the most striking examples of how butterflies have gradually and dramatically vanished in the last century. What are the reasons behind this? If we believe the slogans (that have been created to achieve an impact on the public), industrial estates and road construction have played a role, and *of course* the insect collectors, with their nets and poison gas collection containers. However, this is not the case; it is quite the opposite, in fact. In some locations, road construction has even encouraged the Apollo. In Baden-Wuerttemberg and the Altmühl mountain region, the embankments of major roads, other road pavements and the slag heaps of quarries have become the last, still-suitable habitats for the survival of the Apollo butterfly (Weidemann, 1986). The zeal of butterfly collectors can only constitute a threat to the survival of the Apollo when other reasons have already caused its population to become almost extinct (Ebert and Rennwald, 1991).

The reason behind the disappearance of the Apollo is quite different. In earlier centuries, the slopes of the hills and mountains were still bare, stony and rocky, and erosion took place when they were exposed to heavy rainfall; but in recent decades, the slopes have become totally overgrown with grass and bushes. The Apollo needs bare rock and bare surfaces on hilly slopes; otherwise it cannot warm up enough because there are not enough vertical rays of sunlight. Grassy and scrub-covered mountainsides are no habitat for the Apollo.

Today's excess nitrogen deposition from the air promotes the rapid growth of vegetation and subsequent clogging of the slopes (Lethmate, 2005). Add to this the fact that the mountain slopes are no longer grazed by goats, which for centuries prevented the growth of dense grass and bushes. Even wood and branches are no longer cut and collected as fuel, and this was a common practice which lasted for centuries. On the hill slopes of the Mosel (as opposed to the slopes of most other mountains), the Apollo was only able to survive because the slopes are very steep; this meant that the rain constantly washed away the humus, so no dense vegetation could take hold. Viticulture, which is still widespread, also keeps the slopes free of bushes. When the harvesting of grapes stops on an area of wine-producing slopes, several locations quickly become overgrown – and the Apollo can no longer live there. The Apollo does not only need the protection of nature it also needs human intervention in nature. The suggestion to plant the Apollo caterpillar's food source white stonecrop (*Sedum album*) in order to enable the butterfly to resettle its abandoned habitats and multiply there is quite ridiculous (Ebert and Rennwald, 1991), because there is hardly any evidence that the decline of butterflies in Central Europe has anything at all to do with the shortage of caterpillar food plants.

The Apollo butterfly is almost a textbook example of environmental protection acting against species conservation. In the Swabian Alb of earlier centuries, the crop damage wreaked by goats led to the humus being washed away and erosion setting in. This caused great environmental damage. However, it was just this environmental damage that was a necessary condition for the occurrence of the Apollo. If the slopes are covered with vegetation, erosion is prevented, but it means death for the Apollo.

Many Apollo butterflies can still be seen today around the Mosel River. It is found in habitats which nature lovers regard as being contrary to nature. Running alongside the Mosel riverbank sites which the Apollo inhabits are an extremely busy road, a much-used cycle path and a railway line that sees a passenger or freight train every few minutes. The rocky slopes rise steeply beyond the railway. There is no trace of aesthetic beauty here. The air reeks of vehicle exhaust gases, and it is noisy and dusty because of the traffic.

These, however, are the locations where the Apollo has survived. They fly over the steep slopes, continually descending to the railway lines and the road where the ground is warm and where a coveted nectar plant, knapweed (*Centaurea*) blooms next to the tracks. The ballast of the railway tracks consists of stones that are kept free of grasses and herbs by railway maintenance workers. While many butterflies are killed by the trains and cars every year, the habitat is nevertheless optimal; and that is more important for the conservation of the species than the death of many individuals.

You will find no pristine nature (= original forests in Central Europe) and no aesthetically cultural landscape (= classic orchards on village outskirts, fields with hedgerows, mountain pastures and heaths) in the areas where many Red List species have survived the wave of Central European species extinction over the last 60 years. The habitats where the lost species of the past can still be found today are not 'natural'. Such habitats are, for example, military training areas and the abandoned military runways of the Second World War, the aerial images of which look more like those of Algeria or Egypt than like natural European landscapes. These landscapes are exactly what environmentalists desperately want to avoid (Anonymous, 2008): heaped earthworks, churned-up soil, shell craters, abandoned bunkers, bullet-riddled former airfields and scorched earth. Landscapes like this do not match the classic criteria of nature conservation. These are by no means pristine landscapes untouched by man, which is something that nature conservationists have desired since the nineteenth century as a counterpart to industrialisation; but it is here that the last hoopoes, black grouse, nightjars, tawny pipits and woodlarks live and thrive.

Species do not care about man's ideal images of nature. The edges of major roads and road embankments are refuges for steppe species and salt-loving plant species, railway lines offer the bare stone surfaces that are overgrown elsewhere, urban wastelands are substitutes for the lost barren areas formerly found on fields and pastures where the skylark, the lapwing and the grey partridge used to live, open-cast mining excavations are home many of the Red List species and military training areas simulate the original glacial landscape of 10 000 years ago (Schwägerl, 2003).

References

Anonymous (2008) Ausverkauf des Niederrheins stoppen - NABU kündigt Widerstand gegen Kiesabbaupläne an. *Naturschutz in NRW – Mitteilungen des Naturschutzbund Deutschland*, **19**, 7.

Bauer, H.-G., Heine, G., Peintinger, M., and Zeidler, U. (2002) Veränderungen der Brutvogelbestände am Bodensee – Ergebnisse der halbquantitativen Gitterfeldkartierungen 1980, 1090 und 2000. *Vogelwelt*, **126**, 141–160.

Berthold, P., Kaiser, A., Querner, U., and Schlenker, R. (1993) Analyse von Fangzahlen im Hinblick auf die Bestandsentwicklung von Kleinvögeln nach 20jährigem Betrieb der Station Mettnau,

Süddeuschland. *Journal für Ornithologie*, **134**, 283–299.

Blackbourn, D. (2007) *Die Eroberung der Natur*, Deutsche Verlags-Anstalt (DVA), München.

Ebert, G. and Rennwald, E. (1991) *Die Schmetterlinge Baden-Württembergs: Tagfalter I*, Eugen Ulmer, Stuttgart.

Hesse, E. (1914) Die Vögel der Havelländischen Luchgebiete. *Journal für Ornithologie*, **62**, 334–386.

Hesse, H. (1979) *Schmetterlinge*, 1st edn, Suhrkamp, Frankfurt am Main.

Lemoine, N., Bauer, H.-G., Peintinger, M., and Böhning-Gaese, K. (2007) Effects of climate and land-use change on species' abundance in a Central European bird community. *Conservation Biology*, **21**, 495–503.

Lethmate, J. (2005) Ein globales Eutrophierungsexperiment: Stickstoff-Regen. *Biologie in unserer Zeit*, **35**, 108–117.

Pennisi, E. (2004) Ecology. Naturalists' surveys show that British butterflies are going, going. *Science*, **303**, 1747.

Retzlaff, H. and Robrecht, D. (2011) Die Veränderung der Fauna der tagaktiven Großschmetterlinge im Kreis Lippe im Lauf der vergangenen 140 Jahre. *Melanargia*, **24**, 193–268.

Schulze-Hagen, K. (2005) Allmenden und ihr Vogelreichtum – Wandel von Landschaft, Landwirtschaft und Avifauna in den letzten 250 Jahren. *Charadrius*, **40**, 97–121.

Schwägerl, C. (2003) Jüterbog-West: In einer ausgedehnten brandenburgischen Kunstwüste wird der Naturschutz neu erfunden. *Frankfurter Allgemeine Zeitung*, **190**, 38.

Südbeck, P., Bauer, H.-G., Boschert, M., Boye, P., and Knief, W. (2007) The Red List of breeding birds of Germany, 4th edition. *Berichte zum Vogelschutz*, **44**, 23–81.

Thomas, J.A., Telfer, M.G., Roy, D.B., Preston, C.D., Greenwood, J.J., Asher, J., Fox, R., Clarke, R.T., and Lawton, J.H. (2004) Comparative losses of British butterflies, birds, and plants and the global extinction crisis. *Science*, **303**, 1879–1881.

Weidemann, H.-J. (1986) *Tagfalter*, Neumann-Neudamm, Melsungen.

Weitzel, M. (2012) Verzeichnis der im Jahre 2011 im Naturschutzgebiet "Mainzer Sand" beobachteten Tagfalterarten (Lep., Rhopalocera). *Melanargia*, **24**, 89–92.

Zöckler, C. (2007) Waldsterben und Wiesen-Limikolen: Naturschutzbewegung in Deutschland zwischen Katastrophen und technischem Umweltschutz. *Der Falke*, **54**, 257–263.

3
What Does Nature Conservation Want: Clean Air, Untouched Habitats, to Make Rare Species More Numerous?

3.1
What Is Nature? What Is Nature Conservation?

Why is a mountain pasture with colourful flowers or a blooming heathland not nature?

The term nature only makes sense if nature is seen as a contrast to humans.

What distinguishes man and animals/what is man? – Is man part of nature?

What is the reasoning behind species conservation? It may either be anthropocentric or holistic.

According to the anthropocentric reasoning, species should be protected because they have a special meaning for human beings; according to holistic reasoning, species should primarily be protected for their own sakes. They have an intrinsic value.

In the practice of nature conservation, it does not seem to be possible to ascribe the same value to all living organisms.

Nature and species conservationists would do well to relegate to the background those arguments that point out the environmental and economic benefits of species richness, because they could damage the credibility of species conservation. The aesthetic significance of species is much more important for humans.

In earlier centuries, species richness in Central Europe was the unintended by-product of purely economic objectives.

Many species-rich habitats in Central Europe are man-made habitats that have nothing to do with natural areas or even with wilderness.

Many endangered species in Central Europe live in habitats to which the term a stable state of nature cannot be applied. The habitats are unstable and can only be maintained if humans constantly intervene.

3.1.1
What Is Nature? Nature Is What Has Not Been Created or Formed by Humans

What is nature? A dale in a meadow with a meandering brook is beautiful to the human eye and is perceived as being nature, as is an alpine meadow with colourful flowers or a blooming heathland. However, in Central Europe these are all

Species Conservation in Managed Habitats: The Myth of a Pristine Nature, First Edition. Werner Kunz.

landscapes that man has created, and they would also quickly disappear without continuing human intervention. All these beautiful landscapes will become overgrown and overrun with bushes in the foreseeable future if this growth is not stopped by burning, tree felling, turf cutting or the browsing of cattle. If landscapes like this are defined as being natural, then nature must not be protected from man, but from itself.

If we designate such landscapes as 'natural' (which is often done), although regular human intervention is necessary, then the question arises why a heath (which is declared a nature reserve) is nature and a golf course (which would never be declared a nature park) is not. In terms of animal and plant wealth, and even of the abundance of endangered species, it is now often the case that city parks, gravel pits, military training areas, brownfield sites and open-cast mining excavation sites are taking over from the 'classic' landscapes, such as forests, meadows and pastures. Why is a military training area thought to be less natural than a meadow landscape? It cannot be the 'man-made' aspect that drives this opinion, because both of these landscapes are man-made.

The term *nature conservation* includes the protection of nature. To discuss nature conservation, two questions must first be answered:

1) What is nature?
2) From what should nature be protected?

Many people have a yearning for nature; they consider nature to be something important, valuable and healthy – but they do not know what nature is (Großklaus and Oldemeyer, 1983; Kirchhoff and Trepl, 2009). There is no societal consensus as to what nature is. In line with common parlance, however, we give the names 'nature' and 'natural' to areas that have not been created by man. Nature is the 'left on its own', the non-controlled. The opposite of nature is the artificial. It is what the human mind has conceptually conceived and willed, and what is formed by man. All the objects and processes that man has created or formed are artificial, and that is not nature.

In everyday language, the term *nature* is used for shapes, structures and formations which have arisen without human help and have achieved their forms without any human intervention. No one would describe a church tower in a village as being natural, but a tower-like termite mound in the African plains is, on the other hand, considered as being 'natural'. A flower on a lawn has either settled there 'on its own', that is, it is 'natural', or it was planted there by humans, in which case it got there by 'artificial' means. Medicines are either 'natural products', or they have been synthetically produced by the pharmaceutical industry, in which case they are known as *artificial products*. Charles Darwin called the selection in the title of his famous book the 'natural selection', to distinguish it from the 'artificial selection' of the gardener and animal breeder. 'Natural laws' are laws that have not been made by humans and which cannot be influenced by man.

It is a fact that in our everyday parlance, we say that nature is the opposite of man-made design; but this is a conventional agreement. There is no general theory upon which a definition for nature could be based, and none upon which the concept that nature is the opposite of man-made design could be based. Since man in his evolution has gradually evolved from animal ancestors in seamless transitions, and since there is a wide transition zone between humans and animals, a zone in which organisms that were neither humans nor animals lived, it is difficult to draw a line of distinction between humans and animals. Humans and animals flow into one other, and it is impossible to determine empirically, that is, to read from nature, when man became a 'person'.

Even in the case of the current living human being, however, there is no definition that clearly distinguishes him from the animal and, therefore, from nature. If we wish to distinguish human beings from monkeys, then we must fall back on a group of many characteristics, or we must quantify these characteristics. Man does not differ in any individual qualitative characteristic from his closest relatives, the great apes. There is no 'human gene' which all people have in common and which the apes do not have, and there is no phenotypic trait that every human being has, but apes do not have.

In terms of the biological sciences, man belongs to nature. He is *Homo sapiens*, a species of the genus *Homo* and a member of the family of hominids. However, this is not just a question of biology here; it is also in line with the current zeitgeist to consider the human being as being a part of nature, that is, that he belongs to nature. This modern conception of the world contrasts with the hubris of the past centuries, when man saw himself as the crowning pride of creation and as the centre of the world; the Old Testament commands man to 'subdue the earth'.

Considering man to be part of nature is, however, a standpoint which cannot be applied to the conservation of nature for pragmatic reasons. If man does not distinguish himself from nature, then everything is nature – and it then becomes very difficult to explain why there has to be nature conservation and, in turn, what nature conservation actually is. All the devastation and destruction that man inflicts would then be classed as 'natural' processes, just as it is natural that storm surges flood the land and locusts destroy crops. If man were a part of nature, we could find no rationale for why and for whom nature should be protected at all. Regardless of whether man is actually part of nature or not, all conservation measures must be based on the assumption (even if it is for purely pragmatic reasons) that man sees himself as if he were beyond the pale of nature.

Nature is not what man has done. Nature is what has developed on its own and 100% nature evinces no signs of human intervention. Hence, the question arises whether 100% nature even exists on earth – and the answer is that it does indeed exist, but only at extreme locations to which man previously had no access or almost none; these are the Arctic regions, the great deserts, the high mountain regions and the ocean depths. Almost all other locations on earth bear the signs of human changes. For millennia now, man has encroached upon nature in most locations on earth, even in the supposed primeval forests.

In Central Amazonia thousands of years ago, massive, prehistoric forest fire clearances took place, followed by agriculture. Even in the Xingu National Park in southern Amazonia, where today there are contiguous tropical forests covering the largest rainforest area on earth, the native population thousands of years ago was much denser than it is today, until a catastrophic depopulation occurred around 3500 years ago. The same is true of the Congo Basin, where great areas were deforested by fire about 3000 years ago to make way for oil palms and bananas; however, the native population collapsed around 500 years later. In the rainforests of Papua New Guinea and Thailand about 7000 years ago, there were large banana plantations and a relatively high population density.

All of this is evidenced by archaeological and archaeobotanical investigations (Willis, Gillson and Brncic, 2004). Although we have no exact conception about actual population densities in these prehistoric times, it seems certain that in earth's rainforests, human settlements were much more common in prehistoric times of past millennia than they are today, and that more humans lived there than is often assumed. What remains today of the primitive tribes, such as the Xingu Indians at the headwaters of the Xingu River in the Brazilian Mato Grosso and the pygmies in the Congo Basin, for example, were historically preceded by more civilised and more populous sedentary cultures. We cannot make any conclusions about the state of rainforests millennia ago from the state they are in today. In many locations, our tropical rainforests are not in such a 'virgin' state as we commonly assume.

If we, therefore, define nature as being 'untouched by man' (as this book attempts to do) outside the extreme locations on Earth today, we will find almost no biocenoses that we can call 'nature pure' without any limitations. There mostly exist only landscapes and habitats that are characterised by lesser or greater degrees of human intervention in nature.

3.1.2
What Is Nature Conservation? For Whom Should Nature Be Protected?

Almost everyone thinks they know what nature conservation is. At first glance, everything seems to be clear: it is nature that should be protected. But which nature is to be protected? Why should nature be protected? From whom should nature be protected? And for whom should nature be protected? Is there any objective reasoning which ethical convictions and the resulting practical measures for the implementation of nature conservation can rely upon, a reasoning from which they can derive their justification?

First of all. **For whom** should nature be protected? For humans? For which humans? For the birdwatcher who wants to see a lot of birds? For the hunter who wants to shoot his prey? Or has nature an end in itself, so that man should not think of himself when he carries out nature conservation?

Second. **From whom** should nature be protected? From humans? From which humans? From the hunter, from the farmer and from dog and cat owners? From modern urban and industrial man, or even from primitive peoples, such as the Xingu Indians in the Brazilian Mato Grosso? Should nature have even been protected from the Neanderthals? Should nature be protected from predators and vermin? Or should it be protected against natural disasters and their consequences?

The question for whom nature should be protected has different answers. The zoologist and philosopher Martin Gorke (University of Greifswald, Germany) attributes an intrinsic value to nature and species that is independent of the value that we humans attribute to nature and species (Gorke, 1999a,b, 2010). The botanist Martin Schnittler (also of the University of Greifswald) sees human evaluation as a benchmark for the protection of a species (Schnittler, 2004). Man prefers some species to others, finding criteria that cannot be justified in strictly scientific manner. This includes the so-called *flagship species* of nature conservation, to which humans assign a high aesthetic value, for example, eagles, cranes and the black stork.

What is the reasoning behind species conservation? The answer is that it may be anthropocentric or holistic (Gorke, 1999b). According to anthropocentric reasoning, species should be protected because they are useful to humans, if only because of the aesthetic value they possess for humans. On the other hand, holistic reasoning says that species should be protected primarily for their own sakes, that is, they have an intrinsic value.

Anthropocentric reasoning

We humans have several essentially economic, ecological and aesthetic reasons that lie behind our interest in protecting species from their extinction. **Economic** reasons are on hand, when certain species play major roles in human consumption, livestock and plant breeding, biological pest control, pollination in fruit orchards and the development of drugs. Species are also **ecologically** important for our well-being: marine plankton ensures the global production of oxygen, the forests regulate the climate and microorganisms recycle waste and produce humus. Third, species have a non-material **aesthetic** and emotional significance for us, because they belong to the formative image of a landscape and are, therefore, important for us in that they help us to feel 'at home'. They also bring a breath of life to birdwatchers and insect collectors.

Holistic reasoning

The holistic view attributes an intrinsic value to nature and species, or in other words, the extinction of species should be prevented primarily for the species' sake (Gorke, 2010). We know an ethical, intrinsic value from interpersonal relationships. Many people are not willing to simply evaluate the relationships they have with other people as being emotional or economic needs. They recognise that respect for others has an intrinsic value. Interpersonal relationships are discerned and respected for their own sake. However, when it comes to the value of nature and its species, then not so many of us will stand ready to protect the species because they have an intrinsic value.

The reason why it is difficult to protect species because of their intrinsic value does not lie in the logic of holistic ethics, because this particular logic is very convincing. The real problem here lies in the pragmatics of the actual implementation. Nature and wildlife conservation measures are expensive. Politicians have the task of convincing the population of the importance of spending taxpayers' money on the conservation of species. Most people are hardly convinced by the importance of the intrinsic value of nature. They want to hear anthropocentric reasoning about the 'benefits' of the species, although they intuitively grant the species a 'right' to continued existence, regardless of whether these species benefit anyone or not. However, when it comes to making decisions on practical nature and species conservation measures, most of us simply want to see economic, ecological or aesthetic arguments. Arguments for intrinsic value are indeed 'worthy', but they are ineffective (Gorke, 1999b).

In the case of anthropocentric reasoning, it is not simply about the selfishness of the modern consumer society. There are much deeper, less superficial attitudes here; not all species can be given the same right of protection 'worthiness', although the holistic ethical reasoning stipulates this. Tiny and evolutionarily lower organisms are not even perceived by most people. They are insignificant for us (as long as pathogens do not enter the equation), but this is of course due to the lack of technical means that are used to perceive the tiny organisms. Many of the apparently inconspicuous small butterflies (Microlepidoptera) are more colourful than many large butterflies (Macrolepidoptera) when a powerful magnifying glass is used to view them. However, although every nature lover could have a magnifying glass in his or her pocket, the purple emperor will always be accorded a higher value than that ascribed to the arched marble (*Olethreutes arcuella*) of the Tortricidae Microlepidoptera family. And when a scanning electron microscope is used to display single-celled organisms (protists) or even prokaryotes, it reveals the aesthetic beauty of these organisms that for most people are totally unknown.

It does not seem to be possible to ascribe the same value to all living things. The effect of an eagle circling in the mountains, a crane calling at dawn on a moor or a blackbird singing in the garden is undoubtedly more meaningful for us than a tiny nematode or a protist in a flowerpot. We would complain less if the latter were to become extinct. For those who see the value and aesthetics of species richness as an asset worthy of preservation, inconspicuous worms, protozoa and bacteria have a significantly lower value than birds, mammals and amphibians.

The three anthropocentric reasonings behind species protection, the ecological, economic or aesthetic motive, comprise different levels of importance. Today, the ecological importance of species conservation is classified lower than it was in the 1970s and 1980s of the last century. In the highly popular ecology movement of that time (a movement that was propelled to broad acceptance by the population through successful public relations), the general opinion was that almost every species in the network of mutually ecological dependencies had its own importance and contributed to ecological stability (Bölsche, 1982; Johnson *et al.*, 1996). This position was motivated by ideology and was, therefore, unscientific. The concept of the ecological value of every species in the balance of nature has now been largely revised. Ecosystems seem to possess a kind of 'species redundancy'. Many species appear to have no importance for the functioning of ecological relationships; they are redundant as far as the ecological system is concerned (Walker, 1992).

Neither should the second anthropocentric reasoning behind the protection of species, that is, the economic significance of species, be thrust into the foreground. Many different plants are used in traditional and modern medicine and some substances from the skin of frogs are of significant pharmaceutical importance. Nevertheless, the 'know how' about the industrial production of such substances continues to expand, and production like this is cheaper in the long run and of course much cleaner. The importance of producing pharmaceutical products from substances that are taken directly from nature will increasingly decline in the coming decades.

Nature and species conservationists would do well to push the arguments for the environmental and economic benefits of species richness into the background, because many of these arguments could prove to be wrong – and they would consequently damage the credibility of species conservation. The ethics of nature and species protection would stand on a firmer footing if the aesthetic value of the species were emphasised. This has nothing to do with any reasoning that the species-protective value of the white stork in Germany is important for the balance of nature, nor has it anything to do with the presence of the stork being an indication of a healthy and clean environment. The former is probably not true, and the latter is certainly not true. Anyone who has seen the multitudes of storks on the landfills on the outskirts of big cities in Spain or Turkey will be quickly convinced that the stork is definitely not an indication of a clean environment (Meyer, Eilers and Schnapper, 2003).

The aesthetic significance of the species should instead be pushed into the foreground for people, together with the holistic ethics of the intrinsic value of nature. Natural values and cultural values should be similarly classified. If, for example, the white stork were driven out by economic interests because wetlands had to be converted into cornfields, that would be a violation of a law of ethics; it would be like demolishing Cologne Cathedral because a multi-storey car park was urgently needed for economic reasons. Wildlife conservation is hardly an economic or ecological commitment; it is more like monument preservation (Reichholf, 2010a).

3.1.3
What Is Nature Conservation? From Whom Should Nature Be Protected?

The earlier development of the Central European cultural landscape was not determined by the fact that man wanted to promote nature, but every human intervention in landscaping was based on the self-interest of the land managers. And it was no different in earlier times than it is today. Former agriculture, however, was very imperfect, due to a lack of technology which our farmers all have available today. The problem for species is today's perfect technique of cultivation and harvesting.

Due to the imperfection of former agriculture (and not because of an agriculture that was closer to nature or considerate of it), many species-rich habitats were created through the forming of areas that were not managed on account of species, but for the achievement of purely economic objectives. In the past, there were no species-protective objectives, and yet man improved habitats for many species through his economic encroachments upon nature. Human intervention in the forests and moors, the creation of meadows and pastures and the browsing of sheep and goats on the hillsides created an unintended by-product of totally different objectives, namely, optimal habitats for a whole range of species, almost all of which are threatened today.

Many species-rich habitats in Central Europe were created through the technical forming of the landscape. Abandoned railway stations, former railway lines and railway and road embankments developed into ideal habitats for endangered butterfly species. Many bird and dragonfly-rich ponds have developed from former peat workings, fish breeding ponds and water mills. Sheep and goat grazing on the Central German hillsides was not carried out with the intention of creating butterfly habitats, just as brownfields were not created for butterflies; but in past decades and centuries, those areas were at the heart of the wealth of butterfly populations (Ebert and Rennwald, 1991a). All of these species-rich habitats were man-made – and they had nothing to do with natural areas or even wilderness. Habitats that were less intensively formed by man and were, therefore, more 'natural' could not even begin to offer anything that was comparable to the man-made biotopes.

In Germany, the destroyed military airfields became particularly species-rich natural paradises after the Second World War. The former aircraft runways were breeding areas for little ringed plovers, wet areas with rushes on the airfield were settled by considerable numbers of common snipe (Kunz, 1959), reed warblers hatched their young in the reeds of the bomb craters, and tawny pipits bred on the boundary walls of the former airfield (Kunz, 1959). Many alcon blues (*Phengaris alcon*) fluttered in the nitrogen-poor, sparsely vegetated areas. None of these habitats had been created with the aim of increasing biodiversity in these areas, and none of these areas have anything to do with nature, nor are they even near-natural.

Established on the edges of several cities since the nineteenth century, generously sized sewage farms evolved into species-rich habitats, especially for migrating waders which used them as gathering places. Until just a few decades ago, sewage farms in Central Europe were used to purify urban effluents. The wastewater was spread over large water-permeable surfaces and filtered during seepage into the ground. The filtered materials were slowly biodegraded by microorganisms. Today, most sewage farms have been shut down. The sewage is treated by modern sewage treatment plants that take up less space. In some cases, however, the sewage farms have been preserved in order to protect the species diversity that has built up over the years. However, this has only achieved limited success.

One excellent example of this is the sewage farm of the city of Münster in North Rhine-Westphalia in Germany; it is a European reserve. A technical wastewater treatment plant was constructed in 1975 to purify the effluents and the loss of the former sewage farm, a species-rich habitat of international importance, loomed on the horizon as a result. To preserve the wildlife, the trickling of the areas was continued with treated water. The species richness that previously existed when the water was still dirty and consequently rich in nutrients declined due to this alternative measure; the abundance of migratory waders was reduced, because the clean, nutrient-poor water no longer offered enough food.

The fact that in Central Europe, the 'technical biotopes' such as airfields, military terrain, gravel and open-cast coal mining sites and wastewater treatment areas offer a higher level of species richness, and are also areas where rare species occur, should force us to reconsider our species conservation policy. In the majority of cases, species richness and rare species in the majority of cases cannot be found in untouched nature or even in the wilderness. Most species-rich habitats in Central Europe have been created as artificial biotopes by man himself. There is no need to protect nature here. Instead, the protection of biodiversity requires extensive technical encroachments upon nature by man. The equation of a 'humans out!' ideology with the term *protection of nature* is missing the target, if the objective is supporting species richness in Central Europe.

The preservation of many endangered species today can never be promoted as long as nature conservation bodies take a passive stance and continue to support the preservation of the natural state of a 'reserve'. Species protection must not be equated with the concept of nature conservation. The forces of nature, to which many species have adapted for millennia, often do not preserve the status quo (Flade *et al.*, 2003). The term *nature conservation* does not equate with the measures that are urgently necessary for the conservation of many endangered species in Central Europe.

Many endangered species in Central Europe live in habitats to which the term *a stable state of nature* cannot be applied. The habitats are unstable and can only be maintained if humans constantly intervene. Such habitats do not simply have a long-term future as a matter of course. Ecological succession usually begins after just a few years and initiates a process which is aimed at the restoration of the natural state. Nature itself strives to approximate the natural state; no nature conservation is needed. On the contrary, the task of species protection in Central Europe is to counteract nature in the case of many species and to prevent the habitats in question from reverting to a natural state.

It is the endangered species in Central Europe today that have adapted to habitats which are not in a natural state. They have adapted to habitats that would lose their value as refuges of endangered species if they returned to their natural state. Dry grasslands and rocky hillsides are quickly covered with bushes, areas with shallow water and sand and gravel surfaces become overgrown and ponds begin to silt up if they are not regularly dug out or dredged. In these areas, it would be counterproductive to initiate nature conservation measures that would seal off these habitats, preventing human intervention. Quite the opposite, in fact: continuous human intervention is absolutely necessary to keep nature in check. The term *nature conservation* cannot be applied to species protection in such areas, because the threat to the species comes from nature itself. Without human intervention, ecological succession would be unhindered; it would overgrow everything, and the endangered species would disappear as a result.

While previously deforested areas, cliffs, escarpments and gravel banks were permanently re-created by natural processes such as wildfires, storms and floods, this dynamic process of nature is severely restricted in order to protect people in today's cultural landscape, since fires, windfalls, floods and landslides are nowadays prevented by technical means if possible, or their consequences quickly eliminated. In the last few centuries in Germany, man increasingly 'conquered' and exploited nature (Blackbourn, 2007), and in doing that, he created areas that were just as ruined and destroyed as those that nature had previously created by disasters; so by exploiting nature, man created the conditions for species richness and the occurrence of rare species.

We need such destroyed areas more urgently than nature reserves and national parks; it is the extreme habitats to which many species have adapted and these species are on the Red List today. Nowadays, natural disasters are curbed, forests are no longer thinned and cleared, foliage and firewood are no longer taken out of the woods and the fields and meadows are well fertilised; so the future of species protection does not lie in creating and maintaining pristine nature, but areas must instead be created on which the earlier destruction of nature is simulated, thus restoring the habitats for many endangered species by technical means.

3.2
Which Species Should Be Protected?

Species protection in Central Europe is often quite the opposite of the protection of wilderness. Species protection must often protect a specific biological community from the threat of nature itself.

Which nature and which species should be protected? We cannot protect everything.

Can we justify the standpoint that species should be protected only where they have their main or core populations?

Neozoa are not the focus of species protection, because they are 'non-resident'.

In a broader sense, however, many species in Central Europe are 'non-resident'; there are only a few truly 'native' species in the region.

Installing tit nesting boxes is an encroachment upon the ecological balance.

The winter feeding of garden birds primarily satisfies human needs.

Species protection means the protection of selected species (chosen by humans) through the preservation or creation of the habitats required by these species.

In many cases, the loss of species in Central Europe means that the generalists become commoner, while the specialists decline.

The term *wilderness* is often used in nature conservation. The term is overlaid with emotion. It expresses a certain hostility to technology. Those who rave about wilderness do so without knowing what they are actually saying. Wilderness is nature without human intervention (Kirchhoff and Trepl, 2009). Wilderness is perceived as freedom, although not a single wild animal and not a single plant are truly free under wilderness conditions, since they are subject to predation and competition from neighbouring species and this prevents their free development. In the zoo, all animals are protected from predators, and in the garden many plants are protected from competitors. In the wilderness, however, every animal and every plant is defenceless.

As nature without human intervention, wilderness does not need nature conservation. When an active nature conservationist raves about wilderness, he must be aware that he is pulling the rug from under his own feet because for nature conservationists, wilderness means simply doing nothing. Anyone who wants wilderness in his garden fires his gardener. In many respects, nature conservation is the opposite of wilderness. In many cases, nature conservation requires the protection of a particular state of nature in an area (usually temporary and influenced by humans); but protection from the threat of a different, 'other nature'. This 'other nature' would gain a foothold in the area in question if man did not protect it from the 'other nature'. In extreme cases, nature conservation means the protection of the nature which dominates in one particular area against the penetration and domination of the wilderness.

A key motive why one specific (usually temporary) state of nature should be protected at all is the protection of some particular species that live there or are to live there in the future. This raises the initial question for every species and nature protection body: 'Which nature and which species should be protected?' We cannot protect everything because to protect everything means doing nothing. Although such a standpoint is debatable, it has nothing to do with nature conservation.

Why are moors and heaths important? Why is the removal of a forest to restore a lost heath in north-western Central Europe desired by species conservationists, but often unwanted by nature conservationists? Why is the restoration of a heath desired by species conservationists, while a golf course is seen as being undesirable? Every intervention in nature (there is no difference between a rut left by a tank and a habitat management measure taken by an active nature conservationist) benefits some species, but displaces others. It would be only logical to identify nature conservation areas by lists, on which not only the species to be protected are shown, but also the species that are not explicitly to be protected in this area.

But which species are worth protecting? There are at least four options by means of which 'protection-worthy' species can be selected.

3.2.1
The First Option: Only Those Species Should Be Protected in an Area in Which a Significant Proportion of Their Global Population Is Present

Should we mainly (or even exclusively) protect a species in a country in which that species has its main population distribution, when it only occurs in smaller numbers in other countries? While this at first sounds convincing, a detailed examination shows that this standpoint leads to consequences that most people certainly would not want. This would mean, for example, that in Germany the Elbe cocklebur (*Xanthium albinum* subsp. *albinum*) would have to be protected with more financial and manual effort than any orchid, because the Elbe cocklebur has its main population distribution here; but there are highly endangered orchids in Germany, which are widespread outside Germany.

In earlier centuries, when the German countryside strongly resembled the open areas of the Mediterranean, the hoopoe, European roller, crested lark, wryneck, woodchat shrike, lesser grey shrike, cirl bunting and the rock bunting were still widespread breeding bird species. Today, all of these species have either become extinct or very rare in Germany. All the above-mentioned species, however, live in several Mediterranean countries and in south-eastern Europe, where they are not endangered. In considering the introduction of biotope maintenance measures to prevent these species from further decline, the question arises why such species should be protected in Germany at all, when they are not threatened elsewhere and are even common.

Animal and plant species that only occur in a specific region in sparse numbers and become extinct there are not endangered in their existence *as species*, if they are common in other areas. One could argue that species should be protected only where they have their main or core populations, while protective measures for these species in the areas or countries where these species only occur peripherally and in sparse numbers do not make sense. This standpoint is, 'pure species protection', so to speak; it only pertains to the conservation of the species.

One simple reflection shows that the focus of the nature conservation bodies is not on species conservation alone because that would mean that the smaller the area under consideration is, the number of species worthy of protection there would become fewer. We would ultimately find ourselves with areas where there are hardly any species left to justify their regional protection. Such areas do not necessarily have to be very small. Even a nation such as Germany has only a few species that have a major proportion of their worldwide populations there. Among the (approximately) 250 domestic bird species that regularly breed in Germany, only one single species fulfils this condition and that is the red kite (Südbeck *et al.*, 2007). If we consistently follow the standpoint that protection in an area should only apply to the species that have a significant proportion of their worldwide populations in that area, almost no birds would have to be protected in Germany.

No one can possibly agree with this; and it is apparent here that the species protection authorities in a country cannot primarily pursue a policy that would save a globally threatened species; instead they are concerned with preserving species diversity in a region, irrespective of whether the species is common and not endangered elsewhere. National or regional species protection is actually more like homeland protection. In the majority of cases, it does not concern the rescue of endangered species.

3.2.2
The Second Option: Only Original, 'Homeland' Species Should Be Assisted

With regard to certain species, some argue that they are exotics, the protection of which is not so important, because they are in any case immigrants from other areas. Some conservationists, for example, give the bee-eater (with nearly 1200 pairs breeding in Germany) a lower 'worthy of protection' status, because it is a Mediterranean and south-eastern European bird that occurs only as a guest in

Germany. This standpoint is totally unjustified, because most breeding birds in Central Europe are Mediterranean, south-eastern European, Eastern European or North European species. The bee-eater is only exceptional because it looks exotic and it started immigrating in real numbers during the last decade. However, many other species of birds breeding in Germany are also post-glacial immigrants with their core populations outside of Central Europe. The bee-eater belongs to Germany as much as the brown hare and the rabbit.

Central Europe is characterised by the fact that most animal and plant species are immigrants from the north, east and south, because the drastic changes between the warm and cold eras wiped out most of the residents during the Pleistocene, and Central Europe had to be constantly repopulated (see Section 6.1). Every 100 000–200 000 years, Central Europe was steppe country and then (usually for shorter periods) forested land. During the transition from forest to steppe, the steppe animals migrated from the east, and the transition from steppe to forest brought the forest animals from the south. It is debatable whether or not the forest or the steppe animals are native to Central Europe, and the opinion perhaps closest to the truth is that most animals currently living in Central Europe are not original or native at all. As a result of glacial destruction and anthropogenic deforestation, Central Europe in the recent past has only a few typical Central European or even endemic species.

Even in more recent historical times (and not only as a result of post-glacial warming), new animal species have migrated into Central and Western Europe. Among the bird species, the European serin and the collared dove are well-known immigrants that first arrived in Central Europe in the twentieth century from the south or south-east. The common rosefinch immigrated from the east and the melodious warbler arrived from the south-west only a few decades ago.

It is, therefore, of secondary importance if and/or when certain species immigrated into Central Europe. The term *naturally* is much more important here. This term is used to describe species that came 'on their own' to Central Europe, that is, 'naturally', without any interference from man, in accordance with the definition of the term *nature* (see Section 3.1.1). On the other hand, all the species that man brought to Central Europe (intentionally or unintentionally) fall into another group. They are not 'natural' inhabitants of Central Europe. In a broader sense, they are also described as being 'non-original', 'non-indigenous' or 'non-resident'. These species are the Neobiota (Neozoa and Neophytes). In the 'Red List of breeding birds of Germany', they received their own status category (Status III, see Section 4.2.2) (Südbeck *et al.*, 2007).

The German Federal Agency for Nature Conservation limited the terms *Neozoa* and *Neophytes* to species that were introduced after 1492 in Germany. This year is vital, because from then on animals were introduced into Europe from America, thanks to the discovery of that continent by Christopher Columbus. All introduced species that lived before 1492 in Germany, are classified as 'native', not Neobiota. Exceptions here are the pheasant, which was naturalised from Asia, and the rock dove (*Columba livia*), which has run wild as the domestic pigeon in Central and Eastern Europe and originally lived only on the rocky coasts of

Western Europe and in the mountainous areas of the Mediterranean countries. Both of these species were indeed brought to Central Europe before 1492 and released there, but the pheasant was continuously reintroduced for hunting and the dove was reintroduced for use as a carrier pigeon. The population has thus been supported by human intervention even after 1492; so it cannot be ruled out that without new additions made to their populations, the birds could have died out; but that, however, is very unlikely (Südbeck *et al.*, 2007).

We must, however, understand the terms *naturally* and *immigrated without human involvement* in a very direct sense, as far as human intervention is concerned. In a broader sense, the immigration of many animals is due to 'human involvement'. Man has encroached upon nature almost everywhere in Central Europe. He changed the habitats and landscapes to such an extent that certain species were able to migrate to Central Europe, species which probably would not have come to the region without man's influence, or at least not in such numbers. This is especially true of the open land species, which were offered a new home by man's post-glacial destruction of forests. In Central Europe, the skylark is a steppe bird living on arable land; the crested lark is a bird of the semi-desert, but here it lives on railway tracks and industrial wastelands; the golden plover is a bird of the northern fells (the plateaus above the coniferous treeline), but in Central Europe it lives on heaths and the drained high bogs. The black grouse largely migrated into the anthropogenically deforested, open areas in Central Europe. The 'original' habitats of the black grouse are the treeline regions on higher mountains and in the Asian–European north. In a broader sense, all these species are 'non-resident', because most of them would not be present 'naturally' in Central Europe today without man's help. Are these species 'native' to Central Europe? Or do they not also meet some criteria that are applied to the definition of Neozoa (invasive species)?

The Neobiota (in the narrower sense) have been either intentionally introduced (e.g. by keeping exotic species in parks) or unintentionally introduced (e.g. by importing food): they then escaped into the wild, where they survived independently and propagated for decades. Only a very small proportion of exotic species can survive independently in Central Europe. Most plant and animal species introduced from foreign countries cannot permanently feed and reproduce without the help of man, or they succumb to competition from native species. Alien species that survive only a short time in freedom are not included among the Neobiota. Neobiota must establish themselves successfully in the host country, and it can take two or three decades until we can say for sure that they are established species.

When species from foreign countries do establish themselves in the host country, the question then arises, what changes do they cause among the existing biological communities? Alien species can establish themselves in their new home without any problems, but others can displace certain native species and push them to extinction, or they can cause economic damage to humans, such as the common water hyacinth (*Eichhornia crassipes*), which originates in the Amazon. It has spread throughout African waters, where it causes annual damage of around \$100 million to drinking water, transport and fisheries. Invasive species can wipe out the indigenous species directly by eating them, they can assert themselves as superior competitors, or they can introduce new pathogens or parasites to which the original local residents are not immune. According to the statistics of the IUCN (International Union for Conservation of Nature and Natural Resources), Neobiota are the main threat to 30% of all globally threatened species of birds (326 species), 11% of all globally threatened species of amphibians (212 species) and 8% of all globally threatened species of mammals (61 species) (Baillie, Hilton-Taylor and Stuart, 2004).

Neobiota particularly threaten endemic animal and plant species that inhabit isolated geographic regions. Here, the native species are highly specialised and have not yet had to fight competitors or enemies. On ocean islands, for example, some species of birds have lost their ability to fly, such as the extinct dodo (*Raphus cucullatus*), a relative of the pigeon on the islands of Mauritius and Reunion in the Indian Ocean), and the kakapo (*Strigops habroptilus*), a New Zealand parrot. These species originally had no enemies and consequently developed neither escape nor defensive behaviour. Such animals are defenceless against enemies or competitors, and quickly fall victim to invasive species. Even a large island such as Australia with its highly specialised marsupials suffered great species losses when placental mammals (such as dogs and rabbits) entered their habitats with the help of man. The animals of the large continents (with open borders to the neighbouring areas) have always had to assert themselves to a much greater extent against rivals and enemies, so they are less vulnerable to Neobiota.

The islands of the Atlantic and Pacific have almost invariably lost many indigenous species to invasive Neobiota, either because they were eaten or were not able to assert themselves against the invaders in the struggle for food resources and nesting sites. Sixty seven per cent of the globally endangered bird species living on the sea islands are threatened by invasive alien species, compared with only 8% of the endangered species on the continents (Baillie, Hilton-Taylor and Stuart, 2004). Invasive non-indigenous species do not generally threaten amphibians and mammals as much as birds, because there are more bird species than amphibians and mammals among the endemic islanders. Birds were able to colonise ocean islands more than amphibian and mammalian species.

Central Europe with its open borders and its Pleistocene past as a permanent immigration region is characterised by a majority of particularly stable and less specialised species. There are not so many 'typical Central European' species. Other continents have more regional species, so Central Europe is relatively low-risk as far as Neobiota are concerned. Evidence of Neozoa or Neophytes causing adverse effects on the Central European animal and plant species has only been found in a few cases. In Central Europe, the majority of Neobiota have succeeded in establishing themselves, without forcing out any resident species compositions (Thomas *et al.*, 2004). In Germany, none of the 20 Neoaves which have firmly established themselves cause significant economic or ecological damage.

Rose-ringed parakeets have bred in many Western European cities for the past 50 years. They have fully adapted to their urban habitats and require no human help, neither in winter nor for raising their young. As tree hole nesters, it seems that they hardly compete with the resident jackdaws, stock doves and starlings in the parks. It is highly probable that the appalling decline of the starling in past decades in many parts of Germany (in North Rhine-Westphalia it is on the Red List at the early warning stage: Category V) has nothing to do with parrots commandeering their nesting cavities in the trunks of trees.

Many of the Neoaves breeding in Central Europe today belong to the *Anatidae* family (swans, geese and ducks), for example, the black swan, the Canada, snow, bar-headed and Nile geese, the ruddy shelduck and the wood and mandarin ducks. Numbering among the conspicuous Neoaves are three different species of flamingo and three parrot species. Remarkably, there are absolutely no songbirds among the Neoaves breeding in Germany, although exotic songbirds regularly escape from their cages.

Man is usually quite powerless to stop many alien species (Neobiota) from entering regions. Most of the successfully established Neophytes and Neozoa could not be driven back again, even with the utmost effort. They have to be accepted as new members of our fauna and flora.

Nevertheless, Neozoa are special, and as such, they are not included in many species population surveys. Ornitho.de, the Internet portal for avifaunistic data collection of the Dachverband Deutscher Avifaunisten (umbrella organisation of German Avifaunists: DDA), marks all Neozoa observation data that is entered in a separate colour, thereby separating them from the other bird species. Nature conservation and species protection measures largely ignore the Neobiota. NABU, the 'Naturschutzbund Deutschland' (Nature Conservation Association of Germany) recommends limiting nature conservation activities to only the preservation of native species (http://www.nabu.de/). In 'The Red List of breeding birds of Germany' the Neoaves are listed separately (Südbeck *et al.*, 2007) and data about the extent of their species losses is not entered into the statistics.

This negative attitude towards the Neobiota is also called into question by the fact that among the Neoaves breeding in Central Europe, there are several species which evince declining populations in their original homelands, or which have even been classified by the IUCN as endangered. The introduction and spread of some specific Neozoa and Neophytes in Central Europe must, therefore, be regarded as an opportunity for the global survival of certain species; for example, the population of the mandarin duck, a native of East Asia, has been expanding for decades as a wild breeding bird in Western Europe, while in its original home in north-east China and Japan, the number of breeding pairs is on the decrease (http://www.iucnredlist.org/details/22680107/0). Thousands of Egyptian geese breed in Germany, but the bird has disappeared from its eponymous breeding country, Egypt; on the African continent, it now breeds almost exclusively south of the Sahara (http://www.iucnredlist.org/details/100600396/0). The great yellow-headed amazon (5–8 pairs breed in Stuttgart), has become rare in its home countries of Mexico and Guatemala and is classified as endangered by the IUCN (http://www.iucnredlist.org/details/106001679/0).

The terms *indigenous* and *domestic* automatically associate species with a homeland. In earlier centuries, the nightingale, the skylark and many other creatures have been associated with the homeland by poets and songwriters. Species like this are considered to be 'landscape-typical' and are an expression of regional character. They are seen in the light of a harmonious nature and as the ideal of a harmonious human–nature relationship (Kirchhoff, 2012).

This association of certain species with the feeling for homeland seems to play an important role when endemic creatures are assigned different values than exotic newcomers. The reasoning behind the protection, promotion (and separation) of 'native' species as opposed to 'non-resident' species is apparently based on subjective motives; but actions like this can only be very insufficiently justified by objective criteria. The rose-ringed parakeet does not stir any youthful memories. Apparently, the exclusion of Neozoa also has something to do with nostalgia. Here is an example of this: when European emigrants settled the areas around Melbourne in south-east Australia in the eighteenth and nineteenth centuries, they introduced skylarks, goldfinches and other European species to give the alien landscape a domestic character (and possibly to alleviate their homesickness).

3.2.3 The Third Option: The Species Which Are of Importance for the Ecosystem or for Humans Should Be Protected

In Germany, species protection began with the birds. As predominantly diurnal organisms which often behave conspicuously and have a colourful plumage and a melodious voice, the birds are the most attractive group of creatures for human beings; and many species of birds were deemed to be 'useful' because they destroy insects, which in turn were generally classified as 'harmful'. These purely anthropocentric motives were the main motivation for the introduction of species protection at the beginning of the last century.

In 1899, Lina Hähnle founded the BfV, the Bund für Vogelschutz (Association for Protection of Birds) in Stuttgart/Germany; it was renamed the 'Deutscher Bund für Vogelschutz' (DBV, the German Association for Bird Protection) in 1966, and again renamed the NABU (the Nature Conservation Association of Germany) in 1990. In the first decades of its existence, the Association focused mostly on the protection of birds. Its main activities were concentrated on the provision of nest boxes and feeding the birds in winter. These activities were consciously associated with the accomplishment of a good deed. More birds in the forest and garden would eat more insects and therefore bring benefits and mitigate damage.

Insects were sweepingly classified as pests. Today, no nature conservationist would want to be associated with this standpoint. In north-western Europe, we are now living in the era of the death of the butterflies (Thomas *et al.*, 2004). In recent decades, the Netherlands and Belgium have lost around a quarter of their butterfly species and in north-western Germany, the loss rate is similar. Of the remaining butterflies, the number of individuals has declined by more than half (see Section 5.2.2).

Faced by such dramatic butterfly losses, the protection of tits, for example, cannot be endorsed. The great tit and the blue tit are among the 10 most common birds in Germany – and these 10 account for almost half of all the birds in Germany (Südbeck *et al.*, 2007). There is no reason to continue protecting tits. A blue tit pair has two broods every year. Each brood numbers around 10 fledglings, and the parents feed each brood continuously for 18 days, at a rate of one caterpillar, beetle larva or spider per minute. That amounts to 15 000–25 000 arthropods that a single blue tit pair in the garden or the forest feeds to its young within 1 year. Given the sharp decrease in insects in Germany, these statistics must be viewed with alarm.

It is inappropriate that many nature conservationists consistently blame pollutants and the use of chemicals in the fields for the current loss of insects. When they hang tit and nuthatch nesting boxes every 100 yards or so in the urban woods, they should think very hard about whether they themselves are contributing to the loss of insects with their so-called conservation measures. 'Nesting boxes in the woods and gardens as a substitute for poisonous sprays' is one of the conservation propaganda watchwords that should be put to the test. In Central Europe, we can do without tit boxes in the garden, forest and open countryside without any drawbacks. No friend of the butterfly can support the installation of nest boxes.

Those who feed garden and woodland birds in winter have been similarly misled. This feeding only benefits the sedentary birds that do not migrate to Southern Europe or Africa in winter. Fed by humans through the winter, they have a lower rate of mortality. Winter feeding interferes with the biological balance, because it only benefits the sedentary birds which then survive in an unbalanced high number (Anonymous, 1996). The too-large populations of sedentary birds that have been fed through the winter can then occupy more than the usual number of breeding territories and nesting holes in early spring, leaving no nesting room for the migratory birds which arrive later in the breeding area. A strong instinct underlies the human urge to feed birds; this was one reason why the book 'Vögel füttern, aber richtig' (Feed the birds, but properly) by Berthold and Mohr became a bestseller (Berthold and Mohr, 2012). Winter feeding, however, is not about the protection of the birds; it satisfies human needs, similar to the feeding of pigeons and ducks in parks, also a common custom that is quite simply a bad habit.

It was only in the 1970s and 1980s of the last century that nature conservation organisations began to rethink their policies and abandoned the preferential treatment of birds. Even the reasoning that insect-eating birds and ants should be protected for the benefit of human beings was pushed into the background and replaced by a more comprehensive ecological perspective. The 'German Association for Bird Protection' (DBV) was renamed 'Nature Conservation Association of Germany' (NABU) in 1990.

The new name is a clear sign that the original goal of the Association, the protection of birds, has been expanded to encompass the protection of all nature. This change to the objective of the original bird protection association was in line with the 'green movement' of the 1970s and 1980s of last century, when 'comprehensive thinking' was propagated that embedded the conservation of the species in a 'intact nature' or 'balanced ecosystem', and in a health awareness which also served human beings (Engels, 2006). NABU adopted a revamped objective that linked the future of biodiversity with 'good air, clean water, healthy soil and a considerate exploitation of nature' (http://www.nabu.de/nabu/portrait/).

This resulted in a collection of environmental protection bodies within one task area of an association, the original goal of which was the protection of species. However, environmental protection and species protection are not one and the same. There are many species, which are not helped by environmental protection measures, and there are also several species which are actually harmed by various environmental protection measures (see Section 3.3). Clean air and clean water are not the paths we should tread towards rescuing many Red List species in Central Europe.

In the last decades of the last century, the NABU went through a process of reorientation. It broke away from the 'benefit to humans' perspective, the standpoint that birds and ants control the insect pests and thus replace poisonous sprays in the forest and in gardens. However, with the propagation of an 'intact ecosystem', NABU approached yet another, new benefit perspective. The diversity of species has been made into an indicator for a 'healthy ecosystem'; but it is a trivial ecology that often underlies the promises of a healthy ecosystem. In reality, the so-called balance of nature and the ecosystems do not exist (Reichholf, 2010a). If an ecosystem is to function properly, many animal species are superfluous. They have no ecological value and consequently do not act as indicators that the system is in order (Walker, 1992; Gorke, 1999b).

So if most of the Red List species are not important enough to keep an ecosystem functioning, and they do not have any indicator value that would show that an ecosystem is in order, what value do they actually have? The answer is that they have an intrinsic value like a historical monument, which also has no pragmatic benefit (Gorke, 2010); they also have an aesthetic value and an experience value for humans.

Species protection has distanced itself from its earlier 'benefit to mankind' and 'harmfulness' paths. Today, individual species are no longer seen as competitors for food with humans; they are no longer called harmful because they are 'predators' or fish-eating animals that take away humans' food; but this does not change the realisation that species protection must be carried out from the perspective of the human being, because it is not feasible to allocate the same protection value to all species. We have no choice; we must grant certain species a higher protection value than others. And if certain species have a higher protection value than others, then we need a yardstick. The scale for this yardstick can only be the subjective understanding of the human imagination. We will probably not be able to break away from the notion (which is believed to be obsolete) that nature is there for humans, simply because it is impracticable to protect mosquitoes and pathogens to the same extent as butterflies and cranes. When it comes to investing time and money in our efforts to protect certain species, subjective evaluation always comes into play.

We prefer some species to others, and in doing this, we find criteria that cannot be justified in strictly scientific manner. The protection of the eagle is more important than the protection of the nematode; animals and plants are protected for humans, and there will always be some species that are the 'flagships' of nature conservation. Subjectivity is hazardous, however, and we must always question ourselves about the roles nostalgia and rigid habits play here (Reichholf, 2010a). The population increases of spectacular species such as the black stork, the sea eagle, the crane and the eagle owl during recent decades must also be weighed against the local disappearance of skylarks, swallows and house sparrows.

This anthropocentrism sounds like a relic from the days of the medieval conception of the world, when man saw himself as the central point and lord of the world. What gives man the right to decide what should be protected? Holding back from this decision would match our much more modest conception of the world today compared with that of the Middle Ages. Since we no longer believe that we are the centre of the world as we did in the Middle Ages, it would be more appropriate to shelve our self-centred self-interests and not presume that we can decide which species are to be protected and which are not.

However, in species protection you cannot simply ignore anthropocentrism. Species protection means the protection of species selected by humans through the preservation or creation of the habitats required by these species. But which species should be selected? Deciding which species should be protected is not possible without reference to the subjective wishes of human beings, and this makes nature anthropocentric. It is a matter of course that species whose very existence is threatened have a preferential right; but the term *endangered* does not only apply to vertebrates and insects. Some inconspicuous nematode species in the topsoil are also currently threatened with extinction as a result of the changes in agriculture.

No one likes to admit to anthropocentrism, but it should be openly acknowledged. We feed birds in winter for our own pleasure, but this is not a good deed. The lynx does not kill to control the roe deer population, but population control is something which hunters very much like to claim, rather than admitting that they are following their instincts and pleasures (Reichholf, 2010a). We do not miss the now-rare species such as the turtle dove, the corn bunting and the fritillary because this indicates a disturbance in the ecological balance; no, we miss them because we have lost something that involves experience value. It is in the nature of man to disown some activities that are aimed at obtaining pleasure and to present them as good deeds, comparable with a Champion's League soccer game that we pretend to enjoy because it is good for international understanding.

3.2.4
The Fourth Option: The Objective of Species Protection and Habitat Management Should Be the Achievement of a Maximum Number of Species

It is a characteristic feature of Central European biodiversity loss in recent decades that the generalists are becoming more common in some animal groups, while the specialists are in decline or even disappearing altogether. In the state of North Rhine-Westphalia in Germany today, for example, every second bird belongs to just 10 species, although 194 bird species breed in this state (Grüneberg and Sudmann, 2013; Weiss and Schulze-Hagen, 2014). Biodiversity is declining, even though the total number of birds in some regions is not on the decline. This can give us humans a false impression if only the birds are seen, but not the species to which they belong.

This also applies to butterflies, at least in north-western Europe. At locations where up to 70 species occurred a 100 years ago (Lenz and Schulten, 2005), only around 15 species can now be found; these, however, can be numerous. Some people believe that butterflies are still common, particularly if more than 10 butterflies are sitting on the garden shrub Buddleia at the same time; but these 10 usually belong to only 3 or 4 species. In most regions of the north-western German lowlands, only around 15 species of butterflies can be found today; but it was a different story just 60 years ago. These 15 species are among the whites (Pieridae), the three cabbage whites, the brimstone and the orange tip; among the brush-footed butterflies (Nymphalidae), the tortoiseshell, the peacock, the admiral, the painted lady, the comma and the map; among the browns (Satyridae), the meadow brown and the speckled wood; and among the blues, the icarus blue and the argiolus blue (see Section 5.2.2).

This species decline which has been clearly noticeable for the past 50 years is mainly caused by the disappearance of suitable habitats. In Central Europe, besides marshy stream valley meadows and fen areas, these are especially the nutrient-poor areas, the sandy areas with poor grass growth and those with hardly any bushes or trees. Butterfly species that were dependent on specific micro-climates used to live in areas like these. To minimise the number of endangered species, the species protection authorities must focus on measures to promote areas like this, or even re-engineer them (see Section 6.4). The demand of the conservation organisations for habitats that are as pristine and natural as possible is not the right way to achieve a maximum number of species. Many butterfly species living in Central Europe cannot tolerate a virgin landscape. They do not need forests, but open land and soils that are not excessively compacted and overgrown.

Many species on the Red List in Central Europe are not species which can live in untouched nature (Haupt *et al.*, 2009). The Red Lists document today's serious species decline; and if we would realise that there was a need for action to stop species decline, we would also realise that this goal cannot be achieved by the preservation and creation of natural or semi-natural habitats and landscapes for many species. The majority of endangered species are those of the agricultural and pasture land. The reason the agricultural landscape in Central Europe has become a place of extinction for our species lies in the modern, machine-friendly management of fields, meadows and pastures; such agricultural methods have made these habitats uninhabitable for many species. The cause of this is not a disregard of the near-natural. The agricultural species are dying out because every square yard is used, fields, meadows and pastures are too densely overgrown and less food waste now remains after harvesting has ended. The agricultural species are not dying out because the fields were once more in harmony with nature, but because they were more sloppily farmed and their yield was much poorer.

However, today's agriculture cannot be restored to the state it was in during the nineteenth or the early twentieth centuries, because the poor soils in Central Europe at that time were often associated with poverty and the malnutrition of the population. There can be no agrarian reform which would cause a drastic reduction in crops yield. The return of typical agricultural birds such as the corn bunting and the ortolan cannot be achieved by a 'green' agriculture with biologically controlled cultivation methods (Lanz, 2009) (Plate 9).

The dry grasslands of Transylvania in Romania are particularly rich in species. If we focused the measurement of species diversity on small areas of just a few square yards, even the tropical rainforests which are the record holders would be beaten by some European semi-dry grasslands (at least in terms of vascular plants) (Dengler, 2012). And in terms of butterflies, the grasslands of Romania are at the forefront of Europe (Cremene *et al.*, 2005). The main cause for the high species richness of grasslands in the temperate zone is human use that has continued for centuries. Today, the species richness of the European grasslands is endangered by the abandonment of agricultural use, while in contrast, the species richness of the tropical rainforests is endangered by the onset of agricultural use.

Grasslands are not natural habitats. Regular mowing and grazing is essential for the biodiversity of grasslands in Romania, because it deprives the surface of the earth of a large proportion of biomass. This biomass deprivation, described as 'disturbance ecology' (Dengler, 2012) creates a well-balanced disparity in the ecosystem. The key factor for this imbalance is the fact that strongly competitive species are kept at bay. This imbalance can be quickly disrupted and upset. Too-intensive usage of a meadow means that certain strongly competitive species get a chance to assert themselves even more, with the result that the richness in different species declines. This same consequence, however, arises if the land usage (and with that the ecological disturbance) ceases, resulting in an established balance in the ecosystem. Even under these conditions, strongly competitive, specially adapted species will dominate and richness in different species will also decrease (Connell, 1978).

The species richness of Eastern European grasslands (www.edgg.org) depends on the use of these areas by man; but this use should neither become too intensive nor should it be stopped altogether. Eastern European grasslands provide us with impressive evidence that the widespread longing for unspoiled nature (particularly evident in Germany), and also the self-conception of intact ecosystems (which arose in the 1980s of the last century and was taken up by the conservation organisations) pursue different objectives than saving the dwindling biodiversity in Central Europe.

Airports, open-cast mining sites and military training areas have long become substitute habitats for the former agricultural and meadowland birds of Central Europe. Here, the habitats are similar to the meagre agricultural land of earlier centuries, many of the Red List species can be found here today. We will not achieve the protection of these species through the protection of nature. Species protection is not nature conservation. We cannot achieve a maximum number of species in Central Europe through near-natural or aesthetic landscapes and a biologically oriented agriculture. These are all the goals of the nature and environmental conservation bodies, which do not focus primarily on the protection of species.

3.3 Separation of the Terms Environmental Protection, Nature Conservation, Species Protection and Animal Protection

Ideological history of the term nature.

Nature conservation in Germany possesses ideological traits; any criticism of nature conservation is perceived as an attack on nature conservation as a whole.

In earlier times, nature was perceived very differently than it is today.

The extinction of species in Central Europe is continuing despite the achievements in environmental protection.

The rhetoric of the 'balanced ecosystem' amounts to 'trivial ecology'.

Most Central European areas placed under conservation are not 'nature', they are cultural landscapes.

Bio-tomatoes and apple juice cannot bring back the corn bunting and the wryneck.

Animal protection bodies believe that the animals as individuals have a right to a 'worthy' life; it is not the objective of these bodies to preserve species from extinction.

3.3.1 Nature Conservation, Environmental Protection and Species Protection Are Not One and the Same

The term *nature* has an ideological history. This is a result of the following deliberations: natural products are regarded as being healthy, but industrially produced food is believed to be unhealthy. Medicines as natural products are supposedly healthier than those produced by the pharmaceutical industry, although the latter are usually much purer, with accompanying substances that are known and are, therefore, much more reliable. The fact that many plants naturally contain toxins (if only because they have to protect themselves from herbivores) is hardly perceived, because plants are perceived as being natural and, therefore, healthy. This awareness of health, which is allegedly present in nature, is deeply rooted in the minds of the population; but it is a myth, and myths cannot be disproved with justifications, nor can they be corrected.

In the minds of many Germans, nature conservation possesses the essential features of a religious consciousness; there are even those who embrace trees. For many people, nature conservation is a 'good thing' and therefore from the outset is not debatable. Nature conservation is taboo. A taboo is a 'discussion stopper'.

Modalities are discussed, but the assumption that nature conservation is important is respected from the outset and the reasoning behind this assumption is not questioned. The questioning of some conservation measures triggers rejection and refusal and is perceived as an attack on nature conservation as a whole. This situation is a heavy burden for pragmatic nature conservation measures.

The rescue of an endangered species raises a general feeling of rejection in the consciousness of the population, if the rescue operation requires interventions that are directed against nature. The most pragmatic argument often falls on deaf ears because the general consensus of opinion is that it simply cannot and should not be that we have to act against nature to protect endangered species. For many people, it simply should not be that trees have to be felled, areas burned or predators shot for the purpose of saving endangered species. A veritable taboo-consciousness dominates here, and no one may touch on the topic without being roundly criticised.

The taboo states that nature conservation and species protection are one and the same and that Mother Nature will do everything that is necessary for the good of the species if we leave her in peace. However, this awareness, nourished by positive feelings, is not realistic.

From the outset, the implication is that man is to blame for the current decline of species; one has simply to keep man out of nature, even to the extent of letting nature run riot. Some nature conservationists simply do not believe endangered species can be saved by the targeted shooting of predators and the use of forestry machines and excavators; and it is often pointless to discuss it, because something like this cannot occur, because it must not occur.

Remarkably, the ideological history of the nature conservation movement in other European nations is less widespread than in Germany. In the Netherlands and in England, the need to help endangered species through technical encroachments upon nature is more readily accepted by the populations there than in Germany. In these countries, argumentation about nature conservation is more rational, and the people are not so readily willing to accept that high values are violated.

In Germany, the term *nature conservation* is historically and heavily burdened by the illusion of a pristine landscape untouched by man (see Section 7). Added to this was the eco-ideology of the 'green revolution' in the late 1970s and early 1980s of the last century; this movement adopted nature conservation as its own, distanced itself from the pragmatism of realistic nature management and embedded itself in idealised goals and messages of salvation.

The logical question arises: What is the reasoning behind this 'pristine' nature concept (if it could be achieved), and would pristine nature even be desirable? There are two main reasons to love nature. On the one hand, there is the aesthetic feeling we get when we encounter nature, and second, it is the wealth of animals and plants that makes nature valuable for us. However, neither of these have hardly anything to do with 'untouched' and 'pristine'. So why should nature be 'untouched'?

If we take a close look at the species in Central Europe that are classified as endangered on the Red Lists, it should immediately strike us that the majority of these species live in habitats that have nothing to do with pristine, intact nature. Orchid habitats in Central Europe mostly stem from the direct or indirect influence of man and these habitats consist of areas where he has prevented the natural development of vegetation (Almers *et al.*, 2001). Fifteen of the 20 butterfly habitats in Germany's North Rhine-Westphalia were created by humans, and these habitats would vanish if we stopped intervening in nature; indeed, most of them are disappearing today because man is not doing enough to combat nature (http://www.lanuv.nrw.de/natur/arten/roteliste.htm).

Heaths and nutrient-poor grasslands are home to many endangered animal and plant species, so they are areas that have to be protected; but heaths and nutrient-poor grasslands are not natural, and it has nothing to do with nature conservation when habitat managers annually uproot growing birch and pine trees from those grasslands. It is important that the measures taken by biotope management are justified by the right arguments (Küster, 2008). This is not about the conservation of nature. It is not nature that deserves to be protected here, but the identity of a landscape shaped by human beings.

The forest has a special role in German nature conservation ideology (see Section 7.1). Most Germans believe that the forest, home to wild animals and birds, is synonymous with nature (Küster, 2008). The forest prevents soil erosion, maintains the groundwater level, filters the air and provides a clean environment. The forest, however, is not the embodiment of nature in Central Europe, a role which Germans happily assign to the forest. This is a fundamental misconception, and one which is still adhered to by Germans today. The forest is the only Central European habitat which is not in short supply. We have enough forests. It was the open land that was destroyed in the last 100 years, not the forest.

Land reparcelling, fertilisation and machine-oriented agriculture have made the agricultural and grazing land uninhabitable for most animals. Apart from the agricultural land, what remains of the open country today is largely overgrown with thick grass and bushes. This is why most of the endangered animal and plant species have disappeared. The last surviving individuals of endangered species have scurried or flown to military training areas, airports, open-cast mining sites and industrial wastelands. But such habitats are not accepted as being 'valuable', despite the diversity of species living there. What is the reason for this?

The term *afforestation* is closely associated with a taboo in the German national awareness. This has historical reasons. Afforestation is generally welcomed. Afforestation is a good deed in itself. It is comparable to the term *marsh drainage* at the time of Frederick the Great. No one usually checks whether or not afforestation on a given area or on a specific slope is at all meaningful for nature conservation and species protection: afforestation simply takes place. The same thing happened 250 years ago on German soil: every marshy area provoked revulsion and drainage was requested, whether it was meaningful or not (Blackbourn, 2007), and today every clearcutting operation is rejected and every cliff at an open-cast mine site is regarded as something that must be quickly removed. The phrase 'restoration to a natural state' is used for this process (see Section 3.5.3).

The attitude of man to nature has changed repeatedly over time. From the Middle Ages up to the start of modern times, nature for people was a threat, whereas at the beginning of the nineteenth century during the Romantic period, it became a glorified object of desire; and in the present bio-movement, nature has become a world in which everything is fine and healthy. All points of view here have one thing in common: they are quite simply illusions which have little or nothing to do with reality.

By the end of the eighteenth century, nature was predominantly regarded as being the enemy of man. Forests invoked fear in people, because supposedly (and sometimes actually) dangerous animals or evil robbers hid in them. A stagecoach ride through the forests of the Spessart in Germany risked an attack, like the one described by Wilhelm Hauff in his 1826 novel 'Das Wirtshaus im Spessart' (The Spessart Inn) (Hauff, 1827). This meant that forests were mostly not appreciated and by no means revered (as they are today); instead, they were often shunned.

The progressive industrialisation of the nineteenth century changed the population's attitude to nature. The ugly, inhospitable factory areas and the residential areas of everyday life built in their neighbourhoods produced a longing for nature. A romanticised notion developed of a pristine, untouched landscape which took the role of a stronghold against life in polluted industrial sites, and this role was played by the forest. Poets such as Adalbert Stifter and composers such as Carl Maria von Weber romanticised the forest (Stifter, 1882). In Germany, the forest was the embodiment of nature (see Section 7.1.1). In the present decades in Germany, there is more forest than there has been for centuries and in the eyes of many this equates with an abundance of nature. However, we live in an era of species extinction and Germany is no exception.

In the 1970s and 1980s of the twentieth century, the attitude to nature changed yet again when an awareness of environment and sustainability was adopted by large segments of the population. A growing number of people were convinced that the cleanliness of soil, air and water was a prerequisite for the restoration of a nature that had been destroyed by industry; they believed that if the water were to become clean again, many of the lost rare species would again return. Unfortunately, in terms of many Red List species, this view also has very little to do with the reality of the situation.

The year 1975 saw the launch of the 'Bund für Umwelt und Naturschutz Deutschland' (BUND, Association for Environment and Nature Conservation in Germany) and the 'Green' political party was founded in 1980 in West Germany and West Berlin. An ecological movement was established with the founding of this association and this party. The movement's objective was to tackle the lack of environmental awareness of our industrial society (Engels, 2006). One key concept of this ideology was the embedding of species protection into environmental protection. The embedding of species protection into the bio-movement and into a 'healthy ecology', plus the association of species losses with endangered health and poisoned food were messages that had a powerful effect on the public and brought the associations and political parties a rich influx of members.

The awareness of nature conservation was now oriented on a holistic concept. The popular topics were 'sustainability in the ecosystem' and 'healthy ecosystems'. The monistic or one-tier doctrine was then adopted and it was made known that the environment would become healthy and natural, and this would in turn benefit the diversity of species. However, this concept could not be proved for the simple reason that ecological relationships are fundamentally very complex and can never be brought together in a simple formula; but it was extremely media-effective to associate the disappearance of the white stork in West Germany with a risk to human health.

Three decades ago, there was significantly more environmental damage in the German Democratic Republic (GDR) than there was in West Germany, where the water was cleaner and the air was less polluted. Nevertheless, the GDR was a refuge for many Red List species, which became extinct in West Germany at this time. The beaver, for example, survived in the GDR section of the River Elbe in spite of the filthy state of the water. And you only had to compare the eastern border of Schleswig-Holstein with the western border of Mecklenburg to see that the survival of the corn bunting had nothing at all to do with a clean environment. However, no conclusions were reached from these well-known and highly visible facts.

For decades now, open-cast mining excavations, military training grounds and industrial sites have been refuges for many endangered species, and this is also a well-known fact. Bird watchers, entomologists and orchid friends regularly visit the destroyed nature of brown coal excavations, the remains of the former military airports destroyed in the Second World War, former ammunition depots and military training areas that have been opened to the public. Many nature lovers prefer these locations, even to some protected areas, because they are home to more rare species. Once again, however, no logical conclusions were drawn from these facts.

It is a known fact that environmental protection does not automatically benefit species; but the conclusions drawn from this realisation have not been integrated into the nature conservation movement, which largely ignores them, even today. The fact that rare species occur where tanks churn up the soil is still regarded as being paradoxical; it is not seen as a doctrine from which we must draw conclusions on how to counteract the loss of species. The opinion that the rare species occur there *in spite of* the tanks rolling through still holds sway, because the realisation has not yet got through to the general public that the rare species are present *because* the tanks roll through these areas.

The ideologisation of nature conservation, which has existed since the 1980s of the last century, still continues up to the present time. Right from the start, a high-ranking value is assigned to pristine and untouched nature. Man is seen as the main enemy of many endangered species. Hunting, butterfly netting and flower picking are among the worst on the list of sins. The concept of a balanced, intact nature, which will regulate everything optimally if we only leave it alone, achieved a particularly high priority level during the heady optimism of the eco-movement in the 1980s of the previous century (Bölsche, 1982). Many still frown upon any actions aimed at engineering habitats, whether these actions are performed by humans or human-driven machines. Why do the nature conservation organisations not even consider using some of their membership fees to purchase an excavator? This machine would of course destroy nature, but if this saved even one rare species, would it not be a worthwhile project?

The nature conservation authorities believe that the purpose of nature reserves is to save that which man has not yet destroyed, and to protect this 'residue of nature' against further human intervention. But for many Red List species, it is not about protecting the species from the encroachments of man, it is instead a matter of protecting their habitats from vegetation succession, that is, from nature. In the form of this succession of vegetation, nature threatens many species, among which are those which, from far back in our past, have become the centrepieces of the flora and fauna of Central Europe. Gravel banks along river banks are becoming overgrown with vegetation, mountainsides are becoming increasingly covered by forests and meadows in river valleys are becoming overgrown with bushes. If we continue to give nature a free rein, we will be doing (almost exclusively) the species of the thick forest and thickets a favour; but the numbers of almost all of these species have increased in recent decades, so they do not need protection.

The publicist Horst Stern, a BUND co-founder ('BUND', Association for Environment and Nature Conservation in Germany), provoked public awareness in the 1970s by claiming that the red deer, the 'King of the Forest', was a dangerous pest. He stated that the red deer's population density was much too high and this prevented the growth of young trees in the woods, so the deer were harmful. Today, we could provoke public awareness with precisely the opposite view: we miss the large animals, because they thin out our forests. Today, the forests are too dense and the canopy has closed in many places, resulting in insufficient sunlight getting through to the forest floor. This leads to the decline of almost all butterfly species living in the forest (Ulrich, 2002), and some forest birds are also being increasingly endangered, such as the hazel grouse and the grey-headed woodpecker, because these creatures are dependent on thinned-out sites in the forest; but such sites, like the birds themselves, are becoming rarer (Klaus, 2012) (see Section 6.2).

The widespread belief that a clean and 'ecologically intact' environment almost automatically benefits the species must be questioned in the case of the Central European region, at least for the terrestrial species. The disappearance of many species is continuing despite considerable progress in environmental protection, and most red-listed species are profiting less than expected from the upward trend in environmental protection. The 'untouched nature', the demand that tree species in the forest must be native, and the refusal to accept forestry interventions are value terms of nature conservation; but they are definitely not the right measures to rescue many Red List species. The disappearance of many species in Central Europe will continue if landscapes are left to their own devices and later become overgrown.

The integration of environmental protection into nature conservation movement is not a suitable measure. Whether one and the same association can fight for clean air *and* for the conservation of species is highly questionable, because the population sees that the extinction of species is progressing, although visible successes in environmental protection have meanwhile been achieved, and the ecological awareness of the population has changed. Our environment is much cleaner today than it was a few decades ago, but many of the lost or rare species have still not returned. Almost all of the bird species that are now more frequent in Central Europe than they were in the 1970s have not gained in numbers because water and air are cleaner again (see Section 3.5). Species diversity increased precisely where it was least expected, namely in the large cities. A greater number of different animal and plant species live in cities such as Berlin, Hamburg and Munich with their millions of inhabitants, than in many nature reserves (Reichholf, 2010a). Cities have more or less become islands of biodiversity in the uniformity of the modern cultural landscape (Plate 4).

Environmental protection successes benefit only certain species. There is hardly a butterfly species that has benefited from improved environmental protection. In the last four decades, England, the Netherlands and north-west Germany have lost more than 50% of all their butterflies, and that has occurred in a period during which water, soils and air have become cleaner again (Thomas *et al.*, 2004). Nevertheless, the nature conservation authorities still cling firmly to the ideology that environmental protection equates with the protection of species; but the bio-tomato in the fields will not bring back the corn bunting and the wryneck will not return because apple juice from protected apple orchards is successfully marketed by nature conservation associations, with the promise that it will bring back this bird. The wryneck and the woodchat shrike will never return, as long as the ban on early mowing is also applied to the management of orchards (rather than propagating the partial removal of grass sods from the ground of the orchards) (Coudrain, Arlettaz and Schaub, 2010; Hübner, Rauh and Will, 2013).

The importance of nature conservation is held on high in the nature association journals and at meetings of the nature conservation organisations. Any restriction of nature conservation is seen as a threat, and the conservationists are unanimous about all the underlying objectives. However, some highly praised merits and values, such as the much-vaunted and longed-for 'balanced ecosystem' are nothing more than 'trivial ecology' because the question as to how nature should be conditioned, a nature that is deemed worthy of protection, is not answered (Reichholf, 2010a). Since there has been almost no unspoiled nature in Central Europe for millennia now, and each habitat is more or less shaped by human intervention, both the criteria and the reasoning behind *which* nature should be protected in Central Europe are lacking. If this question is pursued, many inconsistencies in nature conservation emerge; but most people seem to be willing to ignore these inconsistencies because 'it's all in a good cause' (Reichholf, 2010b).

The fundamental flaw here would seem to be in the actual concept of nature conservation itself. The fact that many specially protected species occur on brownfield sites, beside the runways at airports and in gravel pits debunks the phrases beloved by the conservation ideology, for example, 'at peace with nature' or 'sustainability' and rightfully places them in the role of 'sacred cows'. Most Central European areas placed under conservation are not 'nature', but cultural landscapes. Many nature reserves should not be conserved, because they would lose their value as a home for rare species in just a few decades if they were conserved. They should be managed by continued human intervention. Nature is not a permanent structure that has to be 'preserved'; nature conservation should concern itself more with preventing nature from running riot in the nature reserves. Many nature conservationists find this very difficult to hear, to admit and to accept.

3.3.2
Animal Protection and Species Protection Are Not One and the Same

The terms *animal protection* and *species protection* are constantly confused and mixed up with one another. Animal protection is the protection of individuals; it is about the individual, suffering creature. Species protection is the conservation of the species; it is about preventing species from becoming extinct or rare. From an ethical standpoint, animal protection is quite different from species protection. In public (even in the press), however, animal and species protection are often not perceived as being different. This misunderstanding also considerably hampers the financing of necessary species protection measures, since some sponsors are more willing to do something about suffering dogs and cats than to fund support for threatened plants and wildlife (Wirth and Glaw, 2009).

Animal protection bodies believe that animals as individuals (such as human beings) have a right to a 'worthy' life. This objective has nothing to do with preserving a plant or animal species from extinction; and yet, people still link the two goals again and again. The following example makes this clear: a public presentation on the impending extinction of the golden plover in Germany due to the destruction of open heath and moorland provoked the following reaction from someone in the audience who called out, 'OK, where do the poor animals go?' This response clearly expresses that the issue of the disappearance of a species has quite simply not been understood. It is confused with the suffering of some individuals. Even in radio and television programmes, species protection matters are often referred to as *animal protection*.

Seal hunting on the coast of north-eastern Canada is another example that shows that animal protection is not distinguished from species protection. Harp seals (*Pagophilus groenlandicus*) occur in large numbers in the Arctic Ocean and have always been hunted by the Eskimos. The Canadian Ministry of Fisheries estimates the harp seal population to be 5.8 million animals (https://de.wikipedia.org/wiki/Robbenjagd). In the nineteenth and twentieth centuries, the hunting of the harp seal by Europeans greatly intensified, to such an extent that it had a significant impact on the overall number of animals. Under international pressure, commercial hunting was largely restricted by the Canadian government, and today the population of the species is no longer at risk.

What triggered the worldwide protests, however, was not the threat to the harp seal that was threatened with extinction as a species. It was the hunting method, with which the young 'sweet' baby seals were slain. This was done with a hacking tool (hakapik). The French actress Brigitte Bardot exploited her fame in 1976 and publicly protested against this hunting method; the ex-Beatle Paul McCartney also had himself photographed with white baby seals to capture media attention. The success of these protests was enormous, because they aroused the pity of significant sections of the population. Very few of the protesters understood that their protests had nothing to do with species protection, or even with nature conservation.

Yet another example is the so-called *species conservation on buildings*. Birds often are killed when they fly against windows and other glass surfaces of buildings. However, quantitative studies have shown that the number of victims is low. At the LANUV five-storey building complex in Recklinghausen, Germany, only nine dead birds of six species were found during 56 days of observation. Compared with the millions of birds that are killed every year in Europe by domestic cats, this figure is absolutely negligible, and therefore meaningless. In spite of this, it is being debated whether or not a triple-digit amount of euros should be spent on the retrofitting of glass windows with a special film which is bird strike-proof. It is totally misleading to call an act like this species protection or nature conservation, when we consider the sheer insignificance of such a safeguarding measure in comparison to the astronomical number of birds killed by cats. The truth is that this is an animal protection measure, the aim of which is based almost solely on the reasoning that the birds which impact on the window panes are usually not killed immediately and therefore suffer as individuals.

The difference between species protection and animal protection is far greater than the difference between species protection and environmental protection, because species protection and animal protection have absolutely nothing to do with one other, while environmental protection very often benefits species; but in many other cases, environmental protection can collide with species protection, so the frequent equation of the terms *environmental protection* and *species protection* only occasionally leads to wrong conclusions being made, but not always.

The confusion of the terms *species protection* and *animal protection* for the public can have serious consequences, because if the rescue of endangered animal and bird species is placed on the level of animal protection, then the rescue of endangered species is not considered to be so important, since the suffering of animals is pushed into the background on the evaluation scale in relation to unresolved humanitarian problems. However, the protection of biodiversity has nothing at all to do with averting the pain and suffering of any animals.

Animal protection should not be described as an understanding of nature. A proper understanding of nature cannot fade out and conceal how 'cruel' nature often is. The cruelty of nature is one of its rules that cannot be changed. On the other hand, animal protection people cannot reconcile themselves to cruelty, while nature conservationists must accept the cruelty. In line with the guiding principles of animal protection, the tadpoles of the natterjack toad must be rescued from a puddle that is drying up. This acts as a soothing balm for the human conscience (Wirth and Glaw, 2009). According to the guiding principles of species protection, however, an action like this makes no sense at all because a natterjack toad population will remain stable even if only two tadpoles from a spawning string of 1000 eggs attain adulthood and reproduce again. On average, Mother Nature has programmed 99.8% of offspring to die (and an even much larger percentage in many cases). In the case of all species which have many offspring, mass death must occur (and this mass death must be accepted by those who really understand nature), otherwise all the ecological systems would collapse (see Section 3.7.4).

Species protection demands that some alien invasive species, such as feral domestic cats on islands, be killed to preserve the indigenous fauna. Neozoa can threaten the population of some resident species in certain areas, so in these cases it is rational and logical to eliminate them, even if this appears to be cruel from an emotional standpoint. Species protection measures must accept necessary cruelty, but animal protection can never allow this, and this also makes the significant differences between animal protection and nature conservation abundantly clear.

The constant confusion of the terms *species protection* and *animal protection* in public also results in many animals' sensitivity to disturbances being greatly exaggerated. Public awareness that certain birds are greatly disturbed by the presence of man is much more widespread and accepted than the fact that many species of birds are disturbed by the deterioration of their habitats. People usually only disturb animals in the short term, but habitats which have been allowed to become unsuitable have a disturbing influence on many animals over the entire reproductive period, and this will drive the animals away more than humans can by simply startling them with their presence. The fact that the young of the lapwing starve on a nitrogen-fertilised meadow is hardly noticed when compared with pictures of an animal photographer setting up his hide next to a lapwing nest. Lapwings have declined sharply in Central Europe over the past decades and in Germany they are now on the Red List at category level 2 (endangered); but this is not at all the fault of the nature photographers. The disturbance of breeding lapwings by wildlife photographers has certainly not affected lapwing populations; and if this has occurred at any time, then only in locations where it was already on the way out because of other reasons.

3.4
Nature Conservation Misunderstood – Fighting the Wrong Enemy

Many heat-loving species in Central Europe have become rarer in recent decades, although the climate has become warmer.

Current climate warming cannot be the main reason for the disappearance of many species; on the contrary, they should have increased in number.

The eutrophication of the Central and Western European landscapes has resulted in a damp and therefore cold microclimate on the ground that is unfavourable for insects.

The general climate warming and habitat losses through eutrophication are mutually cancelling forces; but habitat losses largely predominate, so the positive effects of climate warming are ultimately suppressed by the negative consequences of habitat loss.

The campaign against climate change is an environmental protection problem; but climate change has so far had little impact on the Central and Western European species.

The proliferation rates of most species are so high that imposing prohibitions on hunting, fishing and specimen collecting does not make sense in the case of many species.

Exaggerated bans on man's access to specific protected areas are evidence of a nature conservation awareness that is wrong; and they also alienate people from nature.

3.4.1 What Is the Significance of the Campaign against Climate Change for the Central European Species?

In Central and Western Europe, almost all the butterfly species have drastically declined within half a century. When we realise that most butterflies are heat-loving and that we are currently living in decades in which a significant climate warming is occurring, then the parallelism of climate warming and the decline of the butterflies is apparently contradictory (Warren *et al.*, 2001; Segerer, 2012; Filz *et al.*, 2013). This means that the disappearance of butterflies is due to other causes that are not compensated by climate warming.

This phenomenon is exacerbated by the fact that it is particularly the thermophilic butterfly species with mainly Mediterranean distribution that have declined, although these species should have benefited from the climate warming. In contrast, the mesophilic species require temperatures that are not so high, yet these are the butterflies which are less affected, although they should have been repressed by climate warming. In Bavaria, the largest number of heat-loving species of butterflies was recorded in the nineteenth and early twentieth centuries, when it was much colder than today (Segerer, 2012).

Heat-loving birds, mainly Mediterranean or Eastern continental birds such as the hoopoe and the wryneck have also declined in number over the past half century (Burton, 1995; Huntley *et al.*, 2007). If we look at the loss of biodiversity in Central and Western Europe, we can clearly see that at least in the case of some birds and butterflies (and against all expectations), many thermophilic species occurred much more frequently during the colder climate phase 100–200 years ago in Central Europe. It follows that the current climate warming cannot be the primary cause for the disappearance of many heat-loving species (although this is oft-lamented by the nature conservation organisations). Temperatures have risen considerably in the last 50 years, but the number of thermophilic species has declined, so the loss of many species must have other causes than climate warming.

Among the many species of breeding birds that have declined sharply in Central Europe over the past decades are many thermophilic species, such as the hoopoe, the lesser grey shrike and the woodchat shrike. It is remarkable that among the many predictions about forthcoming species changes due to climate warming, a positive image is actually emerging (Huntley *et al.*, 2007). The number of species that should have benefited from climate warming is greater than the number of species that should have declined as a result of it. However, these forecasts do not coincide with what really occurred. Since most butterflies are heat-loving and it used to be much colder in the past, there would seem to have been small patches of ground that could heat up strongly despite the generally colder temperatures of previous centuries.

The intensive eutrophication of the Central and Western European landscapes that has lasted for half a century creates a wet and consequently cold microclimate on the ground. The strong vegetative growth on the formerly open sandy or rocky soil surfaces shields these areas from the sun and prevents them from temporarily heating up to any great extent (Plates 1, 3, 4 and 8). Apparently general climate warming does not affect the ground sufficiently to enable a rich insect life after the ground has been covered with too much vegetation. This limits the nutritional options of a number of bird species. The newly hatched chicks of ground-breeding birds mainly need ground with sparse vegetative growth. It is not warm enough for them in dense ground vegetation, despite the current climate being warmer than that of previous decades. Thick grass also prevents them from moving around to hunt for insects, which even their sharp eyes cannot discern (if the insects are there at all) (Südbeck and Krüger, 2004). Hunters which sit on high perches, such as the shrikes and the common redstart can no longer see any insects running around beneath the overgrown vegetation (Martinez, 2010). This seems to be one of the reasons behind the non-fulfilment of the positive predictions made for the heat-loving species after climate warming started raising the temperatures. Eutrophication in Central Europe keeps the sun away from many ground-level regions, shades the earth's surfaces and causes a ground-level cooling in this age of 'climate warming'.

The general climate warming and habitat losses through eutrophication are mutually cancelling forces; but the habitat losses largely predominate, so that in the end the positive effects of climate warming are suppressed by the negative consequences of habitat loss (Warren *et al.*, 2001). Microhabitats which heat up strongly on earthy, sandy and rocky surfaces are a thing of the past in many areas, because they have either had to make way for thick, moist grass growth or they have been overgrown by bush encroachment and afforestation. This caused the loss of warm microhabitats for insects (Segerer, 2012) (Plates 1, 3, 4 and 8).

Nature conservation organisations overestimate the consequences of climate change on Central European species. In the case of the German nature conservation and environmental organisations and several provincial environment ministries, they seem to be fighting the wrong enemy, if their aim is to combat the menacing decline of species with climate protection. The campaign against climate change is a problem of environmental protection; but climate change has so far had little impact on the Central and Western European species.

Higher winter temperatures could be a problem for hibernating caterpillars, as their torpor would not last long enough to maintain their metabolism at a low enough level. The caterpillars could starve as a result of higher temperatures, because in winter they need food that they cannot find. The result would be the death of many insects in warm winters, the exact opposite to the popular assumption that cold winters would reduce the insect populations. However, predictions about the impact of climate change on insects are hardly possible, because the interrelations are too complex. Most forecasts of recent decades about the imminent spread of thermophilic species have not been fulfilled (Burton, 1995; Huntley *et al.*, 2007).

3.4.2
Will Apple Juice Bring Back the Wryneck to Us?

The German nature conservation organisations complain in their association journals about clearcuts in the forest, power lines that cut through the woods, wind turbines erected close to the forest, the planting of tree species that do not correspond to local climatic and soil conditions, the running times of nuclear power plants, CO_2 emissions and the loss of habitats through human habitation, road construction, excavations and leisure activities. Unfortunately, not every member of these associations understands that they are airing problems and grievances which often have very little to do with the causes of the current biodiversity loss. One of the main factors for the current species diversity loss in Central Europe is simply being disregarded, and that is the changed structure of the ground vegetation (Coudrain, Arlettaz and Schaub, 2010; Coudrain, 2011; Hübner, Rauh and Will, 2013). The consistent demand to use technical habitat engineering to create large areas that have sparse and patchy ground vegetation is either not given enough priority by nature conservation – or, sadly, none at all.

Birds of the open agricultural landscape, such as the common redstart, the woodlark, the hoopoe and the wryneck are becoming increasingly rare in Central Europe. Researchers at the University of Bern and the Swiss Ornithological Institute in Sempach have an explanation for this: they say that the primary cause is ground vegetation (Coudrain, Arlettaz and Schaub, 2010). Using telemetry transmitters, they studied the foraging behaviour of the common redstart, the woodlark, the wryneck and the hoopoe in grassland-orchards and vineyards. In all the sites which were studied, the birds had a strong preference for the sites with sparse ground vegetation; they avoided the areas which were densely overgrown with grass and herbs. On all the sites, the main factor in finding food was the structure of the ground vegetation. Only open ground allowed the birds to prey on insects. Overgrown plots of land in the same areas were simply abandoned by the birds. The researchers at the University of Bern stated that throughout Europe, there is a lack of ecological compensatory measures which would actively promote patchy ground vegetation; if such measures existed (in their opinion) the populations of these four species would then recover.

Studies in Germany's Western Upper Franconia (Oberfranken) led to the same results (Hübner, Rauh and Will, 2013). There and in many other locations, the wryneck's numbers have declined sharply as a breeding species, but especially so in Western and south-western Germany. This decline took place although the classic habitat of the bird, the grassland-orchard, still existed over a large area and had in fact been augmented by support measures and special protection. Wrynecks mainly feed on ants, so the researchers investigated whether or not the ant populations had declined in the birds' deserted breeding territories, because this would mean that the wryneck no longer had enough available food; but this was not the case. In the abandoned breeding territories, food was mostly still abundant; but the wryneck could no longer detect it in the thick grass.

It was found that the cause for the decline of the wryneck lay in the structure of the vegetation. The birds had left the grassland-orchards and departed for other habitats where open surfaces and sparse grassland were available, such as disused railway platforms and seldom-used railway lines, because in these areas the ants were running around and their nests could be clearly seen. It is, therefore, clear that the wryneck can no longer target its main food, ants, on 'protected' habitats, that is, on meadows that are uncultivated, unmown, mown later or no longer grazed.

The wryneck abandoned its traditional habitat, the grassland-orchard, because nature conservation measures had ordered a late mowing of meadows. The Bavarian contractual conservation programme (Vertragsnaturschutzprogramm) legally prescribed late mowing for the management of grassland-orchards in order to protect nature (Hübner, Rauh and Will, 2013). Grazing was also stopped. The vegetation in the sage – oatgrass meadows (Salvio – Arrhenatheretum) was far too high throughout the breeding season of the wryneck; so it is obvious that the 'protection' of the grassland-orchards took the wrong direction.

These nature conservation regulations were counter-productive for the preservation of the wryneck and other former, typical grassland-orchard inhabitants such as the woodchat shrike. For wrynecks and shrikes, grassland-orchards only make sense if intensive cultivation provides sufficient areas with bare earth and sandy surfaces (Coudrain, 2011), but such measures require significant human intervention with technical equipment and this can often not be carried out for financial reasons. Nevertheless, grassland-orchards are still held up as a shining icon of nature conservation. They continue to be propagated by the nature conservation organisations and even touted at public events, with effective advertising using grassland-orchard apple juice sales.

3.4.3
Bans on Collecting and Catching Specimens – A Distraction from the Main Causes of Species Loss

Many animal species propagate so strongly that after only a few generations the earth's surface would be covered by the descendants of a single original pair if the vast majority of individuals did not die shortly after birth (Darwin, 1859). If much more than 99% of all descendants of a species did not die prematurely, then one single species would inundate all the others with huge numbers of individuals. This would lead to unchecked overpopulations, and the ecological balance would be destroyed (see Section 3.7.4).

Fortunately, the death rate in almost all animal species is extremely high, because most individuals either die from diseases or have no chance in the struggle for food resources and starve as a result. In most cases, only a few descendants of a species need to survive for the species to achieve a stable and lasting survival rate.

The importance of the enormous propagation rate of most animal species is often underestimated, but it actually leads to some compelling conclusions: it is in most cases useless for the protection and conservation of a species to protect specific individuals, unless the species in question is already on the verge of extinction. Any form of help for sick, injured or otherwise damaged animals does not benefit the conservation of the species, because specific individuals are not of key importance for species conservation. Caring for injured animals in breeding clinics, feeding freezing animals in the severe winter cold and the cleaning of oil-smeared seabirds after tanker accidents are all actions which do not help the species to exist. These are not nature conservation or species protection measures; but they are, however, animal protection measures (see Section 3.3.2). The care of individuals relieves the pain of the animals in question; but isolated actions of caring for animals have mostly nothing to do with preventing a species from becoming rarer.

Accordingly, hunting and catching animals in most cases do not reduce the numbers of individuals of a species, for example, it is not comprehensible (even with logical arguments), that the export of captive butterflies from tropical countries by insect collectors is prohibited, while the rainforests are being cut down at the same time in these countries. Legal measures like this are detrimental to species protection, because they put the credibility of species protection at risk.

Hunting and collecting specimens can only threaten the existence of a species under special circumstances. These circumstances occur

1) if the existence of the species concerned is already threatened for other reasons. This applies, for instance, to the shooting of most wader species by hunters in Western and Southern Europe. Hunting may also affect the existence of a species.
2) if the number of creatures shot is very high. In past centuries in Central Europe, a very high number of crows, magpies, cormorants, herons, raptors and geese were shot, for example. Third, hunting can cause damage to some animals' existence.
3) if the animals are made excessively timid by the hunting and consequently maintain a sufficient escape distance from humans. They can then no longer live in many suitable habitats in populated areas, because they must repeatedly flee from man, although the quality of the habitats in these areas would be suitable for the rearing of their young. This is a very high level of damage

caused by hunting; and it is often much more serious than the fact that some animals are shot by hunters.

In all other cases, however, the numbers of a species cannot be affected by hunting, collecting or capturing specimens. Ten billion birds are killed by domestic cats in Europe and the United States every year, and fowlers kill 100 million annually in Western Europe and in the Mediterranean area (Section 3.7); but even these deeds only seem to affect the individual numbers of species that were already rare beforehand. For most bird species, there is no clear evidence that these decimations reduce population levels, although the number of animals killed is so great. The primary cause for population declines is habitat-destruction in the breeding areas, not shooting. If the breeding areas are in a healthy condition, then propagation rates will be high, and hunting will cause little damage (if it is not exaggerated).

Legal bans on flower plucking and animal collecting and catching were issued by the German nature conservation authorities in the late 1970s and 1980s. However, these measures solved no problems and missed their targets. The numerous catching and collecting prohibitions have not benefited the vast majority of plant and animal species at all; on the contrary, they have caused almost irreversible damage, because they have taken away many of man's possibilities to have contact with nature and have promoted his alienation from nature.

Many of the laws passed reflected an insufficient knowledge about the problems of species diversity protection. Wrong priorities were set here, using bureaucratic and impractical measures. The measures taken were those that had as minimal an effect as possible on the economy, did not upset the taxpayer and were in line with a broad populist public opinion. Instead of providing fundamental protection and the technical engineering of habitats, the delusory impression was given to the public that amphibians, butterflies and flowers could be saved through laws that stopped us taking tadpoles out of ponds, catching butterflies and picking flowers.

These species protection regulations, which make the capture of several hundred species of butterflies in Germany punishable by law, saw to it that people were no longer brought into sufficiently close contact with nature (especially young people) and that enthusiasm for nature could no longer be instilled in them. Young people, who were still unknown in professional circles, but who were interested in butterflies, had no chance of ever obtaining a legal permit to catch and collect butterflies. This has had a devastating impact on the next generation of the entomological societies. The young were kept away from their interest in insects. This contributed to the ageing of the societies, with the result that irreplaceable knowledge is already being lost today.

Genuine interest in nature and its protection usually arises from formative experiences and fascination in early childhood. If direct contact with the creatures is lacking, because it is illegal to catch and touch them, and if educators and teachers must first put in a lot of effort to obtain permits to show children the fascinating development of tadpoles in the aquarium, then we should not be surprised if our young people stay away from natural science associations, and even biology students no longer recognise a peacock or an admiral butterfly. Knowledge is especially easy to remember if people can 'grasp' it, in both senses of the word. If schoolchildren cannot take frogs out of the school pond and actually 'grasp' them, they will not be able to 'grasp' how nature actually is. The legal bans on collecting and catching creatures has succeeded in preventing a whole generation of children from developing more interest in nature and consequently in its protection, because unnecessary laws have prohibited them from having close-up experiences in and with nature.

The first thing that struck passers-by on a nature reserve was the warning sign 'No Entry!' Man was declared an enemy because he might trample plants underfoot or disturb animals. After the departure of the former Allied forces from Germany, several military training areas were opened for the public. These were large open areas covered with only sparse vegetation, because the ground was torn up and rutted by tank and heavy vehicle tracks, which kept it generally open. Plants and animals that had disappeared elsewhere due to the eutrophication of the Central European landscape lived here. When visitors were allowed to enter these areas, they were ordered to walk only in the wheel tracks of the former military vehicles. This would avoid disturbing the animals and trampling the plants; but this was a wrong order, because it encouraged the growth of grass and bushes. The intention was to protect the plants and animals living there – but it achieved exactly the opposite effect: the plants and animals disappeared because they were dependent on the crushed, flattened and track-rutted ground.

When the upland moors of north-western Germany threatened to disappear altogether because of drainage and peat cutting in the 1970s and 1980s, efforts were made to save the last remnants of the remaining moors and their animals. In the Diepholzer Moorniederung in the German state of Lower Saxony, for example, some golden plovers were breeding until well into the 1980s (see Section 6.5.1). To conserve and protect them, the last moorland areas were regularly patrolled and guarded by community service workers and volunteers. The local inhabitants of the surrounding villages reacted with surprise and incomprehension; why should they not be allowed to walk in 'their' moors where they had grown up, not even on the paths?

A few years later, the golden plover had disappeared, but not because the walkers disturbed them; the habitat had changed. The heathlands and bare peat lands disappeared because thick grass covered them. Drastic habitat-conservation action would have been needed to conserve the golden plover as a breeding bird. Prohibiting access to people proved to be a completely useless nature conservation measure, one which also angered the population and alienated them from nature.

It is a grotesque situation that European agriculture and the agriculture of tropical countries, the main causes of species loss, are hardly affected by the species protection laws, because countless special permits nullify the effectiveness of the laws and regulations. Instead, species protection legislation negatively affects those who have the least to do with species loss: the nature lovers. Access to nature is made difficult for them, and any study of the protected species is almost completely denied to them.

Many regulations of nature conservation legislation in Germany and in other countries are more of a hindrance than a help to research and the protection of species biodiversity. Research biologists who want to work in the field must now spend much of their time sitting at their desks in order to obtain various permits for their work and for the legal export and import of specimen copies of animals or plants. Even meaningful agreements such as the Convention on International Trade in Endangered Species of Wild Fauna and Flora (CITES) can greatly impede research by the excessive lengthening of the lists of species. Long CITES lists are neither a proof of successful species protection, nor are they an indication of the 'real' risk of the relevant taxa. This is why only as many species as necessary should be included on CITES lists, because bureaucracy costs a lot of time and consequently money, which could be better spent on concrete nature conservation improvements (Wirth and Glaw, 2009).

The Food and Agriculture Organization of the United Nations (FAO) calculated that in the period from 1990 to 2010, a total of 231 661 square miles of tropical rain forests were destroyed in the catchment area of the Amazon (http://www.faszination-regenwald.de/info-center/zerstoerung/flaechenverluste.htm). That is an area that is almost twice the size of Germany. Brazil is a country which imposes heavy fines on the export of butterflies; but in the face of this astronomical destruction of nature that causes the deaths of billions of butterflies, the border control policies seem downright ridiculous.

3.5 What Are Nature Reserves?

In case of the protection of habitats for rare plants and animals in Central Europe, it is not nature that has to be protected in most cases, but it is the habitats that have to be protected against nature.

Nature reserves in Central Europe harbour biotic communities that arose in habitats which have been shaped by continuous human use since time immemorial.

The flower-filled hay meadows and pastures in Central Europe and the grasslands of south-eastern Europe are home to a significant biodiversity, the species communities of which cannot be found anywhere else in the world.

It is areas like these that must be protected, not nature.

The protection of species richness in Central Europe in several cases requires measures which are completely different from those required for the protection of species richness in most other parts of the world.

The airport in Munich, the second-largest in Germany, is home to the largest breeding population of curlews in Bavaria.

The last areas which were successfully protected from impending afforestation are located predominantly on the sites of brown coal and gravel mining and military training areas.

For the purposes of species protection, it would be best if the large brown coal excavation areas were simply left alone and not recultivated after mining operations have finished. However, this is a naive idea, one which ignores the existing legal system.

The brown coal excavation areas could only be preserved as habitats for rare species if the excavators remained in continuous use.

3.5.1
Many 'Nature Reserves' in Central Europe Are Areas That Must Be Protected *from* Nature

We describe nature as how an area would look if humans did not intervene (see Section 3.1.1). If we take the term *nature protection area* literally, it is a protected area the condition of which is such that it appears as if man had not encroached upon it. A biocoenose should be preserved or redevelop itself in a manner that corresponds to the respective local conditions. This means that animals and plants should develop here in a species composition which is determined by the climate, the soil, the topography and other factors, but not by humans. This is what we mean by 'natural conditions'.

Apart from the high Alps and some seashore locations, however, areas like this only exist outside of Central Europe. All the Central European landscapes, forests, rivers, lakes, ponds, wetlands and grasslands have been influenced or even created by man for millennia. Most bodies of water designated as 'nature protection areas', for example, are man-made, such as former fish breeding ponds, peat workings and dammed waters. The dry grasslands and heaths – designated as nature protection areas – were created by cattle grazing.

The term *nature protection area* can hardly be applied to Central Europe, because the region has almost no more natural landscapes. If we closely pursued the objective of protecting nature and preserving it, then there would be almost nothing to protect in Central Europe, because nature no longer exists there. What is designated as a 'nature protection area' in Central Europe is not a nature protection area; the name is used incorrectly. In reality, nature reserves are cultural landscape components resulting from anthropogenic creation. In attempting to find a correct definition, the term *nature conservation* must in most cases be replaced by the term *cultural landscape protection*.

As far as the so-called nature protection areas are concerned, specific areas must be preserved with all the structures, animals and plants that are important to us (humans). This objective is not based on an unspoiled nature (which is of course subject to perpetual change), it is a clear cultural objective directed against nature, because nature would reconquer the protected areas, with the result that the structures, animals and plants that should be protected would disappear. All ecosystems are subject to change under natural conditions. The purpose of 'nature conservation' is to ensure that this anticipated change is halted. To preserve certain animal and plant species, we protect a habitat from a nature that would alter it (without human protection) and which would therefore drive out the species that have to be protected.

If we expand the term *nature conservation* and *cease striving* to create 'nature protection areas', but try to create 'protected areas for near-natural habitats', these, on terrestrial ground, are almost exclusively forest ecosystems in Central Europe. But are we not tacitly approving a wide-ranging impoverishment of flora and fauna, by predominantly encouraging forest ecosystems in Central Europe? A 'natural' forested region such as Central Europe would no longer have heaths, meadows and dry grasslands and would lose most of its species as a result.

Most parts of Central Europe that are rich in flora and fauna are habitats that have evolved through agricultural traditions that have lasted for millennia. It is here that some globally unique biological communities have developed, such as the flower-rich hay meadows and pastures in Central Europe and the grasslands of south-eastern Europe (Schmitt and Rakosy, 2007; Dengler, 2012). Here, we find the actual value of Central European biodiversity, because there are species communities in compositions that can be found nowhere else in the world. Some of the animals and plants that live here are hemerophiles that immigrated centuries ago or even millennia ago from Southern Europe or from the steppe regions of the East.

However, even if we take a radical stand and strive to attain and preserve predominantly near-natural habitats in Central Europe, such a regression to 'pristine' forest habitats would not lead to what could be called an *original state* or *wilderness*. Gradual ecological succession processes of the vegetation would lead to 'wilderness' forests; but these would not be native forests in a composition of tree species which could be called *pristine*. Not every tree species (which would normally grow there due to the soil and climatic conditions) can prevail, but the tree species which do grow there are those that were able to compete against the species that were already resident in the area. Other tree species that could be called native will not have a chance to grow there. Each biocenosis of tree species not only reflects the soil and climatic conditions of a certain location, it also mirrors the past history of this location (Küster, 2008) (see Section 7.1.2).

Nature reserves that are worth preserving in Central Europe harbour biotic communities which have been created through continuous human use since time immemorial. These areas do not possess the character they undoubtedly would have had if man had not existed. The protection of areas that are worthy of preservation in Central Europe faces the task of almost exclusively protecting a species-rich, rural cultural landscape. The preservation of this ecological heritage is our main task.

The main 'nature conservation' problem in Central Europe is that in the second half of the last century the vast majority of species-rich open land fell victim to the dramatic change in agriculture, which was intensified and industrialised. A nature conservation that is solely focused on the protection of natural processes (Prozessschutz) is totally inadequate if we want to preserve the Central European traditional landscape structure with its unique biodiversity for future generations.

Ecological succession processes would swallow up all the open areas in a relatively short time if we left these areas to their own devices. As soon as open spaces are created by floods, storms, fires or by humans, a natural change is launched. Escarpments caused by floods are levelled by erosion again and vegetation regrows on bleak, bare areas created by storms or fires. In just a few years, ecological succession usually ruins the valuable but short-term habitats that were created by earlier destruction. In the case of habitat diversity and (consequently) biodiversity, nature should not be protected or 'preserved' from environmental disasters, but from natural 'reparation' and from human-initiated re-naturalisation and re-cultivation (Plate 7). The most important protection for the preservation of biodiversity is often to protect a destroyed habitat before the immediate onset of restoration to its original state by nature or by humans. Central Europe's soil and climatic conditions are ideal for strong and fast-growing vegetation, that is, a strong and fast ecological succession. In more extreme regions, such as in the colder high north or in the drier Mediterranean regions of Europe, ecological succession does not possess such qualities.

Many 'areas worth preserving' in Central Europe are those that must be protected *from* nature. Instead of the protection of natural processes, Central European nature conservation needs the preserving protection of the landscape against natural processes. This protection is only possible with sophisticated methods and the associated high costs. Old traditional agricultural usage structures must be imitated in order to preserve at least small fragments of the species-rich Central European rural and cultural landscape. Without a protection system that has preservation characteristics, these ecological treasures made by human hands will disappear in Central Europe – and their loss will be largely irreversible. A strong commitment to the preservation of man-made landscapes should therefore be of primary concern for anyone who is interested in the richness of the Central European world of flora and fauna. The protection of species richness in Central Europe includes measures which are significantly different from most of the protection measures of species richness in several other parts of the world (see Sections 6.1.4 and 1.3).

3.5.2
Airports as Refuges for Threatened Species

For almost 3 million years, Central Europe alternated between steppes and forests; so the species that were native to Central Europe for 3 million years alternated between steppe animals and forest animals. At intervals of a few tens of thousands to hundreds of thousands of years, both of these were repeatedly wiped out by the alternating cold and warm periods; they were then replaced by the immigration of animals from neighbouring regions. Currently, we live in a warm period, that is, in an era of the forest. However, that does not mean that the steppe animals are not Central European species. We cannot think in terms of only a few tens of thousands of years. Steppe animals have consistently returned to Central Europe and are waiting 'on the doorstep' to immigrate once again.

Devastating natural disasters such as wildfires, severe storms and flooding have destroyed parts of the Central European forest landscape again and again in the last millennia, and these destroyed areas repeatedly became habitable for the steppe animals for short periods. When man became sedentary in Central

Europe about 5000 years ago, he did the same thing that natural disasters had done, that is, he partially exposed the old Ice Age landscape again through fires and land clearances, paving the way for the steppe animals to return to Central Europe. With the beginning of the Roman colonisation 2000 years ago, man increasingly deforested Central Europe, and with the beginning of the founding of cities in medieval times, Central Europe was almost completely converted into a steppe land over the centuries, and this in turn enabled many glacial species to re-immigrate.

The increasing population in the Middle Ages had an enormous demand for wood, and this removed most forests from Central Europe. skylarks, brown hares, bustards and many other steppe animals spread out in Central Europe, only decreasing in number when the afforestation of the nineteenth and twentieth centuries led to large-scale reforestation in Central Europe. Only certain animal species found a home in the forest; most of the abundant animal wealth in Central Europe was focused on the open lands that remained and to a large extent these were the agricultural and grazing areas.

After the Second World War, these areas were also lost to flora and fauna, because on the one hand, the machining of the agricultural land left no feeding and breeding places for the creatures, and the intensive nitrogen fertilisation on the remaining open lands that were not used for agriculture produced such dense, high vegetation that most of the animals no longer found the steppe landscape to which they were accustomed. In the past 60 years, this has caused the mass extinctions of most butterflies and many species of birds.

However, unlike most other tracts of land, airports are open areas that are not overgrown by bushes and woods and this is why they became a refuge for some endangered species. Airport areas are only covered with sparse vegetation and are therefore nutrient-poor compared with the other remaining open areas. The areas adjacent to and off the runways do not have unbroken, closed-off greensward like other meadows in today's nitrogen age; there are small earthy and sandy strips between the swards. Exposed to the sun's rays, these small open spaces between the turf swards heat up more than the dense, lush grasslands of today's agriculture, so more insects can live there; and they can be more easily seen and caught by foraging birds thanks to the lack of thick vegetation. This is why animals and plants live on airport areas, and even particularly endangered species are often to be found there, although it is an entirely artificial habitat.

Cologne-Bonn airport in Germany plays host to bog asphodel plants (*Narthecium ossifragum*) and there are orchids on the open spaces next to the runways (Gotzmann and Ferber, 2007). These plant species are classified as 'vulnerable' or 'endangered' on the German Red List. In terms of passenger numbers, Munich Airport is the second largest airport in Germany. It is situated around 17 miles north-east of Munich and covers the 'Erdinger Moos' nature protection area. The airport now hosts the largest breeding population of the Curlew in Bavaria, a bird which is classified as 'critically endangered' (Category 1) on the Red List of breeding birds in Germany. Some of these birds even breed right next to the take-off and landing runways (Reichholf, 2010a).

When Munich Airport was built 20 years ago, there were massive protests from the population because the construction was to encompass parts of the Erdinger Moos nature protection area. The protesters did not take into consideration that the Erdinger Moos had mostly been destroyed long before the construction of the airport and no longer deserved to be called a *moor area*. It was not the construction of the airport that destroyed the Erdinger Moos, but the drainage measures carried out decades earlier, plus of course the ingress of agriculture with its fertilisation methods. Most of the typical moorland species had either declined or disappeared even before the airport was built. Those who mourned the passing of the beautiful Erdinger Moos nature area as it was in earlier centuries should have taken part in protests and demonstrations against agriculture, not against the construction of the airport.

For the Erdinger Moos, the term *moor* was already inapplicable in the 1950s and 1960s, as in almost all locations in Central Europe. Since that time, 'moor' has been used as an old field name, for example, in the northern foothills of the Alps and in north-western Germany. The term was wrongly applied to birch forests; however, birch forests are not moors. True upland moors are extensive sphagnum areas with pools of water and bulten (raised domes of peat and peaty mosses) where an open view to the far horizons is still possible. Birch forests where an open view into the panorama of the landscape is obstructed in every direction by stands of the trees, are nowadays often called *moors*, but this is wrong.

However, in spite of the rare species that breed on them, airports themselves are certainly not aesthetic. They are ugly and noisy, polluting the cleanliness of the air and producing CO_2 emissions. The expansion of airports and the building of new runways means that other habitats have been destroyed (often forests). All these are serious reasons why airports and their expansion projects are rejected by a majority of people. However, these are reasons that involve environmental and nature protection, not species protection.

3.5.3 The Re-naturalisation of Open-Cast Mining Sites – A Destruction of Rare Species' Habitats

Grasslands are among the most important insect habitats in Europe and one of the few habitats in Europe that have achieved worldwide importance (WallisDe-Vries, Poschlod and Willems, 2002). Half of all European butterfly species are inhabitants of the grasslands (Van Swaay and Warren, 1999). However, the steppe habitats of Central Europe are not stable in the current climate and soil conditions. The climate is too warm (and above all too humid), and atmospheric nitrogen fertilisation makes the soil too fertile to maintain a stable steppe landscape; ecological succession immediately overgrows the steppes with bushes. Sand and grass surfaces quickly become shrub landscapes, and at the end of the ecological succession, everything is closed forest, forever destroying irreplaceable habitats. To stop this ecological succession, man must intervene with technical equipment. If we do nothing, all the open land will be overgrown after only a few decades. Central

Europe under its recent climatic conditions does not tolerate deserts, steppes and dunes. These habitats disappear in a natural way, as they did when the steppes of the Ice Age had to give way for the immigrating forest.

For centuries and even millennia, many open spaces with poor soils were kept free of trees due to the enormous demand for wood, cattle browsing and the removal of the herb layers; but these areas are now threatened by scrub growth and forest. The current prosperity in Europe has made the removal of the upper layers of vegetation-covered turf, the pulling up of bushes, the extraction of firewood from the forests and the herding of sheep and goats superfluous. All the centuries-old human interventions in nature that created open land have vanished in the mists of time.

The fight against the current eutrophication of the landscape seems hopeless, because a profit-oriented agriculture cannot be scaled down to the technical level of earlier centuries. The struggle to preserve open land would seem to be futile. However, brown coal open-cast mining excavations, the gravel pits on the land and the establishment of military training areas have proved that the preservation of species which have adapted to open land is still possible, even at the present time, if massive technical landscape engineering is carried out in the countryside. During the last half century, eutrophication has destroyed almost all the open areas, but industry and the military have succeeded in preventing open spaces from being overgrown by a surfeit of lush vegetation and the ever-threatening afforestation (Kunz, 2004, 2013; Kunz *et al.*, 2014).

The mass extinction of the species that had already adapted to barren open areas would have been even more drastic if industry and the military and road building authorities had not carried out large-scale encroachments upon nature at the same time as scrub growth and forestation were destroying open land; these encroachments created new, extended areas that have little or no vegetation. Thanks to brown coal and gravel pit excavations and the creation of military areas, the habitats that had been destroyed by afforestation and eutrophication in the twentieth century were newly created. The open land species were given a new home, in which many rare or vanished species were able to resettle.

The general public, however, is not aware that these encroachments upon nature perpetrated by industry, the military and road building authorities which have created new areas for endangered species by ruining nature. Open-cast mining excavations are regarded as 'ghastly lunar landscapes', and the thought that these could actually be valuable refuges for rare species is completely foreign to most people. The reason for this is that most people equate the terms *nature* and *species* richness with one another; they do not understand that an 'intact nature' is not the sine qua non for the existence of valuable species and that a nature which has been destroyed can make a wonderful habitat for rare species.

In recent decades, no less than 18 orchid species that thrive on nitrogen-poor soils ingressed into the large-scale, open-cast mining excavations in the Rhenish brown coal mining region to the west of Cologne in Germany (Albrecht *et al.*, 2005). This unexpected wealth of rare orchids in the Lower Rhine Basin between the cities of Cologne, Aachen and Mönchengladbach can even compete with that of the dry grasslands in the northern Eifel limestone basins, although the soils on the excavated and re-banked areas of the brown coalfields are not calcareous. The orchids grow instead on soils brought up from the depths by the mining process. These are unfertilized, relatively nitrogen-poor soils on which the orchids can grow without being crushed by the competition of nitrogen-loving species. It is due to the open-cast mining excavation areas ranging from Cologne to Aachen that a number of endangered bird species that had largely or entirely vanished from these areas have returned, such as the quail, the red-backed shrike, the wood lark, the tree and meadow pipit, the wheatear, the corn bunting and the eagle owl. These birds breed on barren land or on the escarpments of the excavation pits (Plate 1).

All these species would not occur in these areas if coal mining had not created open ground. The problem, however, is that there are very few ways to preserve these excavated areas when coal production has ended. Only massive technical engineering will save species richness in the long run. Brown coal open-cast mining led to the surprisingly fast settlement of rare species on the excavated areas; but both the areas and the species diversity will vanish just as quickly if these areas are allowed to 'return to nature'.

For the purposes of species protection, it would be best if the large brown coal excavation areas were simply not recultivated. However, this is a rather naive idea, and one which ignores the existing legal system. Almost every square yard of land taken up by the mines belongs to someone or other and the owners of these areas claim their property as soon as coal production has ended (Kunz *et al.*, 2014). The owners have of course not made their land available for species protection.

The legal framework, which led to the approval of coal mining, dictates that the land must be recultivated immediately after the end of coal production. This primarily means that land will once again be available for agriculture. This form of land recultivation results in the immediate disappearance of almost all species. Other areas of the disused pit area made available for recultivation are usually reserved for afforestation and this results in the disappearance of all the species which have a high protection value, because only species which are already common and widespread will colonise these afforested areas (Plate 7).

However, there are some remaining areas that neither have to be returned to agriculture nor used by the forestry sector. It is here that species protection can find the scope to take measures that will preserve the species richness of the excavated areas. However, these areas are usually also 're-naturalised'. This entails the replanting of the wide-open stretches of land, the filling-in of the depressions and the flattening or levelling of the escarpments that are so valuable for the species.

'Re-naturalisation' is a problematic concept; in most cases it does not support species richness, and the areas that are re-naturalised are of course not 'nature'. These areas are only given the name 'nature'. They are wholly artificial park landscapes, which are regarded by the majority of the population as being aesthetic and pleasant; but the rare species which used to occur on the nitrogen-poor soils and the escarpments cannot live in these 're-naturalised' areas, and they vanish almost without exception (Plates 5 and 7).

We could adopt the standpoint that we should completely dispense with re-naturalisation and simply leave the nutrient-poor mining land that agriculture and forestry has not taken over; we might even believe that this would save the endangered species. However, this is not possible for ecological reasons. There are natural laws that conflict with the idea that certain open-cast mining areas should be preserved by not re-cultivating them. In Central Europe, which is influenced by the Atlantic climate, natural ecological succession will very quickly cause even the most nutrient-poor sand and gravel sites to be overgrown with scrub encroachment and woods and this process occurs fast.

In order to permanently preserve the special site conditions of the mining areas, we must engineer a massive crackdown on the dynamics of nature. Without technical habitat engineering, the specific characteristics of the excavation areas will soon vanish, and the habitat of specialised animal and plant species will be quickly ruined. The rare species of open-cast mining sites only immigrated into these areas a few decades ago because brown coal mining was realised. To preserve these species, the excavated landscape must be 'reset to zero' by technical means at regular intervals – and this means the continued use of the excavator. This would require a lot of funding; but the official nature conservation bodies prefer to use their funds to preserve 'pristine' nature and to found national parks. Species protection does not have enough support among the population and consequently has insufficient backing in the nature conservation organisations; and the result of this is that it will be suppressed by the nature conservation associations for some time to come. From the standpoint of nature conservation, the open-cast mining site will probably continue to be classed as a disruptive intervention, one which should be eliminated with as few consequences as possible. The public is right in saying that the 're-naturalisation' of these mining areas repairs earlier landscape destruction; but the people usually do not possess sufficient insight to understand that it is re-naturalisation which initiates the disappearance of rare species (Kunz, 2004) (Plates 5 and 7).

3.6
The Protection of One Species Means Death for Another

Not every species in an area can be protected at the same time.

Protection of natural processes in a habitat is mostly unsuitable for species protection. The protection of certain species requires the preservation of a habitat in an unchanging target state. This means that a habitat must be constantly manipulated by continued care and maintenance measures.

Animals were classed as 'useful' or 'harmful' in earlier times.

In earlier times, nature conservation organisations used to find it easier to explain to their members why a particular species should be protected just because it was so useful to humans.

The alleged usefulness of tits and ants.

In some cases, nature conservation which targets the protection of endangered species would have to destroy certain other species in a habitat to eliminate them as competitors of other species.

A nature upon which man no longer encroaches will in many cases inevitably become very species-poor in Central Europe.

The general public shows more toleration for encroachments upon plant biological communities than for interventions in animal biological communities.

3.6.1 The Competition Among Species for Resources

Nature conservation and species protection are faced with an almost unsolvable dilemma in Central Europe today. On the one hand, we have almost no more nature, because all the landscapes have been man-made for millennia. This of course means if we wish to preserve many species, nature simply cannot be left to fend for itself, because then the species which have adapted to the anthropogenically formed landscapes will be endangered or even wiped out. On the other hand, the preservation of man-made landscapes requires technical encroachments upon nature, and these have different implications for different species. Some species benefit from such encroachments, while others are suppressed. The species compete against one another for the resources offered by the landscape.

In the case of plants, it immediately becomes clear how quickly one species can overgrow another when certain human interventions such as nitrogen deposition, humidity regulation, mowing and grazing are carried out. Each of these human interventions benefits some species, but weakens others. Most inferior species are not directly harmed by these interventions and would also profit from more nitrogen, more moisture etc., but they lose out to the competition against a few dominant species which profit to a greater degree and actually crush the others.

When a species becomes too numerous, other species suffer. Once a species finds conditions that allow it to proliferate, its proliferation is at the expense of other species in the same habitat. The other species lose their air, water, light or food and are pushed back. Specific living conditions are often optimal only for certain dominant species. Other species can also cope with these living conditions; however, they do not benefit to the same extent as certain dominant species, so they cannot reproduce as efficiently. The result of this is that the disadvantaged species become rarer, because they simply cannot survive as successfully as some other competing species, although they can definitely live under the given soil, temperature, humidity and other conditions.

Every gardener knows that it is not necessary to deploy life-preserving measures to promote certain plant species; in most cases their competitors only have to be eliminated. One important task of the gardener is to destroy the 'weeds', to enable the preferred species to thrive. If we transfer this experience to the necessary management measures in a 'nature protection area', then the initial question we must ask ourselves is: 'Which species should be encouraged to grow in this area through our measures and which species should we more or less ignore?'

Species protection faces the task of having to make a trade-off for every single area, deciding which species need to be protected there at the expense of which other species. This state of affairs is not general knowledge, and the nature conservation associations tend to keep it to themselves. Instead they attempt to give the impression that they can protect all species in a certain habitat at the same time. That is impossible and it contradicts a basic principle of ecology.

Established nature conservation associations opt more for the protection of natural processes (Prozessschutz), meaning that certain areas are left to themselves and natural ecological succession can proceed unhindered there. Protection of natural processes is the non-intervention in the natural processes of ecosystems. As a rule, we cannot foresee the ultimate condition of a habitat if its natural-dynamic processes are left to themselves. The protection of natural processes strategy of nature conservation is not suitable for species protection, because the protection of certain species requires the preservation of a habitat in an unchanging target state. In other words, a biotope must be constantly manipulated by means of continuous care and maintenance measures. Each 'target state' is oriented on very specific species that are to be supported in the habitat in question, and this state deliberately neglects other species which the measures do not benefit.

We cannot protect all the species in an area at the same time. Nature conservation measures necessitate a preliminary decision about the species that are to be protected in an area. A transparent nature conservation approach would be to put up a notice board in every nature protection area, naming each species that is protected in that area, and, equally, naming each species that is not explicitly protected there.

This problem, however, is difficult to solve because it demands underlying guidelines that define exactly what is 'worthy of protection'. The nature conservation associations do not make any serious attempts to formulate such guidelines and adopt them in their statutes. In nature conservation today, it would seem that one of its most difficult tasks is to invoke a binding objective. Instead, the nature conservation associations suggest that it is possible to protect all species in a particular habitat and that it is not the task of nature conservation measures to reduce or to eliminate one or more selected species from that particular area. However, we can never hope to create ideal conditions at the same time in one habitat for all the species living there. The protection of one species often leads to the death of another. This sounds harsh, but it is in the nature of things that species compete with one another for the resources in a habitat.

3.6.2
The Earlier Assessment of Species According to Their Usefulness for Humans

In earlier times, there were in fact guidelines which assigned different 'worthy of protection' values to species. These guidelines were in accordance with anthropocentrism, which stamped the worldview of past centuries on many areas of life and can be traced back to Old Testament sources. It was a worldview in which man was the Lord of Nature. It was the hubris that man was the most important creation of God and that the earth was the centre of the world. It was the concept of a nature that had only been created to benefit man. In books that were published in the first half of the twentieth century, we often find precise information about how useful or how harmful certain animal species were, and the extent to which certain species were to be protected because of their usefulness, and how other species were to be shot or destroyed because of the harm they caused.

The ethics underlying this worldview were simple and rather primitive. As soon as an animal had been classed as a competitor for human food, it was referred to as *harmful* – so all the carnivorous animals that ate the same animals as man were classed as harmful. These were the birds of prey and the carnivorous beasts which were classed as robbers because they stole food from man. The same was true for fish eaters such as cormorants and herons. The animals that hunted man's pets and devoured field and garden plants, such as foxes, rabbits, rooks, cranes and many larvae were also generally classed as being harmful. In contrast, exactly those animals that ate the harmful animals were classed as being useful, such as tits and ants because they ate caterpillars and beetle larvae, and ladybirds because they ate aphids. The cycle of such habit-forming thought was continued by classing the animals that ate useful animals or their broods as harmful, for example, ravens, crows and magpies. This balancing of useful and harmful soon attained mannerist traits when individual species were weighed against one other in percentage terms for their usefulness or harmfulness; the sea eagle, for example, was classed as being mainly harmful (useful to harmful = 8 as against to 12). However, the last few sea eagles in Germany were to be protected as 'natural monuments' (von der Nordmark and Zukowsky, 1936).

This worldview of usefulness, which justified the protection or persecution values of animal species from the standpoint of how useful or harmful they were for man, was even supported by the former 'DBV (now renamed NABU, the Nature Conservation Association of Germany) in the first few decades following its founding. Compared to nowadays, nature conservation organisations used to find it easier in bygone decades to explain to their members why a particular species should be protected just because it was so useful to humans. In the light of our present-day nature conservation and environmental awareness, it has become much more difficult to find any justification for the protection of certain species.

A blinkered fixation on the provision of nest boxes and the protection of anthills specified the objectives of nature conservation for decades, although conflicts had already become evident in the 1960s. The drastic disappearance of butterflies, ongoing since 1960, has been associated with the promotion of insect-eating birds by means of the excessive installation of nest boxes and the exaggerated winter feeding of tits. Excessive protection of anthills resulted in a threat to the caterpillars of many blue butterfly species (Ebert and Rennwald, 1991b).

By the 1960s and 1970s, the worldview of anthropocentrism could no longer be upheld in species protection and nature conservation. The simplicity underlying this worldview faltered and the nature conservation associations had to reorient their objectives. To be on the safe side, the German species name 'Fischreiher' (Fish Heron) was renamed 'Graureiher' (grey heron) and the 'Raubvögel' (raptors) were renamed 'Greifvögel' (birds of prey) to mitigate the age-old notion of competitors and food thieves.

As controversial as it always was and still is today, ecological awareness of the importance of species in the 'balance of nature' replaced the centuries-old belief that nature was there to serve man; so man, the only measuring scale for the importance and the purpose of creatures, was now relegated to the background. This transformation in awareness was indeed a paradigm change. In Germany, it was expressed in the establishment of BUND in 1975, and the 1990 renaming of the DBV (which had existed since 1899) to NABU.

Nature conservation now wanted to make the public aware that species were not there for the benefit of humans, that they fulfilled a role in the 'balance of nature', maintaining the ecological balance. The basic concept was that the most effective means for the protection of species was habitat protection; and as logical and correct as this was, the decision-makers within the nature conservation organisations swiftly commenced to place environmental protection and species protection on an equal footing (see Section 3.3.1). As soon as a species became more common and it evinced a positive development trend, this was seen as an indication of a healthy environment. The reappearance of the stork as a breeding bird on the Lower Rhine in Germany was touted as a success for environmental protection, which had ensured that the water was once again clean.

However, policies which place environmental protection and species promotion on an equal footing are very often ineffective. To verify this, we need to only look at the many examples of animals that live on waste and are dependent on pollution (Meyer, Eilers and Schnapper, 2003; Orros and Fellowes, 2015). There are many species which are neither indicators of a clean environment, nor are they necessary for maintaining an intact ecosystem (see Section 3.2.3).

The problem also lies in the fact that there are different ecosystems. A designated nature protection area can either be left to its own devices or we can employ care and management measures to prevent bushes and trees from gaining the upper hand there. Very different ecosystems develop depending on which route we now choose. If we take the standpoint that the protection of certain species is guaranteed by the ecosystem, the decision as to which ecosystem we should deploy in a specific area is still left open on the table.

Since the 'protection worthiness' of certain species could no longer be substantiated by their usefulness to man, the second argument for the protection worthiness of species can now be examined, namely to understand the meaning and purpose of species in the maintenance of an intact ecosystem. The questions still remain as to what an intact ecosystem actually is, and whether or not we can implement different ecosystems in one location. A comprehensive theory is still lacking, a theory from which we could deduce the reasons why certain species are worthy of protection.

In many cases, it again comes down to the anthropocentric background of species protection, but with a completely different rationale today than in earlier times. We do not need species because they bring us economic benefits; we need them for aesthetic reasons, because they enrich our homeland (see Section 4.3.1).

3.6.3
Regulatory Intervention in Animal Biological Communities – Saving Endangered Species by Eliminating Their Competitors

In many plant identification books, you can read that a large proportion of the Central European orchid species can be found on dry grass, 'mostly on calcareous soils' (Fitter, Fitter and Blamey, 1974). These orchids do not need calcareous soil to survive; they grow mostly on calcareous soils in Central Europe because the competition on these soils is weak. On calcareous soils the orchids have space to live, since they are not forced out by nitrogen-dependent plants.

When other factors ensure that the orchids are not supplanted by other species of plants, many orchid species thrive on soils that are not particularly calcareous. In past decades, on the large open-cast mining excavations in the Rhenish brown coal area to the west of Cologne in West Germany, great masses of earth were brought up from the depths with the coal and spread over the ground. These soils had not been fertilised and therefore had low nitrogen content. The plants which are dominant on nitrogen-rich soils, especially grasses, have not yet been able to settle there in sufficient density, so the orchids have a good chance to grow in a competition-free environment; they would not have had this opportunity on fertilised soils, because they would have been overwhelmed by the grasses. In the Lower Rhine basin between the cities of Cologne, Aachen and Mönchengladbach, 18 orchid species have settled in recent decades and are thriving there (Albrecht *et al.*, 2005). Destruction of nature in the mining areas created these orchid habitats, and they have attained a value which is comparable to the orchid-rich calcareous grassland of the northern Eifel, although the soils on the excavation areas of the former brown coal mines are not calcareous.

The orchids of the brown coalfields in the Lower Rhine Basin are neither jeopardised by plucking hands or trampling feet, but by the fact that nitrogen-dependent plants will stifle them in the foreseeable future. It would not help to keep the locations of these precious orchids secret or to give the land upon which they grow the status of nature protection areas and then prohibit access to them. The only way to preserve these orchids is if we regularly uproot the competing plants. Nature conservation which targets the protection of endangered species must selectively destroy certain species in this habitat to eliminate them as competitors of other species – so species protection can kill or drive out targeted species. However, we should view this approach with caution, because ecological relationships usually are complex. There is a definite lack of knowledge in this respect, so the elimination of competing species by humans is often the wrong approach to conserving endangered species.

The view that man should not interfere in principle cannot stand up against rigorous and consistent reasoning. Of course we can leave nature to its own resources, but then many species will disappear in Central Europe, and that is something that no advocate of species protection can possibly desire. A nature upon which man no longer encroaches will inevitably become a species-poor nature in Central Europe (see Sections 6.2 and 7.1.4). In today's nitrogen-rich era, the long-term preservation of an orchid habitat in Central Europe requires continued human intervention in species composition. The atmospheric nitrogen alone means that we cannot preserve many orchid habitats by leaving them alone. That would be irresponsible.

Every gardener knows that weeds have to be pulled up by the roots to help a desired species thrive. This task is not limited to plant communities. It also arises in the preservation of endangered animal species. Meadow-breeding birds, for example, are endangered almost everywhere in Central Europe today, and a major problem faced by species protection associations is to prevent their young from being eaten by predators. Meadow waders in particular fall prey to foxes (Barkow, 2010). The nests and offspring of the last great bustards in Central Europe are also threatened by ravens (Dornbusch, 1996). We can keep foxes out of breeding areas using fences, but ravens attack the nest and the young bustard chicks from the air. There seems to be no other solution than to shoot the predators if we are to save the great bustards.

This sounds provocative for the species protection of the future. Nevertheless, we should remember that non-intervention on our part will mean that the ecological control of a biological habitat will be left to itself. This is 'natural', but undesirable at least in the cases where the preservation of endangered species in anthropogenic biotopes is at stake.

We must first exercise caution and restraint before making any decision to save endangered species by killing the competing species. This, however, is only an initial recommendation on how we should perhaps proceed, because the consequences of such interventions are complex and because of this complexity, the environmental effects of a targeted killing of animal species can often not be predicted. In principle, however, unwanted creatures should be treated like unwanted plants and removed from nature conservation areas. Species protection ethics do not encompass any ban on driving out or killing competing animal species. A ban on killing animals would probably be a must for the ethics of animal protection, but not for the ethics of species protection.

There is an inherent contradiction in the principles of those who protest against the killing of animals as a regulatory intervention for the purpose of species protection. They do not tolerate any intervention in the animal biocenoses. (Author's note: Animal biocenoses, or animal biological communities, are known as *zoocoenoses*, which are biological communities of different animal species in a defined habitat.) However, these same protesters do tolerate intervention in plant biocoenoses. (Author's note: Plant biocoenoses are phytocoenoses, which are biological communities of different plant species in a defined habitat.) The destruction of plants, such as the removal of bushes from dry slopes to promote butterflies, is accepted as a nature conservation measure. Most remarkably, however, killing animals to promote certain other animal species, such as the shooting of ravens and even sea eagles in the great bustard breeding areas, is generally not accepted by the public. This makes it clear that in cases like this where a conservation measure is defined as species protection, we are actually dealing with animal protection. Otherwise, it would be impossible to explain why any difference was made between animals and plants. In the case of plants, there are no ethics which would be an analogue for animal protection, because most people assume that a plant suffers no pain.

The hunters who are frequently attacked by nature conservationists have always acted according to the 'controlled killing of animals' principle, and the official-sounding word 'gamekeeping' is used to sell this principle to a sensitive public. The gamekeeper has however fallen into disrepute in recent decades – and rightly so. First of all, the much-publicised biological importance of gamekeeping measures (which are supposed to preserve the health of animal populations) withered under the pressure of scientific investigations. Second, gamekeeping was pushed forward too much as a seemingly rational argument, in an attempt to cover up the true motives that lie behind the hunt, namely the sensual pleasure that humans get when they gratify their hunting instinct. The gratification of instincts is often veiled behind the respectable label of 'performance of duty'.

Biologists and ecologists have often condemned gamekeeping measures as being unjustified, denouncing them as 'shotgun ecology' (Bezzel, 1995). In fact, all the attempts to justify the hunting of hawks, crows and magpies as a measure to maintain a good balance among songbird populations failed. The ornithologist Einhard Bezzel aptly expressed this in the following sentences: 'It has never been proven that hunting and shooting is a sustainable regulatory mechanism, although representatives of the shotgun ecology never tire of vastly overestimating the importance of the finger on the trigger of the gun' (Bezzel, 1995).

3.7
Who Kills More Birds – The Fowlers of Western Europe and the Mediterranean or Our House Cats?

Fowlers in the Mediterranean area kill more than 100 million birds every year.

Domestic cats in Europe and the United States kill several 10 billion birds annually.

Free-roaming cats perpetrate major animal cruelty acts on breeding birds, acts which are hardly perceived by the animal protection associations.

There are more than 10 cats for every 100 residents in the industrialised countries of the West.

In the United States, at least 3 billion birds fall prey to the great number of cats every year.

Every 10th bird is killed by cats in the gardens of the industrial countries of the West.

Mediterranean fowlers kill species (en mass) that are under strict protection in Germany and other Central European countries, and this raises the question, why do we protect these species at all?

In France, 1 in 50 inhabitants is a hunter or fowler.

Hunting bans are perceived as a throwback to feudalism, which was eradicated by the French Revolution.

Some species of birds are only so rare, because they are so afraid of humans.

The extremely high rate of reproduction of organisms: every animal and every plant proliferates so intensively that life on earth can only function because the vast majority of organisms die relatively quickly as soon as they are born.

The hunting and catching of specimens can only decimate species if these species are already on the decline or if the numbers shot are particularly high.

The number of birds captured or killed by fowlers and hunters from Spain to France and Italy and from the Adriatic Sea to Egypt is in the hundreds of millions (https://www.nabu.de/tiere-und-pflanzen/voegel/zugvogelschutz/aegypten/15708.html), but the number of birds killed by domestic cats in Europe and the United States is 100 times higher than that. In the affluent western countries, cats kill tens of billions of our feathered friends every year (Lingenhöhl, 2010; Dämmgen, 2013).

These figures are staggering. Many factors have to be considered if we want to estimate the full extent of the billion-fold slaughter of birds. This concerns species protection as well as animal protection. To evaluate the impact of the slaughter of birds on species protection, we must differentiate between the bird species that are bitten to death by cats in the gardens and those that are shot by fowlers or trapped in nets in Western Europe and the Mediterranean countries. There are substantial differences between the species that are killed. It should be noted that cats mostly kill the common species of birds in home gardens, while the rare species fall prey to the fowlers in Southern Europe. The lapwing and the common snipe are shot down in tremendous numbers by French fowlers; the Maltese shooters kill the osprey and the golden orioles, and the red-backed shrike and the wrynecks end up lying on the stalls of the Cairo markets. These are species, the protection of which in their Central European breeding areas costs a lot of time and money. The harm caused by cats is therefore in a different league to the harm caused by fowlers.

The slaughter of birds that is tolerated by most people and takes place every year right before their eyes (in their gardens) is probably less disastrous than the harm fowlers do to the populations of migratory birds that cross the Mediterranean. This is why fowling is denounced in the media. From a species protection standpoint, the only important questions are whether or not this mass slaughter reduces the number of birds in the long run and whether or not the reproduction rate of the birds can compensate for these huge losses.

From an animal protection perspective, however, the situation looks completely different. The cat, thoughtlessly let out of the house by its owners and stalking through gardens ready to pounce and rend, perpetrates a significant act of animal cruelty on breeding birds. Anyone who hears the strident and terror-filled clamour of blackbirds on an early spring morning will know that the adult birds have discovered the cat and are trying (mostly unsuccessfully) to drive it away with their alarm calls. Young blackbirds usually leave their nests a few days before they can fly. They then sit on the ground under the bushes, where they are fed by the parents and are easy prey for the cats.

However, many young blackbirds also die in their nests if the garden is regularly visited by a cat. The parent birds who want to feed their young can sometimes suffer stress and anxiety for hours. During the hours in which the cat stalks its prey in the garden, the young birds cannot be fed and warmed by their parents. Investigations have shown that the adult birds often start feeding their young again normally an hour later, when the cat has long since departed to wreak its carnage in neighbouring gardens. The growth rate of young birds in the nest slows down when the parents are kept from the nest for too long, and the number of young fledgling birds decreases. The presence of a cat also attracts crows, magpies and birds of prey that recognise the blackbirds' alarm calls, which more or less inform them where a nest with young birds can be found (Beckerman, Boots and Gaston, 2007).

The degree of cruelty to animals caused by free-ranging cats is truly enormous. The animal protection associations keep their collective heads down in this respect, to avoid any serious actions against cat ownership. This attitude brings the associations more members; it also goes down much better with the members if the policy is to protest against the animal experiments which are carried out at scientific institutes or which serve commercial purposes, rather than taking measures against the exaggerated cat ownership. The lobby of the animal experiments supporters is significantly less powerful than the lobby of the cat lovers. Approximately 10 million people in Germany have at least one cat in the household; every eighth person in Germany owns a cat (http://de.statista.com/statistik/daten/studie/181168/umfrage/haustier-anzahl-katzen-im-haushalt/).

While the animal cruelty caused by free-roaming cats is indisputable, it is much more difficult to assess whether the many birds killed by cats actually reduce the populations of the species; this may not be the case. However, it is absolutely incomprehensible that the worthy garden owner feeds the birds and provides them with nest boxes, then allows his or her cat to roam freely during the bird breeding season. Such contradictory behaviour can only be explained if it is openly admitted that feeding the birds and keeping cats only serve our own satisfaction and personal pleasure. Any suggestion that it is a good deed to hang nest boxes in the garden and carry out winter feeding of birds is absurd.

3.7.1
How Many Birds Are Killed by Cats?

The cat was domesticated in Egypt as a pet more than 4000 years ago and reached Central Europe in the early Middle Ages. Free-roaming cats are gifted, highly effective bird hunters. While cats left to roam on islands are responsible for the extinction of many species of birds, the damage that cats wreak on the mainland is controversial. Although several tens of billions of birds are killed every year in Europe and the United States by domestic cats, this mass destruction is offset by the enormous reproduction rate at which the birds reproduce. Blackbirds, for example, breed in Central Europe up to three times a year, laying 4–5 eggs each time. A single blackbird pair would therefore normally produce up to 15 young per year. This would very quickly lead to overpopulation if no young birds died before they reached adulthood; and most of the young birds are probably eaten by cats. Perhaps cats are even a counterbalance against the overpopulation of certain garden birds; but there is no experimental evidence that would tell us to what extent the number of blackbirds would actually increase if there were no cats.

These 'death numbers' are frightening. The number of cats in towns and villages and the number of birds caught by them are almost unbelievably high. The United States has a population of 300 million people, and more than 100 million cats live there, of which more than half are ownerless (Lingenhöhl, 2010; Dämmgen, 2013). In Europe, the cats' population density is somewhat lower; the United Kingdom, for example, has 64 million inhabitants and 9 million cats, while the number of cats in Germany, with its 81 million inhabitants, is 8 million. The many cats in the United States kill at least 3 billion birds, while in the United Kingdom around 30 million birds fall victim to the feline killers. These figures are of course very inaccurate and there are many unreported cases, so the actual figures are likely to be higher, quite apart from the many birds that are only injured by cat bites but subsequently die from infections. The (approximately) 3 billion birds killed by cats in the United States contrasts with a total bird population of around 20 billion; so more than 1 in 10 birds fall victim to cats.

Cats are highly motivated predatory animals; even if they are well fed at home and their stomachs are full, they will still kill birds. Their menu includes not only sparrows, tits, finches, blackbirds and starlings, but also larger birds such as pigeons, ducks, woodpeckers and rails. Due to the high reproductive rate of birds and the likelihood that slower and less wary birds are killed rather than vital and healthy birds, it is possible that the numerically common bird species can maintain their populations and not become rarer, despite being killed in great numbers by felines.

However, it must be said that there have been significant changes in recent decades. Due to the fact that most agricultural areas can no longer be inhabited by their species, villages and towns have become increasingly important for the birds. Surveys in British cities have shown that cats kill almost half of all house sparrows, robins and dunnocks (Baker *et al.*, 2005). These statistics can probably no longer be ignored (Beckerman, Boots and Gaston, 2007). House sparrows and starlings are now among the species of birds that are decreasing in number from year to year.

However, cats that travelled to remote islands introduced a whole new dimension of danger to birds. The cats proliferated and exterminated the islands' native species, which were not attuned to seeing cats as mortal enemies from which they should immediately flee; the natives, after all, had lived for millions of years in safe seclusion. In just a few decades on the South Atlantic island of Ascension, cats killed nearly 1 million sooty terns. A total of 33 bird species have been exterminated by cats on the islands of the three oceans in the last few centuries.

3.7.2 The Killing of Migratory Birds through Fowling, from France to Egypt

Many migratory birds of Central, Northern and Eastern Europe and Western and Central Asia migrate to Africa, gathering on the coasts before flying over the straits and resting on the islands of the Mediterranean. They occur in large numbers and are often exhausted by their long journeys. Many human residents of the Mediterranean countries expect them year after year; but most of these residents are not bird watchers with binoculars and cameras, they are fowlers with guns and nets. Fowling is succoured by deep human instincts, but also by traditions that are thousands of years old, laboriously achieved hunting rights and lucrative earnings on the gourmet markets. Tackling old traditions and passionate instincts is difficult, and the authorities and volunteers (who want to cut down on the targeted killing of migratory birds) are faced by an uphill battle; they are even subjected to death threats on occasion.

The largest bird trapping system in the world awaits the millions of migratory birds which have finally managed the exhausting flight from Greece and Turkey across the Mediterranean and are searching for a resting place before flying on. An almost seamless 435 mile-long net now extends along the coast of Egypt from the border with Libya to the Gaza Strip. The birds cannot avoid it and can only fly over it with great difficulty. Every year this giant net catches tens of millions of migratory birds, which are then offered as delicacies in the markets and restaurants of the country.

Bird trapping on the Egyptian coast has always existed. What is new, however, is the enormous size of the trapping equipment, from which the majority of the migratory birds cannot escape. Although Egypt has signed the international agreement on the protection of birds and there are legal requirements for minimum distances between the nets and maximum net heights, no one adheres to these laws, and infringements are not prosecuted.

In the United Kingdom and in Central and Northern Europe, thousands of people (volunteers for the most part) strive to protect the last populations of the wheatear, the wryneck, the golden oriole, the red-backed shrike, the nightjar and the hoopoe, and large sums of money are also earmarked for these species' preservation. These are the species which land on the market stalls in Egypt and ensure the slaughterers a lucrative business. On a single day, tens of thousands of red-backed shrike are sold for €3 each and thousands of quail are sold at a price of €5 per bird (https://tierfilm.wordpress.com/2013/04/25/zugvogelmord-in-agypten/). For Central Europeans, it is a valid question whether or not it is worth preserving the populations of the red-backed shrike, wheatear, common redstart, wood warbler, hoopoe, great reed warbler, oriole, quail, cuckoo and turtle dove in their own countries, if these birds are going to be mercilessly killed in Egypt anyway.

The political chaos in countries such as Egypt also reinforces the trend towards lawless actions. In a country where a slight insult to the Prophet may end in beheading, there are seemingly no pangs of conscience over the slaughter of millions of migratory birds, the origin of which no one cares about.

Although Egypt currently holds the sad record for the slaughter of birds, it is also common practice in other countries, several of which belong to the European Union. France, Spain, Italy, Malta, Croatia and Cyprus are just a few examples of countries where fowling is ongoing, although they have all signed the European Birds Directive, albeit with some exemption clauses. The number of birds killed annually in these countries is as follows: 2.5 million skylarks in France, Italy and Malta, 1.5 million turtle doves in Spain and in France alone, half a million lapwings, more than a quarter of a million common snipe and 1 million woodcocks (http://og-bayern.de/?page_id=3116). In Germany, the common snipe is critically endangered (Category 1 in the Red List), while the lapwing and turtle doves are still in Categories 2 and 3 (as yet), but their numbers have declined fast in recent years.

There are 1.3 million hunters and fowlers in France, more than in any other European country (http://www.komitee.de/content/aktionen-und-projekte/frankreich). While the minimum standards prescribed by the European Union for the protection of migratory birds are increasingly respected in the EU countries, the hunters of France fight doggedly against any restriction. The practice of hunting for everyone is a prominent law for which the citizens fought hard during the French Revolution, because in previous centuries, it had been the privilege of the nobility, and citizens were prohibited from pursuing their passion for hunting. Much like the French who fought for democratic freedom and the right for everyone to hunt, the Germans also prevailed against the rule of the aristocracy, in that all citizens now have a legal right to enter the forest. In earlier centuries, the forests were the domain of the nobility, as was hunting.

Both rights are valuable achievements of democracy; but both are often in conflict with nature conservation and species protection today. The prohibitions are seen as a throwback to earlier authoritarian times, although today these laws have to be made legal for very different reasons. In France, if a law against the killing of the lapwing and the common snipe could be passed it would be a real step in the right direction; but the French citizen sees this as a throwback to the old feudal state, not as a species protection measure that is absolutely necessary. In Germany, it would be ideal if access to the forests where the black stork and the crane breed could be prohibited, but this is also difficult to enforce legally, because it would be a restriction on painstakingly achieved civil rights.

France has classified 64 species as open game (wildlife that may be hunted) and allows hunting seasons up to 7.5 months in the year; the country consequently holds an unenviable record within the European Union. These very long hunting seasons partially overlap with the ongoing breeding seasons of birds. However, France is above all a land of passage and of wintering grounds for many important bird species. A lot of effort went into protecting snipes, plovers and other waders that breed in Germany and in Northern and Eastern Europe and they are indeed well protected in those regions – but they are shot out of hand in France as soon as they leave their breeding grounds and cross the borders to the West. France is home to 10% of the total European winter populations of godwits and 25%

of the lapwings and grey plovers that winter in Europe. In France, fowlers shoot 1.3 million lapwings, more than half a million common snipe and nearly half a million teals every year. France is the only Western European country where the red knot, the oystercatcher, the spotted redshank, the greenshank and the grey plover may be legally shot (Krumenacker, 1998). Most of the French fowlers also have an extremely sketchy knowledge of species; they shoot the red-crested pochard, the pintail and the gadwall willy-nilly, in the ignorant belief that they are mallards.

We can only hope that France will reverse the triste role it plays in the annual slaughter of birds. The bird protection organisation 'Ligue pour la Protection des Oiseaux' (LPO) has fought bitterly for a tightening of the laws and regulatory procedures for many years. Some successes have made themselves felt in recent years, although the going is very tough and slow. The problem is that the lobby of the hunting organisations receives support from all parties, from the Communists on the left to the 'Front National' right-wing party; the only exception is the Green Party. The protests of the hunting associations against any French legislation that would adhere to the requirements of the EU range from intimidating letters to Members of Parliament and aggressive propaganda against the advocates of bird protection to death threats against the regional officers of the LPO.

However, fowling is also a kind of national sport in other Western and Southern European countries. On the Adriatic East coast – and this is contrary to existing legal regulations – fowlers fire intensively at the great numbers of migratory birds swimming on lakes and inland waters until the waters finally lie bereft of avian life. Spanish, Italian and Maltese fowlers use lime twigs, clap nets, decoy birds and even electronic decoy equipment to lure the birds. In these countries, only a small percentage of the captured birds end up in the cooking pot or the frying pan. Fowlers in Malta are true connoisseurs who focus on rare prey. Besides targeting the ortolan bunting, the blue rock thrush, the red-throated pipit and the short-toed lark, the Maltese also focus on the waders, such as the avocet, the stilt, the dotterel, the wood, Green and common sandpipers, the dunlin, the little stint and temminck's stint (http://www.komitee.de/content/aktionen-und-projekte/malta/vogelfang). The wings of the captured rarities are then trimmed in order to sell them as live collectibles.

In Spain, the annual bird hunting bag amounts to more than 10 million birds many of which do not any longer end up in the frying pan, but are now sold as cage specimens (http://www.komitee.de/content/aktionen-und-projekte/spanien). Also in Italy, the birds have largely disappeared from the menus of Italian gourmet restaurants and the counters of butchers. Nevertheless, nearly 2 million Italian fowlers wait for the migratory birds as they wing their way to the southern wintering grounds. Whole flocks of migratory birds are lured into special nets with a variety of expensive decoys that lead their unsuspecting brethren to destruction. The fowlers justify their actions using terms such as *tradition* and *national sport*. The 'victors' are honoured with trophies and other awards for the best catch, much like anglers in open competitions. No pangs of conscience plague Italian fowlers, because priests and higher church dignitaries are also avid fowlers. In one bird trapping association, Saint Francis was even invoked as the patron saint of birds, to help religiously minded (and greedy) fowlers reap a rich bounty of feathered victims (http://www.anti-jagd-demo.de/texteartikel/hansjuergenlutzvogeljagdinitalien.html).

To avoid nourishing disparagements on a national level, we must remember that what is happening today in Western Europe and in the Mediterranean countries was simply business as usual a few generations ago in Germany. The siskin, bullfinch and goldfinch were caught and kept in cages, and this was a tradition that lasted for centuries; it was widespread in Germany and no one objected to it. The father of German ornithology, Johann Friedrich Naumann (1780–1856), was himself a great fowler who praised the virtues of fowling in his writings. As for the Mediterranean countries, they do not have any other tradition pertaining to birds as we in Central Europe had in the past; but even here, a different kind of awareness may hopefully be born after one or two generations.

3.7.3 The Main Damage That Fowlers Wreak Is Not the Fact That They Kill the Creatures – They Make Them Timid, with the Result That They Cannot Settle in Habitats Near Humans

Germany has 81 million inhabitants and approximately 15 million chaffinches; this means that man is five times more common than the most common breeding bird in Germany; and it also means that there are too may human beings around for the creatures. This is a major problem, one which the nature conservation and species protection bodies must tackle. Animals do not have enough space – but they would have more space if they were not so timid.

Today, we live in a time in which we can almost forget about agricultural land being a domicile for species. Fields, meadows and pastures are dead, and no one can change that. Most animals need ruderal land and brownfield sites, but these have almost disappeared between the fields and meadows. Agricultural crops on the fields and grass on the meadows have also become much too dense to provide species with warmth and enough room to move. Many suitable ruderal and brownfield areas are no longer in agricultural areas, but near or in villages and cities, or beside roads, hiking trails and other locations where many people pass by. These plots could be inhabited by rare birds and mammals if the animals were not so afraid of man. Some newly introduced bird species, such as the Canada goose und the Egyptian goose are good examples of birds that are able to live and breed in direct proximity to man.

Fowling occurs everywhere that people live, and this situation has always existed, even from the beginning of mankind. Wherever man has encountered creatures, they have learned to flee from him. The only birds and mammals which do not flee from humans are those which live in areas where man does not 'occur', such as small uninhabited islands or at the North and South Poles. Fear of man has also anchored itself in the genes of many animals, and has prevailed as a hereditary characteristic that has propagated under strong selection pressure. When they arrive as passage migrants in Central Europe in late summer, most migratory birds that breed in the high North are afraid of man from the outset, even if they have not had any negative experience with humans at all.

Fear of humans has evolved differently in the forms of a genetic trait and as a learning success. The whooper swan and Bewick's swan are timid, while their close relative, the wild mute swan, which lives on the Baltic Sea, the Black Sea and the Caspian Sea is much less timid with people. The white stork leaves only a short escape distance between itself and humans, but the black stork is one of the shyest birds of all. It is, however, very unlikely that the different swan and stork species have each had different bad experiences with humans in past centuries. Fear of man is genetically anchored, and such genetic properties have prevailed under selection pressure; however, they may weaken by different selection conditions.

Genes usually only change slowly over time under altered selection conditions; but when fear of man is no longer a survival issue, this may change over long periods and generations. Anyone from Central Europe visiting the Everglades National Park in Florida is amazed how close they can come to the herons, ducks and even the birds of prey that live there; these birds all have significantly shorter escape distances. The European relatives of these species are much more timid. The same applies to East Africa and India: lapwings and even the resident species of sea eagles are much more trusting there than they are in Central Europe. After the Second World War, the white-tailed eagle was only able to settle in Germany's Schleswig-Holstein after humans were barred from entering the forests where the eagles bred and lakes where they looked for food, but the African fish eagle (a species of the same genus) breeds on campsites at Lake Naivasha in Kenya, just above the tents and houses where the tourists are present. The grey herons which breed in Central Europe maintained an escape distance of several hundred yards as recently as half a century ago, but this has changed today. Now they breed in urban parks, and you can almost touch the bird in cities and parks when it is standing at the edge of garden ponds. However, the Eastern birds which migrate from Russia to Central Europe in autumn and winter are still as shy as the domestic Central European grey herons used to be in the past.

These significant differences show the extent to which fowling has affected the timidity of birds. Now, in modern times, when most birds have had to vacate the agriculturally cultivated areas and retreat to industrial areas and the cities, it is much more important to find a consensus between humans and animals. Berlin has 141 breeding bird species, making it one of the richest locations for birds in Central Europe. Man has been a terrifying figure in the minds of birds for millennia – and that must change. The reconciliation of man and bird has already begun in the cities.

We look back on a European history, telling the story of how birds were shot and captured for thousands of years.

> Herr Heinrich sitzt am Vogelherd
> recht froh und wohlgemut …
> In Wies' und Feld, in Wald und Au,
> horch, welch ein süßer Schall;
> der Lerche Sang, der Wachtel Schlag,

die süße Nachtigall.
Herr Heinrich schaut so fröhlich drein:
Wie schön ist heut' die Welt!
Was gilts? Heut gibt's 'nen guten Fang!
Er lugt zum Himmelszelt.

Free Translation; no rhyming; it is only intended to convey the rough idea behind the song, which is the catching, caging and eating of birds:

Mr. Henry repairs his bird-catching nets,
his spirits high when he hears the song,
of the Lark, the quail, and the nightingale:
Cage and pan ... expect a throng,
t'will indeed be a catch to hail!

This is a song by Johann Nepomuk Vogl (1835), set to music by Carl Loewe (1836). The song describes the historical tradition how Henry I (875–936) from the House of Saxony was crowned as the first German king in May of 919. He received the message from a cavalcade of Frankish and Saxon dukes while he was fowling. The song expresses how capturing birds for caging and as a food supplement was an old tradition in Europe. Birds captured were kept alive to be used as decoys in large bird trapping installations or were used in mines to warn of dangerous gases. Nowadays, we should not be surprised that Europe is characterised by particularly timid birds when compared with global avian timidity standards. European wildlife photographers know this and prefer to travel to Florida and Eastern Africa or south-eastern Asia if they intend to photograph birds without having to use a hide.

Many species of birds have declined in number in the last few decades and one way of increasing their chances of survival would be to help them lose their fear of man. They could then colonise the available suitable habitats in the vicinity of humans and not flee from them because too many people were there. The harm done to birds by fowling is not actually because the hunters kill their victims; the greatest danger of fowling is that it makes birds timid, driving them away from the vicinity of humans. Birds would have more territory available for rest and feeding, both in the breeding areas and in the transit and wintering areas, if they did not have to repeatedly learn from one experience to the next that humans want to kill them (Reichholf, 2005, 2010a). This is becoming particularly important in this age, in which many birds are leaving the agricultural areas and find new homes in the cities and industrial areas.

3.7.4
The High Reproduction Rate of the Organisms

It is currently impossible to make any verifiable judgement as to whether the billions of birds that fall victim every year to cats and fowlers are a significant reason for the decline of the number of individuals in many bird species in recent decades. It is possible that the high losses caused by the cats can be at least partially offset by the strong propagation rate of the birds. The fact is that the sharp decline in the populations of most bird species (see Section 5.1) is a phenomenon of modern times, while the damage caused by cats and fowlers already existed in earlier times. It is also a fact that many of the species which are on the decline in their individual numbers are among the bird species that are not attacked and eaten by domestic cats. The decline of many bird species is certainly caused by other reasons as well. The losses caused by cats probably only affect the individuals of those bird species which are already in decline due to other reasons. This would also mean that only the individual numbers of the rarer species is affected by cats, while the more common species can offset these losses; and so we can in turn conclude that the fowlers of the Mediterranean region cause greater harm to birds than domestic cats, because it is the more common bird species that probably fall victim to the cats in our gardens.

The almost astronomically high reproduction rate of creatures is frequently underestimated. It was one of Charles Darwin's greatest insights that he actually recognised the importance of the high reproduction rates of living organisms. Progress in evolution is only possible because an excessive number of offspring is produced, and this gives the selection process the opportunity to select mutants (which are suitable for survival and further propagation) from a large number of descendants. Otherwise evolution would come to a standstill. However, the necessary consequence of the high reproduction rate is that most of the progeny which are born have to die prematurely; otherwise the high reproduction rate would quickly lead to overpopulation. Darwin expressed this in the following sentence: 'There is no exception to the rule that every organic being naturally increases at so high a rate, that, if not destroyed, the earth would soon be covered by the progeny of a single pair' (Darwin, 1859, p. 79).

The conclusion to be drawn is clear: it is a biological necessity that the vast majority of offspring born dies prematurely through enemies, accidents or diseases. Otherwise, the earth would be covered by a single species (often after only a few generations). The female of the great tiger moth, for example, lays 1600 eggs (Hering, 1926). This means that in only one generation, a single mother produces offspring which number more than the currently existing number of individuals in a variety of rare animal species worldwide.

We can also choose examples other than a moth to illustrate the enormous proliferation of most animal species; organisms with much lower reproductive rates would also multiply much too quickly if most of their progeny did not perish relatively quickly. The following sample calculation illustrates the proliferation rate of organisms.

Under natural conditions, a pair of northern lapwings lives about 15 years. In 1 year, a pair produces 4 chicks, which sexually mature after a year of life. Assuming that no animal dies prematurely, this results in the following numbers:

at the end of the first year: 2 lapwings + 4 young = 6 lapwings,
at the end of the second year: 6 = 3 pairs + (3×4) young = 18 lapwings and
finally, at the end of the 15th year: $2 \times 3^{15} = 28\,697\,814$ lapwings.

Both parents then die of natural causes. This means 28 697 812 descendants, all produced by a single original pair. We can derive from this calculation that 28 697 810 descendants of the first lapwing pair must die prematurely due to illness or as victims of enemies, to keep the population constant (= 2 lapwings). This is a very revealing calculation. Every animal and plant proliferates so intensively that life on earth can only function because the vast majority of organisms die relatively quickly, in many cases as soon as they are born.

The sample calculation shows that the death of individuals through fowlers, butterfly netters or an oil spill can in most cases not be held fully responsible for the fact that species become rare or extinct. If the habitats are suitable for breeding, then in most cases, these will alone guarantee such a high reproduction rate that hunting and fowling become irrelevant, because they cannot affect the individual numbers of a species. It is emphasised elsewhere in this book (see Sections 3.4.3 and 5.2), that the netting of butterflies probably has not harmed one single Central European butterfly species. The enormous loss rates that butterflies have suffered in Central Europe in the last few decades are probably due to the change in habitats that reduced the insects' proliferation. The legislation obtained by the nature conservation associations against those who net butterflies has combated the deaths of the butterfly on the wrong front.

Hunting and fowling can only decimate a species if that species is already on the wane due to other causes. However, this is now the case for many species due to the eutrophication of our landscape, because there are only a few nesting and feeding places left among the densely vegetated areas. This applies to most waders. It is highly probable that French fowlers who kill and capture the lapwing, the black-tailed godwit and the common snipe are responsible for slowly but surely driving the populations of these species in Central Europe towards extinction. Hunting and fowling can also decimate a species if the numbers of creatures killed is enormously high. This was the case with the crow, the magpie and some birds of prey during earlier decades in Central Europe; and it is probably the case today due to the shooting of tens of thousands of waterfowl on the Croatian Adriatic coast.

References

Albrecht, C., Dworschak, U.-R., Esser, T., Klein, H., and Weglau, J. (2005) Tiere und Pflanzen in der Rekultivierung – 40 Jahre Freilandforschung im Rheinischen Braunkohlenrevier. *Acta Biologica Benrodis*, Suppl., **10**, 1–238.

Almers, L., Baum, A., Baum, H., Jansen, H., Luwe, M., Singer, R., Thiele, G., Wenker, D., and Westphal, G. (2001) *Die Orchideen Nordrhein-Westfalens*, Arbeitskreis Heimische Orchideen NRW, Steyl.

Anonymous1996) Winterfütterung – mit dem Gewissen vereinbar? *Fotografie draußen*, **27**, 60.

Baillie, J.E.M., Hilton-Taylor, C., and Stuart, S.N. (2004) *2004 IUCN Red List of Threatened Species – A Global Species Assessment*, International Union for Conservation of Nature and Natural Resources (IUCN), Gland and Cambridge.

Baker, P.J., Bentley, A.J., Ansell, R.J., and Harris, S. (2005) Impact of predation by domestic cats *Felis catus* in an urban area. *Mammal Review*, **35**, 301–312.

Barkow, A. (2010) Probleme der Wiesenvögel – Ist die Uferschnepfe noch zu retten? *Der Falke Taschenkalender*, 197–204.

Beckerman, A.P., Boots, M., and Gaston, K.J. (2007) Urban bird declines and the fear of cats. *Animal Conservation*, **10** (3), 320–325.

Berthold, P. and Mohr, G. (2012) *Vögel füttern, aber richtig: das ganze Jahr füttern, schützen und sicher bestimmen*, 3rd edn, Franckh-Kosmos Verlags-GmbH, Stuttgart.

Bezzel, E. (1995) Farbveränderte Stockenten und Entenbastarde in Europa. *Der Falke*, **10**, 291 and 312.

Blackbourn, D. (2007) *Die Eroberung der Natur*, Deutsche Verlags-Anstalt (DVA), München.

Bölsche, J. (1982) *Natur ohne Schutz – neue Ökostrategien gegen die Umweltzerstörung*, Spiegel-Verlag, Hamburg.

Burton, J.F. (1995) *Birds and Climate Change*, A & C Black, London.

Connell, J.H. (1978) Diversity in tropical rain forests and coral reefs. *Science*, **199**, 1302–1310.

Coudrain, V. (2011) Wendehals – Gefährdung durch zu wenig offenen Boden. *Der Falke*, **58**, 2.

Coudrain, V., Arlettaz, R., and Schaub, M. (2010) Food or nesting place? Identifying factors limiting Wryneck populations. *Journal für Ornithologie*, **151**, 867–880.

Cremene, C., Groza, G., Rakosy, L., Schileyko, A.A., Baur, A., Erhardt, A., and Baur, B. (2005) Alterations of steppe-like grasslands in Eastern Europe: a threat to regional biodiversity hotspots. *Conservation Biology*, **19**, 1606–1618.

Dämmgen, J.W. (2013) Verstummt der Frühling: Welche Rolle spielen Hauskatzen? *Der Falke*, **60**, 144–147.

Darwin, C. (1859) *On the Origin of Species by Means of Natural Selection or the Preservation of Favoured Races in the Struggle for Life*, Murray, London.

Dengler, J. (2012) Europäische Trockenrasen schlagen tropische Regenwälder. *Biologie in unserer Zeit*, **42**, 148–149.

Dornbusch, M. (1996) Situation und Schutz der Großtrappe (*Otis t. tarda* L., 1758) in Sachsen-Anhalt. *Naturschutz und Landschaftspflege in Brandenburg*, **1/2**, 28–29.

Ebert, G. and Rennwald, E. (1991a) *Die Schmetterlinge Baden-Württembergs: Tagfalter I*, Eugen Ulmer, Stuttgart.

Ebert, G. and Rennwald, E. (1991b) *Die Schmetterlinge Baden-Württembergs: Tagfalter II*, Eugen Ulmer, Stuttgart.

Engels, J.I. (2006) *Naturpolitik in der Bundesrepublik: Ideenwelt und politische Verhaltensstile in Naturschutz und Umweltbewegung 1950–1980*, Ferdinand Schöningh Verlag, Paderborn.

Filz, K.J., Engler, J.O., Stoffels, J., Weitzel, M., and Schmitt, T. (2013) Missing the target? A critical view on butterfly conservation efforts on calcareous grasslands in south-western Germany. *Biodiversity and Conservation*, **22**, 2223–2241.

Fitter, R., Fitter, A., and Blamey, M. (1974) *Pareys Blumenbuch*, Verlag Paul Parey, Hamburg und Berlin.

Flade, M., Plachter, H., Henne, E., and Anders, K. (2003) *Naturschutz in der Agrarlandschaft*, Quelle und Meyer, Wiebelsheim.

Gorke, M. (1999a) *Artensterben: Von der ökologischen Theorie zum Eigenwert der Natur*, Klett Cotta, Stuttgart.

Gorke, M. (1999b) Wie kann und soll man Artenschutz begründen? *Artenschutzreport*, **9**, 1–5.

Gorke, M. (2010) *Eigenwert der Natur. Ethische Begründung und Konsequenzen*, Hirzel Verlag, Stuttgart.

Gotzmann, I.H. and Ferber, D. (2007) Die Wahner Heide. *Biologie in unserer Zeit*, **37**, 396–403.

Großklaus, G. and Oldemeyer, E. (1983) *Natur als Gegenwelt – Beiträge zur Kulturgeschichte der Natur*, Loeper, Karlsruhe.

Grüneberg, C. and Sudmann, S.R. (2013) *Die Brutvögel Nordrhein-Westfalens*, LWL-Museum für Naturkunde, Münster.

Hauff, W. (1827) *Das Wirtshaus Im Spessart*, Bibliographisches Institut, Leipzig und Wien.

Haupt, H., Ludwig, G., Gruttke, H., Binot-Hafke, M., Otto, C., and Pauly, A. (2009) *Rote Liste Gefährdeter Tiere, Pflanzen Und Pilze Deutschlands – Band 1: Wirbeltiere*, Bundesamt für Naturschutz, Bonn-Bad Godesberg.

Hering, M. (1926) *Biologie Der Schmetterlinge*, Julius Springer, Berlin.

Hübner, G., Rauh, M., and Will, D. (2013) Wendehals und Landschaftspflegepraxis – Erkenntnisse einer Lebensraumanalyse im westlichen

Oberfranken. (Wrynecks and landscape management – lessons from habitat analysis in western Upper Franconia). *Natur und Landschaft*, **79**, 118–123.

Huntley, B., Green, R.E., Collingham, Y.C., and Willis, S.G. (2007) *A Climatic Atlas of European Breeding Birds*, Durham University, Barcelona.

Johnson, K.H., Vogt, K.A., Clark, H.J., Schmitz, O.J., and Vogt, D.J. (1996) Biodiversity and the productivity and stability of ecosystems. *Trends in Ecology & Evolution*, **11**, 372–377.

Kirchhoff, T. (2012) Urban grasslands: between nostalgia, control of nature, and wildness. *Local Land and Soil*, **42/43** (2), 19–21.

Kirchhoff, T. and Trepl, L. (2009) in *Vieldeutige Natur. Landschaft, Wildnis und Ökosystem als kulturgeschichtliche Phänomene* (eds T. Kirchhoff and L. Trepl), Transcript Verlag, Bielefeld, pp. 13–66.

Klaus, S. (2012) Lückenbewohner im Natur- und Wirtschaftswald: Auerhuhn und Haselhuhn. *Der Falke*, **59**, 26–29.

Krumenacker, T. (1998) Frankreichs Jäger: Gegen Zugvögel und europäisches Recht. *Limicola*, **12**, 268.

Kunz, W. (1959) Die Vogelwelt des Kreises Bersenbrück. *Schriftenreihe Kreisheimatbund Bersenbrück*, **6**, 1–159.

Kunz, W. (2004) Der Braunkohletagebau als Ort der Wiederansiedlung seltener Tagfalter und anderer Organismen – Was wird durch Rekultivierung zerstört? (The brown coal open-cast mining as a place for resettlement of rare butterflies and other organisms: what will be destroyed by recultivation?). *Entomologie heute*, **16**, 245–255 (in German).

Kunz, W. (2013) Artenförderung durch technische Gestaltung der Habitate – Neue Wege für den Artenschutz (Species promotion by technical shaping of habitats – new ways in species protection). *Entomologie heute*, **25**, 161–192 (in German).

Kunz, W., Schmitt, T., Cölln, K., Dworschak, U.-R., Tumbrinck, J., and Jelinek, K.-H. (2014) Artenschutz durch Schaffung von Offenland – Podiumsdiskussion. *Entomologie heute*, **26**, 161–176.

Küster, H. (2008) *Geschichte des Waldes. Von der Urzeit bis zur Gegenwart*, 2nd edn, C. H. Beck, München.

Lanz, U. (2009) Vage Erfolgsaussichten im Ortolanschutz. *Der Falke*, **56**, 426–431.

Lenz, N. and Schulten, D. (2005) Tagfalter (Lep., *Hesperioidea* et *Papilionoidea*) im Gebiet der Landeshauptstadt Düsseldorf um 1900 und um 2000 – ein Beispiel für alarmierende Artenverarmung im 20. Jahrhundert. *Melanargia*, **17**, 19–29.

Lingenhöhl, D. (2010) *Vogelwelt Im Wandel – Trends Und Perspektiven*, Wiley-VCH Verlag GmbH, Weinheim.

Martinez, N. (2010) Gartenrotschwanz – Wiesenstruktur entscheidet. *Der Falke*, **57**, 388.

Meyer, W., Eilers, G., and Schnapper, A. (2003) *Müll als Nahrungsquelle für Vögel und Säugetiere*, Westarp Wissenschaften, Hohenwarsleben.

von der Nordmark, H. and Zukowsky, L. (1936) *Aus Deutschlands Vogelwelt*, Cigaretten-Bilderdienst, Hamburg-Bahrenfeld.

Orros, M.E. and Fellowes, M.D.R. (2015) Widespread supplementary feeding in domestic gardens explains the return of reintroduced Red Kites *Milvus milvus* to an urban area. *Ibis*, **157**, 230–238.

Reichholf, J. (2005) *Die Zukunft der Arten*, C.H. Beck, München.

Reichholf, J. (2010a) *Naturschutz. Krise und Zukunft*, Suhrkamp Verlag, Berlin.

Reichholf, J. (2010b) Nur fürs Auge – Der Naturschutz setzt falsche Schwerpunkte. *Focus*, **44**, 92–94.

Schmitt, T. and Rakosy, L. (2007) Changes of traditional agrarian landscapes and their conservation implications: a case study of butterflies in Romania. *Diversity and Distributions*, **13**, 855–862.

Schnittler, M. (2004) *Naturschutz und biologische Vielfalt*, Bundesamt für Naturschutz, Bonn, pp. 39–52.

Segerer, A.H. (2012) Die physikalisch-geochemischen Grundlagen des planetaren Klimas und die Auswirkungen auf die öffentliche Diskussion – potenzielle Fallstricke für Ökofaunisten. *Nachrichtenblatt bayerischer Entomologen*, **61**, 32–45.

Stifter, A. (1882) *Der Hochwald*, C.F. Amelang's Verlag, Leipzig.

Südbeck, P., Bauer, H.-G., Boschert, M., Boye, P., and Knief, W. (2007) *The Red List of Breeding Birds of Germany*, 4th edn, vol. **44**, Berichte zum Vogelschutz, pp. 23–81.

Südbeck, P. and Krüger, T. (2004) Erhaltungssituation und erforderliche Schutzmaßnahmen für Wiesenvögel in Niedersachsen – Bilanz und Ausblick. *Naturschutz und Landschaftspflege in Niedersachsen*, **41**, 106–123.

Thomas, J.A., Telfer, M.G., Roy, D.B., Preston, C.D., Greenwood, J.J., Asher, J., Fox, R., Clarke, R.T., and Lawton, J.H. (2004) Comparative losses of British butterflies, birds, and plants and the global extinction crisis. *Science*, **303**, 1879–1881.

Ulrich, R. (2002) Vom Naturschutz vergessen: Die Lichtwaldarten. *Naturschutz im Saarland*, **3**, 22–25.

Van Swaay, C. and Warren, M. (1999) *Red Data Book of European Butterflies (Rhopalocera)*, Council of Europe, Strasbourg.

Walker, B.H. (1992) Biodiversity and ecological redundancy. *Conservation Biology*, **6**, 18–23.

WallisDeVries, M.F., Poschlod, P., and Willems, J.H. (2002) Challenges for the conservation of calcareous grasslands in northwestern Europe: integrating the requirements of flora and fauna. *Biological Conservation*, **104**, 265–273.

Warren, M.S., Hill, J.K., Thomas, J.A., Asher, J., Fox, R., Huntley, B., Roy, D.B., Telfer, M.G., Jeffcoate, S., Harding, P., Jefcoate, G., Willis, S.G., Greatorex-Davies, J.N., Moss, D., and Thomas, C.D. (2001) Rapid responses of British butterflies to opposing forces of climate and habitat change. *Nature*, **414**, 65–69.

Weiss, J. and Schulze-Hagen, K. (2014) Viele Menschen – viele Vögel? Ornithologie und Vogelschutz in Nordrhein-Westfalen. *Charadrius*, **50**, 3–22.

Willis, K.J., Gillson, L., and Brncic, T.M. (2004) How 'virgin' is virgin rainforest? *Science*, **304**, 402–403.

Wirth, R. and Glaw, F. (2009) Globales Unwissen contra Biodiversität: Gedanken zum Schutz der biologischen Vielfalt. *Biologie in unserer Zeit*, **39**, 42–48.

4
Objectives, Content and Limits of the Red Lists of Endangered Species

4.1
The Objectives and the Origin of the Red Lists of Endangered Species

Red Lists are often difficult to understand.

The aim of the International Red Lists is to publish statements about the threat to the existence of species.

The Regional Red Lists, on the other hand, pursue the conservation of the wealth of the home country rather than the conservation of certain species from extinction.

Who creates the Red Lists?

The IUCN List covering the global threat to species mostly pertains to higher vertebrates and only small percentages or thousandth parts of most other species groups.

Should a species not be protected in a nation where it only has a marginal presence?

The IUCN List and National Red Lists are not comparable with one another, simply because different criteria are used to classify the endangerment levels.

The Red Lists of endangered animals and plants are often difficult to understand. This is because there is no 'current state'. The decisive factor here is not the number of living individuals of a species, but whether the population trend is increasing or decreasing. There are species that still exist in large numbers, but display a significantly downward trend in population. Species like this may be more vulnerable than similar species whose total number of individuals is fewer, but whose population is stable. Eagles, for example, are found only singly or in pairs wherever they occur, because they need large territories – they are not found in greater numbers except during their migration periods. Despite their sparse numbers, however, the species is not endangered, as long as their population remains stable.

Even more difficult to understand are the National and Regional Red Lists of Endangered Species, which are based only on defined geographical regions. Here, the authorities differentiate between native species and sporadic and introduced species (Neobiota). Anyone who leafs through the National List of Endangered Breeding Birds of Germany (Südbeck *et al.*, 2007) to find out how many species of breeding birds can be found in Germany will not find that information very

Species Conservation in Managed Habitats: The Myth of a Pristine Nature, First Edition. Werner Kunz.

quickly. The searcher must also decide whether or not the sporadic breeders and Neoaves are to be included – and he or she will also learn that even extinct species are included among the breeding birds in Germany, because these may well return one day. However, any search for a simple number like how many breeding bird species there are in Germany will first necessitate painstaking research and the mastering of several problems of comprehension before that number is found. The current Red List has 244 native species that breed regularly in Germany (Südbeck *et al.*, 2007).

If, however, we also include the species which have only occasionally been brooding in the last 20 years in Germany, like the stilt, the booted eagle and the saker falcon, the number of breeding birds in Germany increases to 269. If we also consider introduced species (Neozoa) like Nile geese and pheasants to be German breeding birds, the number of species rises to 298 – and although the stone curlew, the European roller and the lesser grey shrike are now extinct, they also belong to the list of breeding birds of Germany. Adding these extinct species increases the number of breeding bird species in Germany to 314. This is around 25% more than the initial number of 244 breeding birds. So it is clear that we must first define what 'breeding birds' actually are before their number can be accurately determined in a particular country.

There are also other reasons why Red Lists are often misunderstood. At first they are understood to be directories listing the species that are endangered or even threatened with extinction – but this is not correct, because there is a global Red List and there are Regional Red Lists – and both of these not only differ in that they comprise a larger or smaller area, but the International and National Red Lists also have a different significance and purpose. The global Red Lists contain the global population numbers and trends (increase/decrease) of endangered species (Baillie, Hilton-Taylor and Stuart, 2004). Each particular species placed in the 'Endangered' category of the global Red List is endangered worldwide – or even threatened with extinction. This is why it is only the global Red List which can tell us the current status regarding the existence and future of a species.

The National or Regional Red Lists make a very different statement, because theirs is related to the survival chances of species in limited geographic regions. It is a fact that many species – not at all threatened on a global scale – are endangered or even threatened with extinction in limited geographic regions. The conservation priorities of the National Red Lists are often misunderstood. The black grouse, for example, is critically endangered or even threatened with extinction in most Central European countries, but it is not at risk in Scandinavia, nor in some countries in Eastern Europe and Asia. Only in endemic species is the national threat level equivalent to the degree of endangerment of the species itself. Most other species classified as endangered in the Regional Red List are not globally threatened. This discrepancy between national and global degrees of endangerment particularly affects Central Europe – because the Ice Ages and the subsequent anthropogenic destruction of nature resulted in Central Europe losing many endemic species and being secondarily colonised by species that have their core occurrence elsewhere (see Section 6.1.1). The following statement makes this clear: if all the breeding bird species in Germany were killed today, there would hardly be a bird species the existence of which would be globally endangered (Gruttke, 2004).

This in effect means that the purpose of the Regional Red Lists cannot be the conservation of species. The regional lists predominantly follow a different objective – namely the conservation of the wealth and diversity of a region – but *not* the conservation of a species. Since the Central European countries have very few endemic species, the Red Lists of the Central European countries have more to do with homeland consciousness than the protection of specific animal or plant species from extinction. Among the roughly 10 000 bird species in the world, 12% are endangered (Baillie, Hilton-Taylor and Stuart, 2004). However, among the 244 native bird species that regularly breed in Germany, 31% are endangered (Südbeck *et al.*, 2007) (Table 4.1). The percentage of endangered bird species in Germany is therefore more than twice that of globally threatened bird species. This high percentage, which indicates that Germany contains a particularly high

Table 4.1 The numbers of breeding birds and butterflies in Germany have hardly changed over the last 100 years.

Breeding birds in Germany: species numbers and population trends	
Species numbers	
since 1850 (Status I, no Neozoa or sporadic breeders)	**255**
Extinct 1850–1985	6
Extinct 1985–2005	5
Species number today	**244**
Endangered plus extinct since 1850 (Categories 0–3)	79 of 255 = 31%
Long-term population trend	
Population decline plus extinct since 1850	113 of 255 = 44%
Population increase since 1850	82 von 255 = 32%
Short-term population trend	
Population decline plus extinct since 1985	45 of 249 = 18%
Population increase since 1985	77 of 249 = 31%
Butterflies in Germany: species numbers and population trends	
Species numbers	
since 100 years (Status I, no Neozoa or sporadic occurrences)	**186**
Extinct in the last 100 years	5
Species number today	**181**
Endangered plus extinct in the last 100 years (Categories 0–3)	77 of 186 = 41%
Long-term population trend	
Population decline plus extinct in the last 100 years	113 of 186 = 61%
Population increase in the last 100 years	4 of 186 = 2%
Short-term population trend	
Population decline plus extinct in the last 20–30 years	81 von 182 = 45%
Population increase in the last 20–30 years	10 von 182 = 5%

If we focus solely on the number of species, the true picture of the decline of species is obscured. If the population trends are considered, there is a significant difference between birds and butterflies. While a remarkable percentage of bird species increased in number during the last decades, almost all species of butterflies became continuously rarer.

number of endangered species, must not be misinterpreted; in fact, only 2 of the 79 endangered species in Germany are globally endangered. These two are the great bustard and the aquatic warbler, both of which are classified as 'vulnerable' (VU) in the global list of the 'International Union for Conservation of Nature and Natural Resources' (IUCN) of endangered species (Table 4.2). It is therefore important to understand that most bird species that are endangered in Germany

Table 4.2 Comparison of endangerment categories and classification criteria between the global Red List of IUCN and the Red Lists of Germany.

Endangerment categories, classification criteria/population trends and status categories		
	IUCN	**Germany**
Endangerment categories	EX = Extinct CR = Critically endangered EN = Endangered VU = Vulnerable	0 = Ausgestorben 1 = vom Aussterben bedroht 2 = Stark gefährdet 3 = Gefährdet R = Extrem selten
Classification criteria/ population trends	CR = Total population declined by at least 80% in the past 10 years; geographical distribution area less than 100 km^2; population consists of fewer than 250 sexually mature individuals	1 = This species will die out in Germany in the foreseeable future, unless the causes of endangerment can be eradicated
	EN = Total population declined by at least 50% in the past 10 years; geographical distribution area less than 5000 km^2; population consists of fewer than 2500 sexually mature individuals	2 = This species in Germany has declined significantly due to the visible deterioration of its living conditions
	VU = Total population declined by at least 20% in the past 10 years; geographical distribution area less than 20 000 km^2; population consists of fewer than 10 000 sexually mature individuals	3 = This species is in continuous decline, although some of these Category 3 species can currently boast tens of thousands of breeding pairs in Germany
Status categories		I = Native species that occur regularly in the region II = Native species that occur only temporarily in the region III = Neobiota = species introduced by humans from foreign countries

While the endangerment categories are roughly the same, the crucial difference between the IUCN list and the German lists lies in the selection of criteria that must be met for a species to be classified in one of the endangerment categories. The IUCN classification requires quantitative numerical values, while the German Lists use predominantly qualitative criteria. In addition, the German Red Lists use Status categories to differentiate between native and sporadic species occurrences as well as introduced species.

are not globally endangered. If they were to disappear from Germany, they would continue to exist in other parts of the world.

The processing and the updating of the Red Lists is done by *nature* conservation organisations, although the objective of these lists is the conservation of *species.* Wildlife species conservation does not equate with nature conservation in Central Europe, because we often find the endangered species in Central Europe living in destroyed and/or extreme habitats – and the conservation of these habitats is not the goal of nature conservation (see Sections 3.3.1 and 3.5.3).

The first organisation to create a Red List was the 'IUCN'. The IUCN is an international NGO (non-governmental organisation) that has taken on the task of sensitising human societies to nature and species protection (http://de.wikipedia.org/wiki/IUCN). The organisation has more than 1000 members (from ~140 countries) who are either governmental/NGOs (e.g. government ministries and nature conservation organisations), or salaried persons and volunteers. They include the German Federal Agency for Nature Conservation (BfN = Bundesamt für Naturschutz) and the WWF (World Wide Fund for Nature). The IUCN, founded in 1948 in Fontainebleau in France, now has its headquarters in Gland, Switzerland and has numerous offices in 62 countries.

A commission was established within the IUCN to deal specifically with the conservation of species and the assessment of their degree of endangerment. This commission is called the *Species Survival Commission* (SSC) and it has been publishing the International Red Lists of endangered animal and plant species at irregular intervals since 1966. These lists are made publicly available under the registered trademark 'IUCN Red List of Threatened Species™'. The full list of species in the IUCN Red List is available as an online searchable database at www.iucnredlist.org. A 217-page explanation provides an overview of the current state of the surveys, the number of globally endangered animal and plant species and the causes of endangerment, from habitat destruction to climate change (Baillie, Hilton-Taylor and Stuart, 2004).

The IUCN List includes the species of many vertebrate and invertebrate animals, plants and lichens. The current coverage of almost all of the animal and plant groups, however, is sketchy at best. Among the animals, only the mammals, birds and amphibians are more or less covered, while the gymnosperms are the only plants whose species numbers are registered more fully. Individual countries use the IUCN Red List as a guideline for species protection measures – and the official bodies of the Red List are obliged to be objective in order to prevent political or economic objectives from upgrading or downgrading the endangerment classification of certain species.

The Red List is an electronic medium and as such is kept permanently up-to-date; in the case of many species, however, new information is only added occasionally. One major hindrance for the updating procedure is the lack of knowledge about most animal and plant groups. The IUCN Red List only refers to a tiny fraction of all species – at the time of writing, it 'only' included around 70 000 species, amounting to just 3% of all known species worldwide.

Only these 70 000 are well-documented as regards ecology, population size, the degree of endangerment and the necessary conservation measures. The degree of endangerment affecting 97% of all known species has not yet been researched; therefore these species cannot be classified – and many of the species which have already been classified cannot be updated, so the targeted revision times of 10 or 5 years, respectively, must be delayed, while the Red Lists for several animal and plant groups become obsolete (De Iongh and Bal, 2007). Almost a third of the 70 000 species which have been evaluated worldwide are considered to be endangered – and 7% of these are threatened with extinction. This high number of endangered species mainly consists of endemic species, and these include many species occurring on islands.

The IUCN has only performed fairly comprehensive, global endangerment research on the almost complete numbers of species of three classes of animals – these are the mammals (about 5420 species worldwide), birds (around 9920 species worldwide) and amphibians (roughly 5750 species worldwide). This amounts to a total of just over 21 000 species. In the case of known species of reptiles and fish, only about 6% has been researched – and the data acquired on invertebrates are even more incomplete, because only freshwater and terrestrial molluscs and freshwater crustaceans have been evaluated for endangerment potential (Baillie, Hilton-Taylor and Stuart, 2004). Considering that the species of several insect groups are well-known, we might expect that there would be more information about the potential threat they are facing worldwide; but this is not the case, since there is a distinct lack of knowledge in this sector. We are only aware of the degree of global endangerment for less than 1 in 1000 of all insect species; and we practically have data only on dragonflies and on the Papilionidae (swallowtails) – otherwise wide gaps in our knowledge prevail. From the other invertebrate phyla, only the Echinodermata (sea urchins, starfish), the Annelida (segmented worms), the Onychophora (velvet worms), the Nemertea (unsegmented worms), the Platyhelmintha (flatworms) and the Cnidaria (cnidarians, such as sea anemones) are represented by some species in the IUCN Red Lists, while the Nematoda (roundworms, e.g. *Ascaris*) and the sponges are not included at all. Among the plants, only the gymnosperms are dealt with (almost) comprehensively (Baillie, Hilton-Taylor and Stuart, 2004).

You will usually come up empty-handed if you look for animal names of invertebrates in the search window of http://www.iucnredlist.org. You will be referred to the 'Catalogue of Life' (http://www.catalogueoflife.org/), a collection of 70% of all the animal and plant species that are currently known worldwide. This 70% currently includes more than 1.3 million species (as of 2012). The 'Catalogue of Life' is a very valuable and useful Internet site which can be used to access knowledge about most animals and plants and to learn about their names and the systematic classification. The Catalogue comprises animals, plants, fungi and microorganisms. It does not, however, list the degree of endangerment for any species – only the IUCN List (http://www.iucnredlist.org) is responsible for publishing this information.

There are two main reasons why only a few animal classes are assessed (to the detriment of all the other classes of animals): first, people are only familiar with a few species groups and none more so than birds and mammals; second, the number of species of mammals and birds is small compared to the number of species of most groups of invertebrates – and the sheer numbers of the invertebrates also make it much more difficult to obtain a clear picture of them. These invertebrates comprise around 95% of all known species, in addition to around 5500 mammal species and 10 000 species of birds on earth (Baillie, Hilton-Taylor and Stuart, 2004). The Lepidoptera alone (1 of more than 370 orders of invertebrates, comprise some 160 000 species worldwide according to current knowledge – and that is 16 times the number of species compared with the class of birds (Zhang, 2011). We can also safely assume that the majority of the butterfly species has still not been discovered, whereas the bird species are largely well-known all over the world. The much higher number of species alone creates gaps in the IUCN's endangerment assessment list.

There are also a number of other difficulties that lead to an uneven distribution of the data among the different animal groups. The distribution and species numbers of bats and mice are much harder to ascertain than those of birds, because bats are nocturnal animals that remain hidden during the day. This is also one of the reasons why birds have been more fully researched than the mammals.

In addition to the lists of the world's threatened animal and plant species published by the IUCN, there are also corresponding lists issued by individual nations and their regional states. Many nations have even followed the example of the IUCN and issued their own National Red Lists since the 1970s. At an international conference in November 2002 in Leiden, Netherlands, it was determined that there are 3700 different Red Lists (De Iongh *et al.*, 2003), and of this number, more than 90% applied to only European countries, whereas less than 10% of the Red Lists covered the animals and plants of non-European countries. The various European countries were also very unevenly represented.

In Germany, the National Red Lists are created by an independent panel that works together with national professional associations and bird conservation observatories (Vogelschutzwarten) of individual states. The panel also cooperates with the German Ornithologists' Society (DOG), the umbrella organisation of German Avifaunists (DDA), German bird observatories (Vogelwarten) and the Federal Agency for Nature Conservation in Bonn (BfN). In addition to the all-German Red List of breeding birds of Germany (Südbeck *et al.*, 2007), there are also separate Red Lists for individual federal states, in which the number of endangered species is of course much greater than it is in the all-German Red List. As is the case in other countries, most of the information is about the class of birds. The Red List of breeding birds in Germany has been regularly published since 1971 and four updated Red Lists covering all of Germany appeared after the German reunification. 2014 saw the last version of the fourth edition of the Red List of breeding birds in Germany – it was based on data acquired in 2007 (Südbeck *et al.*, 2007).

The creation of the Regional Red Lists also created problems that still have to be solved, because the categories and criteria of the IUCN List were originally designed to evaluate the global degree of endangerment to species. A global threat, however, poses a different problem than does the extinction of a species in one particular country, if that species is common in other countries. If a species occurs over several political boundaries, then different degrees of endangerment apply for each of these individual countries. If a country's area covers only the peripheral population of a species, then that species can be endangered or even threatened with extinction there, although it may be common and not endangered in other countries. The National Red Lists contain many threatened species that are not threatened on a global scale. Approximately 13% of all the 10 000 bird species on earth is endangered (Baillie, Hilton-Taylor and Stuart, 2004), for instance, but 31% of the 255 species of breeding birds in Germany is threatened (Südbeck *et al.*, 2007). The picture painted by the Regional Red Lists therefore conveys a false impression about the real threat to a species (Schnittler, 2004).

Most species have a geographical core population and peripheral occurrences (Gruttke, 2004). The organisms of a species that lives in a core population area have a different population-genetic and evolutionary significance than organisms of the same species in a peripheral and often isolated population area. In many cases, the peripheral occurrence of a species is meaningless, because the individuals of the species in that area live in less than optimal conditions, so they often only occur there in small numbers. One example here is the bee-eater in north-western Central Europe. In some cases, the population of a species in these border zones would disappear (even without any deterioration in the living conditions there) if the individuals were not regularly replaced by immigrants from the core areas. Here, we could dispense with conservation measures for a species in a region, because its occurrence in the border zone is only significant for that particular region – the population here has no effect on the overall conservation of the species. The question becomes more critical in cases like this is whether or not a species should be protected in an entire nation in which it only has a peripheral occurrence – should we therefore ignore species like the hoopoe and the woodchat shrike in Germany?

However, cases also occur where isolated peripheral populations attain a special population-genetic level of importance. This is because allelic forms of certain genes prevail there, which would normally have no chance of competing in core populations. This 'founder effect' (Mayr, 1963) has a species–protective value, because it could lead to the origin of a new species, thereby enhancing biodiversity. This is possibly true of the western Central European population of the violet copper butterfly (*Lycaena helle*) for instance (Habel *et al.*, 2010).

Migratory species such as birds are yet another problem. One species may be breeding in one country, yet be migratory or wintering birds in another. The criteria of endangerment thus differ in nature – they may affect the breeding populations of one species, or 'only' the populations of migratory or wintering birds. Another difference between the Regional Red Lists and the IUCN Red List lies in the newly introduced exotic species. The National Red Lists know the term

Neobiota, which are non-native species introduced by humans. Some duck and goose species plus some parrot species live in Central Europe, for example – they were introduced from America, Asia or Africa. However, the National Red List officials and authorities have decided not to include these species in the Catalogue of species which are worthy of conservation. Only the native species are considered to be worthy; but what if an exotic species imported by humans is threatened with extinction in its home country? In contrast to the National Red Lists, the global IUCN Red List does not recognise any 'native' species because all species are considered in their worldwide occurrences and all species on earth are indigenous.

One major problem is that the National Red Lists can be compared neither with one another nor with the global IUCN List. The classification of species into different levels of endangerment is carried out by the IUCN and by national bodies in accordance with different criteria. For a species to be classified as 'Endangered' (EN) in the IUCN, certain criteria must be met – but these criteria are not the same as those which would classify a species in the corresponding German Red List's 'stark gefährdet' category (Level 2) (Table 4.2). The IUCN endangerment categories, for example, use the threshold principle if the habitat of a formerly widespread species has been reduced to a small area. This value cannot be used by the National Red Lists, because the threshold value used by the IUCN can already be reached if the country in question has only a small area. If a species is found in only one area of less than 20 000 km^2, the IUCN already includes it in its endangerment categories; but we simply cannot say that all species are endangered in Luxemburg, for example, just because the country only has an area of 2600 km^2?

These issues (together with other problems) are the reasons why the endangerment categories of the IUCN Red List cannot simply be transferred to the Regional Red Lists. The Regional Red Lists cannot be compared with the IUCN Red List, and it is also difficult to compare the National Red Lists of individual European countries with one another, so much so that in some cases, comparisons cannot even be drawn between the lists of neighbouring countries. The IUCN has therefore issued guidelines to determine the endangerment levels for endemic and non-endemic species at the regional level. You can download the guidelines in PDF format from http://www.iucn.org/themes/ssc/redlists/regionalguidelines.htm.

However, the implementation of these guidelines at national level is fraught with difficulty. At an international conference in November 2002 in Leiden, Netherlands, an attempt was made to merge the Red Lists of several European countries and create a comparable evaluation with the IUCN lists. The regionalisation criteria suggested by the IUCN was tested to harmonise the Red Lists of countries with one other and with the IUCN Red List (De Iongh *et al.*, 2003) and were intended to enable all countries to create their Red Lists in accordance with one standard procedure.

This standardised approach, however, would have resulted in many endangered species losing their endangered status in the countries in question – and, inversely, many species that were not classified as endangered before the regionalisation criteria were implemented would then have been classified as endangered. In the case of many countries, the harmonisation of the Red Lists could not be achieved and attempting it proved to be a difficult and tedious task, not least because the Red Lists are mandatory for the national legislations in some countries (De Iongh and Bal, 2007).

4.2 The Classification of Threatened Species in Endangerment Categories

The Red Lists are not lists of rare species, but lists of endangered species.

What are Endangerment Categories, Classification Criteria and Status Categories?

The classification criteria of the IUCN and the German Red Lists.

Which species are 'native'? – The Status Categories.

A rare species is not necessarily endangered – the decisive factor is the population trend.

Can the endangerment criteria of the IUCN simply be transferred to the National Red Lists for classification purposes?

4.2.1 Endangerment Categories and Classification Criteria

As explained in the previous chapter, the IUCN's International Red List criteria for the classification of endangered species in the individual endangerment categories differ from those of the National Red Lists – and this is why the lists are incomparable. This section explains the endangerment categories and the criteria for classification in detail.

One common misunderstanding regarding the Red Lists is that they only deal with rare species – but that is not why the lists were created, because there are species which are inherently rare. There are many more rare species than common species. Eagles, for example, are not found in any great numbers anywhere in the world and yet many species of eagle are not classified as being endangered; on the other hand, no matter how intensive eagle conservation efforts may be, eagles could never occur as frequently as sparrows (Reichholf, 2010). So it is not only the rare species that are included in the Red Lists, but the populations of species which have declined sharply over a longer or shorter period of time, even if the species in question are (still) common. If the causes of endangerment to a species are known, we can say with certainty that its population is declining – and a species is only entered into the Red List if its population declines noticeably. The Red Lists are therefore not lists of *rare* species – they are lists of *endangered* species.

One exception is the Category R (extrem selten) in the German Red Lists. Extremely rare species are added to this category, even if their populations are not endangered.

To understand the Red Lists, we must first distinguish between three separate terms which should not be confused with one other: (i) the Endangerment

Category, (ii) the Criterion for Classification in one of the endangerment categories and (iii) the Status Category (Table 4.2).

1) When it is established that a plant or animal species population is in a state of significant decline, then that species is considered to be endangered and is classified in one of three **endangerment categories** depending on the scale of the risk of extinction.
2) No species population, neither plant nor animal, remains constant. The frequency of occurrence of individuals varies from year to year due to natural conditions, but these swings in numbers should not be confused with a population decline. When we speak of a population decline, certain criteria must be met and these enable a lower or higher risk of extinction to be predicted. These are the **classification criteria** – but the IUCN Red List criteria differ considerably from those of the National Red Lists.
3) Besides classification in endangerment categories, there is also classification in **Status Categories**. Status categories are found only in the National Red Lists – they are not in the IUCN Red List, because it makes no sense to use them at the international level. At the national level, however, the question arises whether or not a species really belongs to a country. Species that occur regularly in a particular country are designated as Status I species, and that country bears the responsibility for their local conservation. Status II species are those that occur only occasionally in a country but have not settled there, neither in the past nor the present. Status III species do not belong to the national fauna or flora of the particular country into which they were introduced with human help; these are the Neobiota (Neozoa and Neophyta). The Neobiota are introduced species and should not be confused with immigrant species.

It is the differences in the defined classification criteria that make the International IUCN List incomparable with the National Red Lists; the endangerment categories, however, do correspond to one another. The IUCN Red List endangerment categories can simply be transferred into the endangerment categories of National Red Lists, because they are similar to one other (Table 4.2). In this part of the book, the endangerment categories of the IUCN Red List are explained first, followed by the corresponding categories of the German Red Lists.

In the IUCN Red List, endangered species are classified in three levels: 'critically endangered' (CR), 'endangered' (EN) and 'vulnerable' (VU). The corresponding three levels in the German Red Lists (Haupt *et al.*, 2009) are 'vom Aussterben bedroht' (Category 1), 'stark gefährdet' (Category 2) and 'gefährdet' (Category 3). Today, however, several European countries such as Switzerland (http://www.bafu.admin.ch/tiere/07964/index.html?lang=de) and Austria (http://www.umweltbundesamt.at/umwelt/naturschutz/artenschutz/rl_tiere/) have adopted the IUCN designations for endangerment categories.

The crucial difference between the IUCN list and the National lists lies in the selection of criteria that must be met for a species to be classified in one of the categories. In Germany, for example, these criteria are predominantly qualitative, while the IUCN classification requires quantitative numerical values. This means that the IUCN threshold values that can be expressed in numbers are defined – and the actual numbers must fall below these defined values for a species to be classified in a particular endangerment category (critically endangered, endangered or vulnerable).

Terms such as *negative population trend, significantly declined, significantly threatened* and *low reproductive success level* are sufficient for classification in the endangerment categories of the German Red Lists, but the IUCN requires numerical values such as population size, the size of the distribution area and the extent of the population decline over a given period of time for its corresponding classifications. The classification criteria of the IUCN are as follows (Table 4.2):

1) Highest risk of extinction – 'Critically endangered', Category 1:
   For a species to be classified in this category as per the criteria of the IUCN,
   a. the total population must have declined by at least 80% in the past 10 years, or
   b. the size of the geographical distribution area must be less than 100 km^2, with simultaneous geographical fragmentation, continuous decline or strong fluctuations or
   c. the population must consist of fewer than 250 sexually mature individuals.
2) Low risk of extinction – 'Endangered', Category 2:
   For a species to be classified in this category as per the criteria of the IUCN,
   a. the total population must have declined by at least 50% in the past 10 years, or
   b. the size of the geographical distribution area must be less than 5000 km^2, with simultaneous geographical fragmentation, continuous decline or strong fluctuations or
   c. the population must consist of fewer than 2500 sexually mature individuals.
3) Still lower risk of extinction – 'Vulnerable', Category 3:
   For a species to be classified in this category as per the criteria of the IUCN,
   a. the total population must have declined by at least 20% in the past 10 years, or
   b. the size of the geographical distribution area must be less than 20 000 km^2, with simultaneous geographical fragmentation, continuous decline or strong fluctuations or
   c. the population must consist of fewer than 10 000 sexually mature individuals.

We can see from these three categories that the IUCN only uses objective standards for the classification of species endangerment – and the organisation wishes to avoid subjective assessments. However, the quantitative criteria defined in (a), (b) and (c) cannot be linked with each other. If a species has lost more than 80% of its individuals in the last 10 years, for example, then that would be a criterion for classification in the 'critically endangered' category, because that particular defined threshold had been exceeded. However, if this same species inhabits a large geographical distribution area (b) and the number of sexually mature individuals amounts to more than 250 (c), this species still has to be classified as 'critically endangered', although these criteria do not support the classification in the category 'critically endangered'.

In Germany and several other European countries, the criteria for classification in the endangerment categories are much more flexible. Here, no fixed numerical values are used in assessing the risk of extinction – instead, the population trend (the change in the number of individuals of a species) is assessed, using qualitative to semi-quantitative standards (Table 4.2). This provides an evaluation of the risk factors for the extinction of a species (Haupt *et al.*, 2009). The evaluation is subjective, so it does not meet the strict scientific requirements, but there are no incompatible numerical values to create conflicts, as in the case of the IUCN criteria.

In contrast to the population trend criteria of the IUCN Red Lists, the German Red Lists differentiate between **long-term** and **short-term population trends**. The long-term population trend looks at the past 100 years and more, while the short-term trend covers the past 10 years up to a maximum of 25 years. Including long-term population trends enables a better forecast to be made about the fate of a species, because there are several examples which illustrate that the decline in the number of individuals of a species can persist (although short-term upward trends can also occur and these can awake false hopes).

Classification in the endangerment categories of the German Red Lists is illustrated by the following examples of breeding birds in Germany:

0) The first category is 'ausgestorben' (Category 0) and this corresponds to the 'Extinct' (EX) category in the global IUCN list. In the last half millennium, 16 breeding bird species have become extinct in Germany. Five of these 16 died out before 1850 and the other 11 became extinct after 1850 – the last European roller disappeared in 1991 and the lesser grey shrike became extinct in 2005.

   It is perhaps confusing that Category 0 includes the bald ibis and the griffon vulture – species which have not occurred in Germany for a long time. We would hardly expect to find the bald ibis (a bird which became extinct in Germany at the beginning of the seventeenth century) or the griffon vulture (which died out even before the seventeenth century) in the list of 'regularly breeding native bird species'. Nevertheless, these types are classified in Status Category I (regularly breeding native species), and this classification is based on a limit set on a deadline date in 1492, the year America was discovered. The time between 1492 and today is regarded as a 'historical period' and all the species that have been breeding regularly in Germany during this period are said to be native species (Südbeck *et al.*, 2007); so the bald ibis and the griffon vulture are classified as native German breeding birds. Conversely, the same deadline date also applies to Neobiota (Neozoa and Neophyta); all the species introduced in Germany by humans after 1492 are classified as Neobiota in Status Category III.

1) The German Red List Category 1, 'vom Aussterben bedroht' corresponds to the 'Critically Endangered' (CR) category in the global IUCN list – and species in this category will die out in Germany in the foreseeable future, unless the causes of endangerment can be eradicated. Thirty species of breeding birds in Germany are included in this category, and the great bustard and the aquatic warbler are also classified as being internationally endangered, so Germany bears a special responsibility for the conservation of these two species. However, the great bustard can only survive in Germany with intensive human management in semi-artificial biotopes (see Section 6.5.3) and the population of the aquatic warbler has declined so dramatically in the last half century that fewer than 10 pairs survive, and it will be an almost impossible task to save them. Category 1 also includes the three typical, former common meadow waders, the curlew, the black-tailed godwit and the common snipe. There are currently several thousand pairs of these species breeding in Germany, but their numbers are rapidly dwindling.
2) Category 2, 'stark gefährdet' corresponds to the 'Endangered' (EN) category in the IUCN list and it signifies that the population of the species in Germany has declined significantly due to the visible deterioration of its living conditions. This list includes 24 breeding bird species, among which are those that were still common in Germany half a century ago, such as the grey partridge, the black grouse and the lapwing.
3) 'Gefährdet' (Category 3) corresponds to the 'Vulnerable' (VU) category in the IUCN list. It indicates that the species is in continuous decline, although some of these Category 3 species can currently boast tens of thousands of breeding pairs in Germany. The German list includes 14 species of breeding birds, among them the turtle dove and typical species that have adapted to sparse ground cover, such as the skylark, the whinchat and the corn and ortolan buntings.

4.2.2
Status Categories – Regularly Occurring Species, Irregularly Occurring Species and Neobiota

The Regional Red Lists differ from the IUCN Red List in that they use the category term *Status* – but the use of this in the IUCN list would make no sense. There are regularly occurring species, irregularly occurring species and introduced species (Neobiota) within the limited geographical area of a country or state and we must differentiate between these three groups. *Status* categories should not be confused with *endangerment* categories; only the endangerment categories state that existing causes of endangerment have led to population losses and can lead to the extinction of the species in a particular country, if the causes are not eradicated. In contrast, the status categories mean that the regularly occurring species are separated from the irregularly occurring species, whose irregular occurrence has nothing to do with the causes of endangerment existing in a particular country.

Only regularly occurring species can have a native species status. Irregularly occurring species and introduced species do not belong to the native fauna and flora and therefore have a special status – so the German Red Lists classify the regularly occurring species, the irregularly occurring species and the introduced species in three Status Categories – I–III. This is illustrated by the following examples of species occurring in Germany:

Status Category I. All the species that have occurred in a region for a period of several consecutive years can be given the status of regularly occurring species in that region. Only these species can be assessed in their degree of endangerment, because fluctuating, sporadically occurring species are not affected by the endangerment factors to which the Red Lists are meant to draw attention. Only the Status I species actually belong to a country or region. Strictly speaking, the birds which only occasionally breed in Germany are not German breeding birds, if there is no reason to assume that they would immigrate and establish a stable population in Germany.

It is at first a little confusing when we see that extinct species from historical times are also included among the regularly breeding native bird species in Germany and given the Status I. The result of this is that the number of regularly breeding species in Germany is greater than the number of species that actually occur there today. For example, Germany has 260 Status I breeding species, but currently (as of 2007) only 244 of these actually breed in Germany (Südbeck *et al.*, 2007). The other 16 species became historically extinct in Germany after 1492.

Status Category II. All the species that only temporarily or irregularly occur in a country or region are classified in this category, in the absence of any criteria that would point to immigration. The areas of distribution of the species concerned are not in Germany and the occasional occurrence must be regarded as an exception, because there are no reasons that would point to a repetition of the occurrence.

As startling as these rarities are and as exciting as they may be for rarity hunters, they hardly represent the state of the domestic nature and landscape. Sporadically occurring species are not suitable indicators of the changes in our landscape over the past decades and centuries – and they are therefore meaningless for the conservation or propagation of native species' biodiversity.

The Red List of breeding birds in Germany classifies 25 species in Status Category II. These include the greenshank, a breeding bird in Scandinavia and northern Russia (seen nesting in 1997 on a sandy area beside a pine forest in northern Bavaria) and the bean goose and white-fronted goose, both of which winter in Germany. It is possible that the isolated broods of these two species of geese are remnants of the winter population that were left behind in Germany during the last few years.

The declaration of some species as Status II species is a rather uncertain issue, because on the one hand, species that may have been native in earlier centuries (and consequently belonged to the Status I species) could be involved, such as the white-winged tern and the saker falcon, which are assumed to be two such species

(Barthel, 2011). On the other hand, there are possibly also Status II species (e.g. the little egret) that may have stably immigrated into Germany in stable fashion during the last few decades and thus become Status I species.

Status Category III (Neobiota). All the species that have been introduced to Germany from foreign countries by humans are included in this category; and these species are able to feed and reproduce independently without direct human interaction. They include the Canada goose, the Egyptian goose, the ruddy shelduck, the wood duck, the mandarin duck, three flamingo species, three species of parrots and the pheasant, which had already been introduced by the Romans to Central and Western Europe, but which was continually imported from Central Asia after 1492.

4.2.3
The Meaning of Population Trends in the German Red Lists

There are species of animals with a large number of individuals (e.g. some sea bird species) and there are species that have never occurred frequently due to their natural characteristics – and these simply cannot occur frequently, because they cannot live in breeding colonies, for example, and they require large ranges to find sufficient food; so the number of individuals a species has actually provides little insight into whether or not that species is endangered. In order to assess the level of endangerment, the change in the number of individuals within a given period (the population trend) is much more important than the species' frequency of occurrence.

The grey partridge, the lapwing and the black grouse are all classified in endangerment level 2 (stark gefährdet) of the Red List of German breeding birds, although the grey partridge (with 90 000 breeding pairs) and the lapwing (80 000 breeding pairs) occur almost 100 times more frequently than the black grouse (only slightly more than 1000 breeding females) (Südbeck *et al.*, 2007). The snipe and the curlew, with 6000 and 3300 breeding pairs, respectively, in Germany are classified in endangerment level 1 (vom Aussterben bedroht), while the green sandpiper – with only 800 breeding pairs – is not included in the German Red List, because its long-term and short-term population is increasing. It is clear from these figures that the number of individuals says very little about the level of endangerment to a species. Certain types *cannot* occur with any frequency in Germany, because of the restricted number of suitable habitats. Only the *trend* in population is crucial in predicting the future population of a species, *not* the current population.

When first published, the Red Lists were lists of rare species – nowadays, however, the Red Lists are increasingly replacing the term *rare* with the term *population change*. This is not feasible in every case, however, because the assessment of population trends requires that the pertinent data remains available over a lengthy period of time. Efforts are now being made, however, to include the populations of significantly declining species in the Red List and not just rare species. Rare species are only included in the Red Lists in extreme cases; they are given the category level R (extrem selten, extremely rare) in the German Red Lists. Several species, such as the rock thrush, the scaup and the Slavonian grebe, which only breed very locally in a few places in Germany and often number just 2–5 breeding pairs are also included in the quantitative statistics and that also increases the total number of German breeding birds. This is especially true of species that Germany has regained in recent years, but these species do not play a very important role in overall biodiversity.

In accordance with the quantitative criteria of the IUCN, species are classified in the 'critically endangered', 'endangered' or 'vulnerable' risk categories if their total population has declined by at least 80%, 50% or 20% in the last 10 years (Table 4.2). In the German Red Lists, the population trends are only consistently taken into account in the latest lists of breeding birds (Südbeck *et al.*, 2007) and butterflies (Reinhardt and Bolz, 2011; Binot-Hafke *et al.*, 2012). In contrast to the IUCN list, however, the decline in populations is based on qualitative assessments in the German lists, which use terms such as *zurückgegangen* (declined) or *erheblich zurückgegangen* (significantly declined). However, the German breeding bird lists enable a far better prognosis to be made about the future fate of a species, since they take the short-term population decline over the last 10–25 years and the long-term decline over the past 100 or even 150 years into account; in the IUCN list only the changes over the last 10 years are recorded, and this means that a short-term recovery phase could effectively obscure a true population decline.

The problem here is that the loss of individuals takes precedence over the loss of species. Before a species becomes extinct, it first of all becomes increasingly rarer. The current species decline in Central Europe should not be measured by looking at the reduction in the number of species, but by comparing the number of species that have become rare with the number of species that have become more frequent in their occurrence. To ascertain the true state of biodiversity, we should not simply look at the *number* of species – the *short-term and long-term population trends of each species* must be taken into account. A species decline can be clearly identified, but the loss of individuals is much more difficult to measure.

A look at the numbers of breeding birds and butterflies in Germany shows that the numbers of their species have hardly changed over the last 100 years (Table 4.1), but these data lead to an incorrect assessment of the true situation regarding current biodiversity loss. Focusing solely on the number of species obscures the true picture of declining German bird numbers and – in particular – declining butterfly numbers in Germany. Only when we consider the population trends within the species does it become clear that the number of species which have become rare is greater than the number of species which have become more common in the long-term trend period over a hundred years and more. In the case of breeding birds in Germany, 44% of all species have become rarer, but only 32% have become more common, and the butterfly population changes in Germany are even more drastic; two-thirds of all butterflies in Germany have become rarer, while only 2% have become more common (Table 4.1).

4.2.4
The Harmonisation of the IUCN Criteria with the Classification Criteria of the Regional Red Lists

Different criteria for endangerment category classification make the Red Lists incomparable with one another, but the data situation also varies from country to country and this is also the case in Europe. Some countries have their data updating system better organised and have surveyed a greater number of animal and plant groups than other countries – although the species diversity of some groups, such as the butterflies, is often better in countries which have not yet surveyed these groups of animals or have surveyed them inadequately at best. In Europe, the endangerment levels of almost all butterfly species have been researched, but from a global perspective hardly 0.4% of all known butterfly species have been surveyed.

There exist almost 4000 Red Lists worldwide. In Europe, roughly half of the lists are from Germany; Britain, however, lags far behind (De Iongh and Bal, 2007). This is surprising, because field biological interest in animals and plants in the United Kingdom is greater than it is in Germany, but this difference in numbers has presumably more to do with the bureaucratisation zeal of the Germans than Germany attaching more importance to the conservation of species.

In 2002, an international seminar was held in Leiden, Netherlands. Its ultimate goal was to harmonise the Red Lists and make them comparable. New IUCN criteria for endangerment category classification (Categories and Criteria C&C version 3.1) were tested for their applicability to the National Red Lists of European countries (De Iongh *et al.*, 2003; De Iongh and Bal, 2007). The new IUCN criteria were 'regionalisation criteria' that had been specifically modified to match the regional requirements of individual countries to one other and to the criteria of the IUCN – but the attempt to use the new regionalisation lists as tools that would enable all countries to create their Red Lists in accordance with one standard procedure proved to be more or less wishful thinking, because the harmonisation of the Red Lists was not achieved.

The harmonisation of the Red Lists was made more difficult by the fact that the lists are mandatory for national legislation in some countries. Only a few European countries, such as Finland, Sweden and the United Kingdom adopted the IUCN's modified regionalisation criteria. Other countries, however – Germany included – are still using their own criteria for the classification of endangered species in the endangerment categories of the Red Lists.

The discrepancy between the use of the IUCN criteria and the national criteria can be partly explained by the fact that geographically small countries, for example, cannot adopt the endangerment criterion which limits the occurrence of a species to a small range. While the IUCN regards the limitation of a species population to a small geographic area as a risk of extinction for the species, this approach naturally cannot be applied to a list system that is only limited to a small area due to political reasons, because the extinction of a species in a politically limited space need not be a threat to the global survival of that species. The golden eagle in Switzerland is one example of this: if the IUCN criteria were used in this case, this bird would be included in the Red List of endangered species, although it is not at all endangered as a breeding bird in Switzerland.

When the IUCN criteria approach was applied to the Netherlands, the result was that a considerable number of species there would have lost their National Red List status if the IUCN criteria had been implemented; but some species would have had to be newly included in the Red Lists. Only around half of the Dutch species would have overlapped one another in their endangerment category classifications if the previous classification criteria of the Dutch Red Lists had been replaced by the IUCN criteria. The Dutch government subsequently decided to retain its own Dutch Red List classification criteria, but to use the IUCN quantitative thresholds to create a European Red List.

4.3 The National Responsibility for Certain Species

The IUCN List and the National Red Lists have different objectives.

A protected rarity in one country is often common elsewhere.

Species colonise both a core area and also border zones, which have suboptimal living conditions.

A nation bears a special responsibility for the protection of a species if its core area is within the country's territorial boundaries.

4.3.1 A Protected Rarity in One Country Is Often Common Elsewhere

Germany and Switzerland are trying desperately to protect the last woodchat shrikes. These birds bred frequently for centuries in both countries, but today there are only a few pairs left – and these few are threatened with extinction. To protect these last breeding pairs, public tax money is being used to fund the planting and cultivation of meadow orchards, the bird's ideal habitat. In Spain, hardly anyone would understand these efforts, because in this country the woodchat shrike occurs frequently and is definitely not endangered. The Spanish would prefer to see the Swiss using their tax funds to promote the conservation of the snowfinch – one-third of its European population occurs in Switzerland. However, the snowfinch does not need any conservation efforts, since it is not at risk in Switzerland (Keller, 2002).

In this case, we are also subject to our own prejudices – it all depends on the perspective from which we look at a country. We tend to see our own countries in a different light. While the Germans and the Swiss strive to conserve the woodchat shrike and willingly raise funds to this end, the same people would naturally not be willing to donate large sums to conserve the chaffinch in Tunisia or the jays in Japan: both of these species are endangered in those countries, because the individuals of these species live there in less than optimal conditions at the periphery of their distribution ranges (Flade, 1998). Germans see the chaffinch and the jay from their perspective – and the chaffinch is the most common breeding bird in Germany with roughly 10 million breeding pairs, while the jay numbers half a million breeding pairs (Südbeck *et al.*, 2007).

Endangerment to a species can thus be viewed from a regional perspective *and* from a global perspective. When we realise that the populations of almost a third of all German breeding birds are endangered, any person who truly cares about biodiversity will be shocked. That shock is tinged with relief, however, when we realise that these catastrophic numbers only refer to a limited region. When a plant or animal species is listed as endangered or threatened with extinction in one of the German Red Lists, this is usually only a regional classification – but this kind of endangerment can look completely different when seen globally.

Several German ornithologists (even the experienced) do not know which species has a high percentage of its total global population in Germany (Flade, 1998). When this question is asked, most name only the red kite, since 60% of its global population breeds in Germany – but a quarter of the world's firecrest population also breeds in Germany and it is hardly ever mentioned. It is also a little-known fact that the breeding area of the wood pigeon is almost completely restricted to Europe – but the wood pigeon is actually the most common non-songbird in Germany (around 2 million breeding pairs).

So in the case of most of the species classified as endangered in the Regional Red Lists, the real issue is not the preservation of a species – it is the depletion of diversity in a given region. It is not the objective of most of the Red Lists of a country or of a state to save species that are declining or even dying out, it has more to do with the home country itself. It is the responsibility of the IUCN to draw attention to the global extinction of a plant or animal species (Baillie, Hilton-Taylor and Stuart, 2004). The purpose of the global IUCN Red List is to counteract the irreversible disappearance of species that cannot be recreated. The Regional Red Lists, however, follow a completely different objective: to preserve the wealth of the homeland by counteracting the depletion of regions and landscapes through the conservation of species. These are very different ethical reasons.

The global IUCN Red List makes statements which differ from those of the Regional Red Lists of individual countries – and this is illustrated by the fact that the percentages of individual endangered plant and animal species are much higher in the Regional Red Lists than those in the global IUCN Red List. The larger a region (surveyed for the Red List), the lower the percentage of endangered species there – so nearly a third of all the 244 native, living breeding bird species in Germany is endangered; but only 12% of the almost 10 000 bird species on Earth are threatened (Baillie, Hilton-Taylor and Stuart, 2004).

If, for example, all the bird species breeding in Germany were exterminated, the global existence of only a few species would be endangered. This applies even more drastically to the butterflies – most of the butterfly species occurring in Germany have the core areas of their distribution in Eastern or Southern European countries. With probably the single exception of the dusky large blue

(*Phengaris nausithous*) no butterfly species in Germany amounts to a large percentage of its world population.

The important difference between the Global and the National Red Lists (mentioned earlier) only applies to species that are not endangered worldwide. Only a few species occurring in Germany are endangered Europe-wide or even worldwide – so these species have a very different status and must be considered separately. Unfortunately, we are only aware of the Europe-wide threat for most animal groups. The worldwide threat to various species is mostly known for mammals, birds and amphibians, since the non-European database for almost all other animal groups (and most plants) is very patchy (Baillie, Hilton-Taylor and Stuart, 2004).

4.3.2
The Occurrence of a Species Is Divided into Core Areas and Border Zones

Limited resources mean that a country cannot protect all species equally. This is why priority lists must be intensively discussed and enforced if necessary. However, this problem does not exist for the global IUCN Red List, because here the simple principle rules and that is that protection must be provided for species that are very rare and for those species the populations of which are in decline.

This principle cannot be applied at national level for most species, however, because most species which are rare and in decline in one country are often common elsewhere. It is only in the case of a country's endemic species that worldwide distribution need not be taken into consideration, since these species do not occur anywhere else. The Central European nations, however, have almost no endemic species, probably due to their decimation through glacial destruction. For example, only about 25 German endemic animal species are known to exist (Gruttke, 2004); but the question for some of these species is whether they are species or 'only' subspecies.

For all other species, the issue remains, which criteria should be used to set priorities? These priorities define the so-called *responsibility principle*, in accordance with which certain countries bear a special responsibility for the protection of certain species (Gruttke, 2004). Each nation should protect certain preferred species and no others. This responsibility is based on the realisation that most species do not occur uniformly in their distribution areas. A species does not simply have one particular area of distribution in purely quantitative terms, it occurs in different degrees of quality in this area. There are one or more focal centres of distribution (core areas, usually in the centre of the range) – and on the borders, there are zones with a low population density and populations which are unstable.

A species has the highest population density in the core area, because it has optimal ecological, climatic and geographical conditions there. This core area is surrounded by a border zone which effectively marks the distribution limits of the individuals of the species. Ecological and climatic conditions are usually less than optimal in this border zone, so a species, weakened competitively and having to live 'on the breadline' because of these conditions, becomes particularly vulnerable (Gruttke, 2004). Species in border zones are often very unstable; they do not reproduce sufficiently and their populations can collapse quickly if ecological or climatic disturbances occur, so they are dependent on immigration from the core area. Such collapses are not always caused by the deterioration of living conditions in the border zone itself, but by insufficient immigration of individuals from the core area. North-western Central Europe is under the influence of the Atlantic climate and as such it is a border zone, in which the survival of several species is dependent on immigrations from the east and the south. Conservation measures are therefore needed more in core areas than in the border zones. One example of this is possibly the white stork in north-western Germany, where the 1970s saw a population collapse in the Lower Rhine and Ems areas – and this collapse may not have been due to local environmental conditions, but to lack of immigration due to low reproduction rates of the populations in the core areas. The white stork is a thermal glider and therefore adapted to continental climates. Despite adequate food supplies, the bird has never occurred frequently in Atlantic climate-influenced north-western European countries such as Holland – but in Central Europe it becomes more frequent the farther east we travel.

Isolated border zone populations, however, can also achieve a high species protection value, which in itself is contrary to their value as 'unimportant border zones of the species'. This is the case if the genetic depletion of allelic diversity in peripherally isolated populations leads to new evolutionary achievements. Some individual alleles, which had no chance to compete with other alleles in the populations in the core area of the species may dominate the individuals of isolated populations and individuals with novel characteristics may emerge as a result. Such populations are known as *evolutionary significant units* (ESUs) (see Section 4.4.3). ESUs can lead to speciation and therefore are especially significant for the protection of species (Crandall *et al.*, 2000).

The responsibility principle is theoretically well founded, but in practice it is often difficult to apply, because the boundaries between the core area and the boundary zones are difficult to define. This is why no distinction can often be made between core area and border zone – and the responsibility a nation bears for a particular species is measured by the proportion of individuals living in that nation. If a high proportion of a species' total global population occurs in a country, it is highly probable that the core area of that species is located in that country, and this is why the country in question bears a responsibility for the conservation of this species. One example of this is the red kite in Germany – around 12 000 pairs in that country represent almost two-thirds of the bird's world population.

The responsibility principle is undoubtedly of great value for setting priorities in species conservation. In the past, little thought was given to the international obligations of responsibility that countries actually bear for the global conservation of a given species. The responsibility principle has played a minor role in German nature protection during the last few decades, for example, and – in contrast to the rarer species – insufficient attention has been paid to the species that have a high proportion of their total populations in Germany. Nature conservation has focused too intensively on the 'rehabilitation' of the flagship species, the crane and the sea eagle but neglected its duties to other species (Flade, 1998).

However, the sole use of the responsibility principle is controversial, since it can often be determined only in Europe, but not worldwide. For example, with probably only one exception (discussed earlier) no butterfly species which has a high proportion of its global population in Germany even occurs in that country. Germany is only a border population zone for most butterfly species and not a core area – so does Germany bear only a minimal responsibility for the protection of its butterflies?

One-third of the European population of snowfinches breeds in Switzerland. This means that Switzerland would bear a national responsibility to give special priority to the protection of the snowfinch within the European zone. However, any possible endangerment which could threaten the snowfinch population cannot be currently identified in Switzerland. On the other hand, if one considers countries outside of Europe, the snowfinch is a very common species. It occurs all over the entire Himalayan region. This reduces the global percentage of snowfinches in Switzerland and also decreases that country's national responsibility for the bird proportionately – but can we say that the Swiss snowfinch population is not really important, because we can always travel to the Himalayas to see them if the birds should disappear from Switzerland?

The expansion of core areas and border zones may shift over time when ecological and climatic conditions change. It is only in recent decades that man has destroyed open areas in Central European landscapes through afforestation and eutrophication. Apart from agricultural land, Central Europe now has hardly any of the open-ground areas, which characterised this landscape for thousands of years. The open-ground-areas became overgrown or 'over-forested', because of declining grazing areas and increasing afforestation.

One example of a species affected by these changes is the black grouse. The black grouse is a boreal–subarctic bird of the tundra–taiga transition zone. The highest populations of black grouse occur in Scandinavia, northern Asia and the treelines of the high mountains, all of which match its living conditions perfectly. Thousands of years ago, postglacial tillers and livestock farmers encouraged the black grouse's immigration to the Central European lowlands by creating an artificial tundra/taiga-like habitat through deforestation, large-area fires and intensive livestock grazing. This expanded the core areas of the species. Up until 50 years ago, the black grouse was a common breeding bird in Germany. It lived on the heaths and the partially drained high moors in northern Germany and in the areas north of the Alps in Baden-Wuerttemberg and Bavaria, as well as on the deforested high plateaus of the German Central Uplands in anthropogenically created, optimal biotopes. These biotopes cannot be regarded as being border zones, but as an expanded core population area which stretched into Central Europe from the north and east. In recent decades, scrub encroachment and afforestation have made the Central European habitats uninhabitable for the black grouse.

This and many other examples clearly illustrate the situation in Central Europe. It is a region of dramatic landscape change – and in recent decades it has lost its status as a core area for many species, which have been pushed back to the north, east and south. This has caused Central Europe to become a border zone for many species, one in which they now have to live under suboptimal conditions and where they are at risk of regional extinction. In earlier centuries, however, these regions were still part of the core area for these species. Should we tamely accept that today we live in a wooded landscape where birds remain scarce – and without beautiful species such as the black grouse, the hoopoe and the European roller? Can we come to terms with the fact that Germany, for example, is no longer a core area for most of the shrike, skylark and bunting species, but only a border zone? And that the conservation of these species is no longer the main focus of species conservation?

4.4
Species or Genetically Isolated Populations – What Should Be Protected?

The term species is not uniformly defined and is dependent on subjective decisions.

Increase of species numbers through taxonomic reclassifications.

Barcode taxonomy only defines a species according to genetic distance.

Since there is no objectively justifiable definition of what a species actually is, there are no numbers of species which can be justified by purely scientific criteria.

What does the term species mean in the protection of biodiversity?

How important are deviations in subpopulations in the protection of biodiversity?

The concept of 'Evolutionary Significant Units' (ESUs).

Do apomictic clones in plants have an impact on the Red Lists?

4.4.1
What Are 'Species'? The Final Decision Is Made by Man

The global IUCN Red list is titled '2004 IUCN Red List of threatened species – a global species assessment' (Baillie, Hilton-Taylor and Stuart, 2004). The term *species* crops up twice in this title – but what exactly is a species? Species are groups of individuals; however, which of these groups exist in reality and how they are segregated from each other, is disputed (Kunz, 2012a). Scientists still argue today about whether a species is an entity which is tangible, with an existing counterpart in nature, or whether species are merely abstractions to classify biological diversity, which in reality is interflowing. It is neither a matter of course for everyone, nor does everyone have the same concept of what a species actually is and of how the genetic and phenotypic abundance of organisms is divided into groups, which are called *species*. Many bio-philosophers believe that species are 'concoctions', stating that only collections of dead animals formed clearly separated groups in their drawers, whereas organisms in nature change in time and range over wide geographical areas – and they are not so well segregated that they can fit into the virtual drawers of the constructed taxons. Nature itself is not as rigorous as this ideal of the classifying taxonomists.

Species are defined, but not without controversy and are often only limited by subjective criteria. The tendency of some taxonomists ('splitters') to intensively subdivide certain genera and families into species and the inclination of other taxonomists ('lumpers') to classify the various groups of other genera and families 'more generously' as species led to different measuring standards being used to segregate the plants and animals into different taxonomic groups and to subdivide these into species (Kunz, 2012a). The number of species between different families from different animal phyla strains often cannot be compared with one another because they relate to different entities.

Species are anything but genetically or phenotypically uniform groups. Many species are genetically and phenotypically heterogeneous. None of the criteria used by taxonomists to identify and differentiate between species is based on a scientifically rigorous explanation of why this criterion in particular should suffice for justifying species segregation. This requires a decision made by human beings. Even if two populations can be segregated by unique characteristics, they could merely be intraspecific variants. In the phenotypical distinguishing of species and the segregation of species according to differences in their DNA sequences, it is ultimately the subjective discretionary benchmarks that decide when a species should be declared as such. Nature provides us with no observable criteria for the final decision (Baum, 2009).

The official bodies which prepare the Red Lists are thus not only faced with the dilemma that species are discovered every day, but they also face a problem in that already long-known species have also been newly subdivided. This results in ongoing changes in the numbers of species while the Red Lists are being reviewed and updated. An increase in the number of species in an endangerment category of the Red List is not necessarily due to the fact that the number of endangered species has meanwhile increased. It may also be because populations or races that were previously regarded as being conspecific have been subdivided into several new species. This could in turn mean that the growth in the number of species in one of the endangerment categories of the Red Lists is only an apparent increase. It is important to distinguish changes in species numbers which are caused by the reclassification of previous species descriptions from those caused by an actual change in the level of endangerment.

This results in a problem for the Red Lists and that is that the numbers of endangered species are not directly comparable with one another after a certain period of time – and this makes it difficult to assess decreasing or increasing population trends. Between 1993 and 2004, the number of mammal species increased from 4629 to 5416 – and this was hardly due to new discoveries, but rather to the subdividing of new species (Baillie, Hilton-Taylor and Stuart, 2004). Between 2000 and 2004, the number of reptiles increased from 7970 to 8163 and the number of amphibians and fish increased from 4950 to 5743 and from 25 000 to 28 500, respectively, during the same timeframe. Only the generally accepted number of species of birds – about 9900 species, researched much more thoroughly – has not changed much since 1996. There are, however, new taxonomic subdivisions involved here as well. European species such as the carrion crow (*Corvus corone*), white wagtail (*Motacilla alba*), yellow wagtail (*Motacilla flava*), black-eared wheatear (*Oenanthe hispanica*) and orphean warbler (*Sylvia hortensis*) were classified as species a few decades ago, but today these are classified in separate species: *Corvus corone* in *C. corone* and *C. cornix*, *Motacilla alba* in *M. alba* and *M. yarrellii*, *Motacilla flava* in *M. lutea*, *M. feldegg* and other *Motacilla* species, *Oenanthe hispanica* in *O. hispanica* and *O. melanoleuca* and *Sylvia hortensis* in *S. hortensis* and *S. crassirostris*. However, during the last few years even the well-known bird species have been radically split into several new species, which has increased the total number of birds in the world from 10 000 to roughly 12 000 (del Hoyo and Collar, 2014).

4.4.2
Barcode Taxonomy – A Conflict between Application and the Theoretical Basis

Today's taxonomy specialists are trying their utmost to use a species concept that approximates nature's entities as closely as possible – a concept which contains a minimum of human-made order and segregation criteria. This goal will most likely be achieved through the 'Barcode' taxonomy concept (Hebert *et al.*, 2004). This taxonomy was launched only 15 years ago – for species classification, it uses only those criteria of organism coherence or organism segregation that can be seen in nature and these can be expressed in numbers, as opposed to a taxonomy based on morphological characteristics, where the usability of a characteristic depends on subjective considerations. Barcode taxonomy has experienced an enormous upswing in the last decade and is considered by many taxonomists to be the 'future of taxonomy' (Tautz *et al.*, 2002). The species definition underlying barcode technology is, however, one-sided, in that it does not take many criteria of biodiversity into account (Ebach and Holdrege, 2005; Will, Mishler and Wheeler, 2005; Kunz, 2006, 2012b, 2015; Dasmahapatra *et al.*, 2010).

Barcode taxonomy is based on the 'molecular clock'. This refers to an (albeit very rough) proportionality between the progression of time and the occurrence of new mutations. For example, two populations that were separated phylogenetically 100 000 years ago are distinguished by a lower number of mutations than two populations that were separated 1 million years ago. If we select two homologous DNA sequences from the individuals of two populations, the number of base differences between these sequences tells us how long ago the two lineages separated from one another – so a DNA sequence comparison gives us a gauge for the relationship between two lineages. Many base exchanges mean that there is only a minimal relationship; fewer base exchanges mean a close relationship.

Barcode taxonomy now refers to all populations that have exceeded a certain threshold of differences in DNA sequences (i.e. those that have exceeded a certain evolutionary age as living populations segregated from one another) as species. If the threshold is not exceeded, then the two populations still belong to the same species. Barcode taxonomy defines each group of organisms as being its own species, one which has a sufficient phylogenetic distance between it and a neighbouring group. This puts species definition on a verifiable and reproducible basis.

The barcode species thus has the advantage of being objectively definable because the evolutionary distance used as a criterion for species definition is from its very nature an empirically readable, quantitative date of the divergence of two groups of organisms. Differences in DNA sequences can be measured by technical means and are therefore not influenced by the subjective discretionary powers of human beings. In contrast, phenotypic species identification characteristics must almost always be initially assessed for validity, while differences in DNA sequences are primarily objectively definable, quantitative differences. In addition, two similar DNA sequences must always have been created monophyletically; they cannot have subsequently become similar to one another by convergence, thus simulating a close phylogenetic proximity that does not exist in reality. In some cases, the application of the barcoding method has recently led to the segregation of species which had previously been considered as conspecific (Hebert *et al.*, 2004); and this has led to an increase in the number of species.

However, the problem with the barcode species is that the limits between species must be subjectively established. Where is this limit between species? At which point are two populations genetically different enough that they can be regarded as being independent, evolutionarily significant entities? When are a certain number of differences sufficient to distinguish two groups of individuals from one another and classify these groups as being 'real' species? There is no biological law which can answer these questions, only subjective human discretion can do so. The genetic distance threshold above which we classify two populations as species and below which we classify two other populations as being in the same species must be determined and established. No empirical rule associated with nature states the evolutionary distance from which two populations are established as separate entities.

Another weak point of barcode taxonomy is that species which have been created in evolutionarily short periods of time cannot be surveyed. In sympatric speciation, new species can originate when two populations colonise different ecological niches and then become subject to different selection conditions. They then acquire different identifying characteristics and are reproductively segregated. In sympatric speciation, these processes happen relatively quickly – they are not reflected in measurable DNA differences somewhere in the genome.

However, it would be unfounded to deny a species level to such recently created species and to devalue them as 'populations undergoing speciation', because in principle the speciation process is constantly in motion in all species. As a matter of fact, these evolutionarily young and newly created species are no longer functionally interchangeable, because they would evince a reduced capability for survival in the other habitat. They therefore are undoubtedly 'true' species; they are not 'populations undergoing speciation'. Nonetheless, their genomes remain almost the same, because the functional changes that have led to the species-specific adaptations in the individuals of both populations are anchored in very few genes, namely only in those that underpin the adaptations to the new habitat (Schliewen *et al.*, 2001). The problem with barcode taxonomy is that these differences cannot be detected by DNA differences; and in accordance with the criteria of the barcode species definition, the individuals of two evolutionary young species must therefore be grouped together as conspecific individuals, because the DNA sequences used for the barcode are identical to one another. In practice, barcode taxonomy solves the problem by using other species concepts to performing species classification – but the barcoders do not take into account that some of the fundamentals of different species concepts are in contradiction with the barcode species concept.

On the other hand, if old species are considered, the same problem arises for barcode taxonomy, especially if these ancient species have a wide geographical distribution. Where barcode technology recognises different young species to be conspecific, old species (in turn) are split into a number of different species. In the long term, two populations that are geographically distant from one another can be subject to identical selection conditions. As a result, the few genes that are responsible for adapting to the environmental conditions do not change, but other parts of the genomes of the distant populations differ substantially, because the 'molecular clock' has worked for long periods. It is difficult to find a convincing argument for subdividing these distant populations into different species because of the genome differences, simply because the geographically distant populations have remained functionally interchangeable with one another. If we placed the individuals of a population into the territory of the other population, they would show the same capabilities for survival there. However, since the number of neutral mutations that significantly distinguish the two populations will have increased during the long evolutionary period in which the geographically distant populations have been segregated, barcode taxonomy is forced to classify the individuals of both populations as two separate species, because the threshold of the DNA sequences which are measured as barcodes for subdivision into separate species has long been exceeded.

Barcode taxonomy must therefore define species borderlines which – in extreme cases – are incongruent with the species borderline definition decisions that have to be taken using other species concepts which are equally justified evolution-wise (Kunz, 2006, 2012a,b, 2015). Phylogenetically young and therefore genetically homogeneous species with minimal genomic differences must be combined into one common species, although they represent groups which function differently – and phylogenetically old and therefore extremely heterogeneous species (genetically) with a high number of intraspecific genomic differences must be subdivided into many different species in accordance with the barcode method. Barcode taxonomy cannot address the fact that both young and old species exist. In critical cases, barcode taxonomy must also make decisions that are not based on strict scientific principles. As objective entities, species are not clearly defined in nature and are therefore not countable. Specifying the number of species is – as a matter of principle – a matter of human judgement.

Barcode taxonomy has nevertheless prevailed over other taxonomies in practice, more or less ignoring all theoretical objections (www.barcodeoflife.org/). Pragmatic reasoning and not theoretical concepts are the reason behind this victory, simply because species identifications can be carried out at breakneck pace using barcode taxonomy. In creating the Red Lists, it is simply not feasible to use traditional methods to find all the species, before these have become extinct. Given the fact that species are fast becoming extinct in many regions – and increasingly so – traditional taxonomy is faced with a dilemma. To classify species, barcode taxonomy uses technologies that can be increasingly performed by machines; so one result of the barcode method is the chance that the classification of species could keep pace with their extinction rates. Barcode taxonomy is thus given a prominent role in species conservation.

4.4.3
The Protection of Evolutionarily Significant Units Instead of Species

Since we do not have a definition of what a species actually is – that is, a definition which is objective and therefore recognised and accepted by all authorities – there are no numbers of species which can be justified by purely scientific criteria. A statement made about an existing number of species initially sounds so objective, but it is ultimately based on a subjective evaluation; and the experts are in total disagreement over the number of species that are scientifically described. Approximately 1.6–1.9 million plant and animal species have been recorded and scientifically described until now – but even in the case of known animal groups such as mammals, the knowledge gained continually results in taxonomic reassessments. Apart from these changes in the 'known species pool', the number of species not yet recorded is estimated to be between an additional 1.5 and 30 million. Some authors even believe the number of currently existing species is as high as 100 million (Wirth and Glaw, 2009). The disunity of science about what a species actually is proves to be a major stumbling block in the effort to provide quantitative data on the number of endangered species.

It is simply not enough that the Red Lists are limited to the protection and conservation of species. Why? Because first of all, species are vaguely defined and there is a risk that a subgroup set could be neglected, one which might be classified as a separate species a few years later. Second, even 'safe' species are not the 'last instance' of biodiversity. Species are anything but genetically or phenotypically uniform groups. Many species are genetically and phenotypically heterogeneous. If we were to search for an entity or 'unit' of biodiversity, we would not find the species.

There are species which are extremely diverse, both genetically and phenotypically. This is due to their advanced evolutionary age, their wide geographical distribution and the fact that they have not gone through a genetic bottleneck. Species can contain subpopulations that significantly differ in their genes from the rest of the populations, although they are classified as belonging to the same species. If we were to protect only a subpopulation of one species and expose other subpopulations of the same species to the risk of extinction, we would reveal the real goal, namely the conservation of biodiversity. The responsibility for the conservation of biodiversity forces us to also address the genetically autonomous subgroups within the species. Intraspecific morphs and geographical races increase genetic and phenotypic diversity in many species. The extinction of a race would be a loss of biodiversity, although the species itself would be conserved.

Some races of higher animals are included in the IUCN list, such as the Siberian tiger, also called the *Amur tiger* (*Panthera tigris altaica*). It lives in the Amur and Ussuri region in the Russian Far East and differs from the South East Asian tiger in many of its characteristics. In 1930s of last century, the total population of the Siberian tiger was reduced to 20–30 individuals. There are about 350 of these tigers today, so the Amur tiger was downgraded from the 'Critically Endangered' category to the 'Endangered' category in the IUCN Red List of 2007 (http://www.iucnredlist.org/details/15956/0).

The handling of intraspecific genetic diversity is even more difficult than the consideration of races. What degree of consideration should be given to subpopulations on the Red Lists, if these subpopulations are neither recognised as species nor as races? It could be argued that an increased degree of responsibility is necessary for the protection of subpopulations if these are genetically very different from the remaining populations (Schnittler, 2004), but there is one problem in this case and that is that genetic individualities are usually found in the geographical outposts of a species. Such outposts live in peripheral, insular border zones of a species and do not affect the core occurrences of the species (see Section 4.3.2). This alone means that these outposts are not relevant for the survival of the species since they do not meet the criteria of a specific regional responsibility for protection, which is necessary where one species has a high proportion of its distribution in a region (Gruttke, 2004).

On the other hand, just a few isolated occurrences of a species can attain a particular value that requires greater protection. The term *erosion zone* is therefore not appropriate for such isolated outposts, because the exact opposite of species 'erosion' exists there. Isolated outposts can attain a high level of evolutionary significance due to various population-genetic reasons. It may be the new, evolutionary-biological birth of a species. The institutions that have made it their mission to prevent the extinction of species cannot ignore the birth of species.

The question of whether or not a recognisably different subpopulation can be considered as a separate unit is a major accountability analysis problem for the creators of the Red Lists. In the context of species protection, isolated populations like this must be granted a high value for conservation of biodiversity, even if they are not totally independent species. However, we cannot protect everything that displays a certain degree of diversity – so the concept of the 'ESU' becomes more important (Moritz, 1994; Crandall *et al.*, 2000). An ESU is a group of organisms which sets itself significantly apart from another similar group, both genetically and phenotypically (http://www.biology-online.org/articles/systematics_endangered_species_conservation/evolutionarily_significant_units.html). This unit would be more clearly defined and not as complex as the species (Moritz, 1994). The ESU would be the minimum unit for the protection of biodiversity. If this concept were applied consistently, it could help to find a workable definition for deciding which units should be included in the Red List.

The call to also include endangered genetic populations in the protection of species and not just species and races is not new. The subject was first broached in Ryder (1986). He propagated the idea that we should use a less complex unit rather than the species for the protection of biodiversity. The problem of the ESU, however, is the need to define what is evolutionarily significant. This is controversial and not so easy to recognise (Pennock and Dimmick, 1997). Isolated outposts can then be most clearly described as being evolutionarily significant if a 'founder effect' can be perceived (Mayr, 1963).

All in all, it is very complicated. Allelic depletion occurs in small separated populations due to genetic drift and many alleles will disappear as a result. The replenishment of lost allelic variants from the core area of the species is missing due to the fragmentation of the species' habitat and this causes many genes to become homozygous in isolated populations. This usually reduces fitness and threatens the survival of the population because many homozygous genes affect vitality and fertility. The result is that the population finds itself in an evolutionary dead end, resulting in the extinction of the species. Protecting this population would then be worthless.

If, however, the species is able to maintain its existence in this genetically depleted state, it will only be in the short term and the population will collapse when the environment changes. In the long term, the loss of allelic diversity can lead to genetically homogeneous populations losing their resistance to habitat changes. Genetically impoverished populations have no allelic reserves to preadaptively prepare themselves for environmental changes.

However, passing through a bottleneck does not always have a negative effect on the remaining population. The greatly reduced size of a population does not necessarily have an impact on the survival chances of a species. There are several examples of a species surviving, even when only a few individuals remain. In contrast to what was previously believed, the lower limit of 5000 individuals may be considerably fewer in some cases (Flather *et al.*, 2011). This means hope for the survival of species that are on the verge of extinction and which could still be saved despite a minimal number of remaining individuals.

Examples of species or subspecies that have recovered despite sharp declines in their populations are the Siberian tiger (*Panthera tigris altaica*) (http://www.iucnredlist.org/details/15956/0), the northern elephant seal in California (*Mirounga angustirostris*) (King, 1983) and the Moselle Apollo butterfly (*Parnassius apollo vinningensis*) in Germany (Habel, Meyer and Schmitt, 2011). All these species or subspecies are today the descendants of an extreme inbreeding and are homozygous for most genes. They document that there is no generally applicable theoretical or practical lower limit to individuals for the survival of a species and that the concept of a 'minimum viable population' (MVP) for species protection is not generally applicable (Flather *et al.*, 2011).

The passage of a species through a bottleneck can also be an opportunity for the 'genetic cleansing' of a population because detrimental, recessive alleles are eliminated by selection. If enough individuals still survive, however, the population is 'cleansed' of harmful alleles and can establish a new, vital and fertile population.

The main population-genetic consequence of habitat fragmentation, however, lies in the opportunity for certain alleles to prevail – and thus establish a new population. Populations at the borders of fragmented areas contain alleles which could not prevail in competition with the alleles in the core area of the original population's occurrence and were therefore of minimal importance. In a new environment and with new competition in the isolated subpopulation, however, these alleles can get another opportunity. They can prevail and give the entire subpopulation a new phenotypic and genotypic character. Such populations may constitute an evolutionary new beginning and lead to the emergence of new species. Such populations meet the conditions for an 'ESU' within the meaning of Crandall *et al.* (2000).

One example of this is the violet copper butterfly (*Lycaena helle*). The core occurrence areas of this North Eurasian species range from Scandinavia to the Baikal and Amur region. This butterfly was widespread across much of the Central European lowlands towards the end of the Würm glacial period. It retreated to Northern Europe during the postglacial warming, but survived the last millennia as a boreal–montane relic at a few locations in Central Europe which are isolated from one another. Today, the remaining individuals survive in small, cold and moist habitats, all of which are threatened by afforestation and the species is now found in all the Red Lists of Germany as a result.

The Central European populations of the violet copper butterfly are typical, isolated outposts of a large North Eurasian core occurrence area. These peripheral currencies are not relevant for the conservation of *L. helle* as a species because it is highly probable that the species would not be jeopardised if the entire central European population were to become extinct. However, isolated populations in Central Europe have become genetically widely separated during the millennia of isolation (Habel *et al.*, 2010). Although individual organisms have become genetically very homogeneous within each population due to the small number of individuals, significant genetic differences between populations exist. This makes the populations on the distribution area's western border very interesting from an evolutionary and biogeographical point of view. An increased degree of responsibility should be allocated to these populations in the Red Lists, even if they are not endemic species (Schnittler, 2004).

The inclusion of genetically different subpopulations during the preparation of the Red Lists remains a major problem. Not everything that has a certain degree of diversity can be protected. This ultimately affects the question of the justification for conservation. The answer to this question does not lie within the realm of science. It is important to distinguish whether the occurrence of a species is only interesting for mankind, or whether that occurrence has an intrinsic value in nature (Gorke, 1999).

In giving an intrinsic value to genetically isolated subpopulations of a species, it is simply not enough to say that the population has been isolated long enough, a fact which can be detected based on deviations in the DNA sequences. Genetic differences between two isolated populations are often based only on neutral mutations and are therefore unimportant in an evolutionary sense, provided that both isolated populations are subject to identical selection criteria. It is not the genetic differences between two isolated populations that are important, rather the functional differences between populations – differences that have been enforced through different selection criteria after the separation of the two populations. It is not enough to simply determine genetic differences in the DNA sequences of two populations. The differences between the individuals of two populations in two different habitats only become particularly significant when the individuals of the two populations are no longer interchangeable with one another, because each has less chance of survival in the habitat of the other population (Schnittler, 2004).

One problematic example of a genetic and phenotypic diversity is the asexual clones of some plant species that result by apomictic propagation from originally sexual plant species. Such clone swarms are distinguished by a high degree of variability. They are common to many composite plants (Compositae), such as the dandelion *Taraxacum*, the hawkweed *Hieracium*, the cinquefoils (*Potentilla* species) and the blackberries (*Rubus* species). Apomictic plant clones colonise different habitats that are usually small. They have a high degree of biodiversity and thus simulate taxonomic individualities. This resulted in the descriptions of a great number of species, for example, the hawkweeds and blackberries.

Apomictic plant clones are not created through fertilisation. Instead, they emerge parthenogenetically from female germ line cells without the fusion of gametes, or they are created from somatic cells of the female flower. In both cases, the products of this reproduction are clearly distinguished groups of numerous individuals which are all clones, that is, they are genetically identical to their respective parent. Since the clones propagate and multiply without sexually mixing together, mutations quickly accumulate that make the clones very different from one another, often after only a few generations. The plant species that are capable of apomixis thus produce many small groups and these attain a high level of genetic and phenotypic diversity.

Since apomictic plants mostly colonise different areas that are usually small, they can quickly attain high levels of endangerment to their existence. However, these are not 'ESUs' within the meaning of Crandall *et al.* (2000). Apomictic clones are often short-lived and they may constitute dead ends in evolution due to their asexual genetic isolation. Apomictic plant clones evince no capabilities for evolutionary progress. They do not have the potential for a separate phylogenetic development.

To justify the evolutionarily significant unit, it should also be noted that there are sometimes misleading discrepancies between phenotypically conspicuous subpopulations and their genotypic independence. The phenotype can often be misleading; here is an example of this: in the early 1980s of the last century, the dusky seaside sparrow (*Ammodramus maritimus nigrescens*) became extinct in Florida. This bird, classified as a race, differed significantly from others of its species by having black colouring in almost all plumage patterns – the other members of the seaside sparrow species were drawn as being light brown or almost unmarked. It was very clear that this race was phenotypically distinct from the other populations of the species. This resulted in an effort to regard the dusky seaside sparrow as being worthy of conservation and to include it in the Red List (http://en.wikipedia.org/wiki/Dusky_Seaside_Sparrow).

The DNA analysis, however, produced a different result. The mitochondrial DNA of the last male survivor was preserved – and a comparative analysis of many different haplotypes showed that the dusky seaside sparrow's genetics were not distinctly different from those of many other seaside sparrows. The genotypic analysis showed a completely different division into races than the phenotype would have provided. Instead of separating the dusky seaside sparrow, the seaside sparrows of the Atlantic coast had to be separated from those of the Gulf Coast and this had been hardly discernible in the phenotypic plumage drawing (Avise and Nelson, 1989).

References

Avise, J.C. and Nelson, W.S. (1989) Molecular genetic relationships of the extinct dusky seaside sparrow. *Science*, **243**, 646–648.

Baillie, J.E.M., Hilton-Taylor, C., and Stuart, S.N. (2004) *2004 IUCN Red List of Threatened Species – a Global Species Assessment*, International Union for Conservation of Nature and Natural Resources (IUCN), Gland and Cambridge.

Barthel, P.H. (2011) Zwischen Freiland und Gesetz – der Würgfalke Falco cherrug als heimische Vogelart. *Limicola*, **25**, 284–316.

Baum, D.A. (2009) Species as ranked taxa. *Systematic Biology*, **58**, 74–86.

Binot-Hafke, M., Belzer, S., Becker, N., Gruttke, H., Haupt, H., Hofbauer, N., Ludwig, G., Matzke-Hajek, G., and Stauch, M. (2012) *Rote Liste gefährdeter Tiere, Pflanzen und Pilze Deutschlands – Band 3: Wirbellose Tiere (Teil 1)*, Bundesamt für Naturschutz, Bonn-Bad Godesberg, pp. 1–716.

Crandall, K.A., Bininda-Emonds, O.R., Mace, G.M., and Wayne, R.K. (2000) Considering evolutionary processes in conservation biology. *Trends in Ecology and Evolution*, **15**, 290–295.

Dasmahapatra, K.K., Elias, M., Hill, R.I., Hofmann, J.I., and Mallet, J. (2010) Mitochondrial DNA barcoding detects some species that are real, and some that are not. *Molecular Ecology Resources*, **10**, 264–273.

De Iongh, H.H. and Bal, D. (2007) Harmonization of Red Lists in Europe: some lessons learned in the Netherlands when applying the new IUCN Red List categories and criteria, version 3.1. *Endangered Species Research*, **3**, 53–60.

De Iongh, H.H., Banki, O.S., Bergmans, W., and van der Werff ten Bosch, M.J. (2003) The harmonization of red lists for threatened species in Europe, in *Proceedings of an International Seminar November 2002*, The Netherlands Commission for International Protection, Leiden, Mededelingen, pp. 1–332.

del Hoyo, J. and Collar, N.J. (2014) *Illustrated Checklist of the Birds of the World – Non-Passerines*, Lynx Edicions, Barcelona.

Ebach, M.C. and Holdrege, C. (2005) More taxonomy, not DNA barcoding. *BioScience*, **55**, 822–823.

Flade, M. (1998) Was will der Naturschutz? *Naturschutz heute*, **31**, 50–51.

Flather, C.H., Hayward, G.D., Beissinger, S.R., and Stephens, P.A. (2011) Minimum viable populations: is there a 'magic number' for conservation practitioners? *Trends in Ecology and Evolution*, **26**, 307–316.

Gorke, M. (1999) *Artensterben: Von der ökologischen Theorie zum Eigenwert der Natur*, Klett Cotta, Stuttgart.

Gruttke, H. (2004) Grundüberlegungen, Modelle und Kriterien zur Ermittlung der Verantwortlichkeit für die Erhaltung von Arten mit Vorkommen in Mitteleuropa - eine Einführung [in German]) Fundamental ideas, conceptional models and criteria of assessing responsibility for the conservation of species occurring in middle Europe - an introduction, in *Naturschutz Und Biologische Vielfalt*, Bundesamt für Naturschutz, Bonn, pp. 7–23.

Habel, J.C., Meyer, M., and Schmitt, T. (2011) Der biologische Wert peripherer Populationen der Tagfalterarten *Lycaena helle* und *Parnassius apollo*. *Entomologie heute*, **23**, 93–102.

Habel, J.C., Schmitz, J., Meyer, M., Finger, A., Rödder, D., Assmann, T., and Zachos, F.E. (2010) Biogeography meets conservation biology: the life history of the highly endangered postglacial relict butterfly *Lycaena helle. Biological Journal of the Linnean Society*, **101**, 155–168.

Haupt, H., Ludwig, G., Gruttke, H., Binot-Hafke, M., Otto, C., and Pauly, A. (2009) *Rote Liste gefährdeter Tiere, Pflanzen und Pilze Deutschlands – Band 1: Wirbeltiere*, Bundesamt für Naturschutz, Bonn-Bad Godesberg.

Hebert, P.D.N., Penton, E.H., Burns, J.M., Janzen, D.H., and Hallwachs, W. (2004) Ten species in one: DNA barcoding reveals cryptic species in the neotropical skipper butterfly *Astraptes fulgerator. Proceedings of the National Academy of Sciences of the United States of America*, **101**, 14812–14817.

Keller, V. (2002) Vögel unter besonderer Verantwortung: Wie die Schweizer ihren Vogelschutz planen. *Der Falke*, **49**, 356.

King, J.E. (1983) *Seals of the World*, Cornell University Press, New York.

Kunz, W. (2006) DNA barcoding in taxonomy and the perception of species in nature. *BioScience*, **56**, 93.

Kunz, W. (2012a) *Do Species Exist? – Principles of Taxonomic Classification*, Weinheim, Wiley-VCH Verlag GmbH/Blackwell.

Kunz, W. (2012bGenetic distance and delimitation of species – barcode taxonomy has its own species concept [in German]) Genetische Distanz und Artabgrenzung – die Barcode-Taxonomie vertritt ihren eigenen Artbegriff. *Entomologie heute*, **24**, 277–286.

Kunz, W. (2015Integrative species or barcode-species – What is the correct species? – Additional note to the article "Integrative taxonomy, a new approach to answer the questions in the biosystematics" by H. Rajaei [in German]) Integrative Art oder Barcode-Art – welche ist die richtige Art? – Ergänzende Bemerkung zum Artikel "Integrative Taxonomie, ein neuer Ansatz, Fragen der Biosystematik zu beantworten" von H. Rajaei. *Entomologie heute*, **27**, 177–181.

Mayr, E. (1963) *Animal Species and Evolution*, Harvard University Press, Cambridge, MA.

Moritz, C. (1994) Defining "evolutionarily significant units" for conservation. *Trends in Ecology and Evolution*, **9**, 373–375.

Pennock, D.S. and Dimmick, W.W. (1997) Critique of the evolutionarily significant unit as a definition for "distinct population segments" under the U.S. Endangered Species Act. *Conservation Biology*, **11**, 611–619.

Reichholf, J. (2010) *Naturschutz. Krise und Zukunft*, Suhrkamp Verlag, Berlin.

Reinhardt, R. and Bolz, R. (2011) Rote Liste und Gesamtliste der Tagfalter (Rhopalocera) (Lepidoptera: Papilionoidea et Hesperioidea) Deutschlands. *Naturschutz und Biologische Vielfalt – Bundesamt für Naturschutz*, **70**, 167–194.

Ryder, O.A. (1986) Species conservation and systematics: the dilemma of subspecies. *Trends in Ecology and Evolution*, **1**, 9–10.

Schliewen, U.K., Rassmann, K., Markmann, M., Markert, J., Kocher, T., and Tautz, D. (2001) Genetic and ecological divergence of a monophyletic cichlid species pair under fully sympatric conditions in Lake Ejagham, Cameroon. *Molecular Ecology*, **10**, 1471–1488.

Schnittler, M. (2004) *Naturschutz und Biologische Vielfalt*, Bundesamt für Naturschutz, Bonn, pp. 39–52.

Südbeck, P., Bauer, H.-G., Boschert, M., Boye, P., and Knief, W. (2007) The Red List of breeding birds of Germany, 4th edition. *Berichte zum Vogelschutz*, **44**, 23–81.

Tautz, D., Arctander, P., Minelli, A., Thomas, R.H., and Vogler, A.P. (2002) DNA points the way ahead in taxonomy. *Nature*, **418**, 479.

Will, K.W., Mishler, B.D., and Wheeler, Q.D. (2005) The perils of DNA barcoding and the need for integrative taxonomy. *Systematic Biology*, **54**, 844–851.

Wirth, R. and Glaw, F. (2009) Globales Unwissen contra Biodiversität: Gedanken zum Schutz der biologischen Vielfalt. *Biologie in unserer Zeit*, **39**, 42–48.

Zhang, Z.-Q. (2011) Animal biodiversity: an outline of higher-level classification and survey of taxonomic richness. *Zootaxa*, **3148**, 1–237.

5
Changes in the Population Sizes of Birds and Butterflies in Central Europe and Germany

5.1
Changes in the Population Sizes of Birds in Germany

Among the Red List breeding birds in Germany, most species show a long-term negative population trend; in the short term, however, the species with positive population trends predominate.

This positive trend is not due to an improvement of habitats.

The recent destruction of many habitats is illustrated more clearly through the example of butterflies than that of birds.

Most red-listed species which are in decline are open landscape species; the majority of forest species are on the increase.

Will the future of birds in Central Europe almost exclusively consist of generalists and no more specialists?

The little bustard, the short-toed eagle, the stone curlew and other species provide us with information about the state of Central Europe's landscape in earlier centuries.

5.1.1
The Increase in Individual Numbers of Some Bird Species in the Last Decades Is Based on the Cessation of Chase, Not on an Improvement of Habitats

The Red Lists only classify the threatened species in 'endangerment' categories 1–3. Although the numbers of all non-endangered species which have declined or increased in the short or long terms are not classified in the categories, their short-term population trend since 1980 and their long-term trend over a period going back from 50 to 150 years can as well be found in the Red Lists (Südbeck *et al.*, 2007). This means that information about increasing and declining populations of breeding birds in Germany is also available for all species that are not deemed to be threatened. Two major questions can thus be answered: (i) whether or not the number of species of breeding birds in a limited region of Central Europe (as shown here by the example of Germany) has either increased or decreased in the long term or short term and (ii) whether or not the individual numbers of breeding species have either increased or decreased during these periods.

Species Conservation in Managed Habitats: The Myth of a Pristine Nature, First Edition. Werner Kunz.

The number of Status I breeding birds, namely the number of species without Neozoa and sporadic breeders amounts to 255 species in Germany since 1850 (Table 4.1). Of these, 11 species are now extinct (Category 0), so 244 native species regularly breed in Germany today. Sixty eight of these 244 species (28%) can be found in the three 'endangered' categories, namely 'vom Aussterben bedroht (1)', 'stark gefährdet (2)' and 'gefährdet (3)'. If we include the 11 extinct species, then 31% of all bird species that breed in Germany are either endangered or extinct. The 26 extremely rare species of Category R have not been incorporated in the 31% of endangered species. For these very rare species, there are no criteria of acute endangerment that would justify classification into Categories 1–3; but they *are* at risk simply because of their rarity. These include the spoonbill, the purple heron, the alpine swift and others. If we add the category R species, the number of endangered species would be further increased.

However, this 31% of German birds being endangered or already extinct does not give us a complete picture of the state of nature, because there are also birds that are numerically common and their populations decline or increase, indicating improvement or deterioration of the environmental conditions; yet such birds are not seriously at risk because they are (still) numerically common. What matters is whether or not and how this number has changed in recent years.

The decisive question is: 'Are birds in Germany in decline or are they on the increase?' There is no simple answer to this, because we must first decide whether or not it is the numbers of species that are meant, or the frequency of occurrence of individual species, that is, changes in the number of individuals of certain species. Second, we must distinguish between long-term changes in population trend (in the last 50–150 years) and short-term changes (in the last 10–25 years) (see Section 4.2.3). In contrast to the International Union for Conservation of Nature and Natural Resources (IUCN) list, the real advantage of the German National Red Lists lies in the distinction that can be made between long-term and short-term population trends. It is thanks to this comparison that the success of conservation efforts can be thoroughly scrutinised and we can then determine whether or not a long-term population decline could be stopped.

The answer to the question 'Has the number of birds in Germany declined or have their numbers increased?' is that the total number of species has remained about the same in the last 150 years. Over the past 150 years ('long-term' period), however, significantly more species have become rarer than species whose individual numbers have become more numerous. If we go back for the 'short-term' period of 25 years, we can see that more bird species have become more common than rarer (Südbeck *et al.*, 2007). The overall balance is therefore negative; but there *is* evidence of a slight 'glimmer of hope' in the last two decades.

The total number of 255 regularly breeding bird species native to Germany since 1850 (i.e. Status I types, neither sporadic guest breeders nor Neozoa) has remained approximately the same for the past 150 years, because the number of species that have become extinct during this period is roughly the same as the number of species which have immigrated. Of these 255 species, 11 species have become extinct and 102 have become rarer. This means that almost half (44%) of all German breeding bird species in the past 150 years have either become extinct or that their individual numbers have been reduced (Table 4.1). Among the 11 extinct species are the short-toed eagle, the little bustard and the stone curlew. Included among the 102 species which have become rare are the grey partridge and all shrikes, larks and meadow waders.

During the same period since 1850, 82 breeding bird species in Germany have become more common, or are newly immigrated. That means 32% of all German breeding bird species in the past 150 years. The species that have increased in the past 150 years include almost all kinds of ducks, geese and swans (the Anatidae duck family), several birds of prey, most gull species, many owl and woodpecker species, as well as nuthatches and many species of crow (Corvidae). The new immigrants in Germany (not to be confused with Neozoa, introduced species) include the spoonbill, the Caspian and Mediterranean gull, the collared dove, the melodious warbler, the common rosefinch and the serin. Despite the increases in numbers and immigrations, the balance in the long-term period is nevertheless negative; this is because the native breeding bird species that have become rarer and those which have become extinct constitute 44% of all 255 German breeding bird species, while only 32% of these 255 have become more common or are defined as new immigrants over the past 150 years.

The short-term balance of the changes in the frequency of occurrence of breeding birds in Germany is in total contrast to the long-term balance, because since roughly 1985, a greater number of breeding bird species has evinced a population increase, while the species with negative population trends are in the minority. 77 species have become more common, while only 45 species have become extinct or rarer (Table 4.1). After 1985, the European roller and the lesser grey shrike became extinct as breeding birds in Germany. This balance includes neither the non-native species introduced by humans (Neoaves), nor the irregularly occurring sporadic breeders (Status II species).

This positive balance which has become apparent in breeding birds in Germany is also supported by more detailed observation. An increase in the frequency of occurrence is a criterion for classifying a species in a lower Red List risk category; for example, from the 'stark gefährdet' Category 2 to the 'gefährdet' category 3; the equivalent in the IUCN categories would be a downgrading from EN ('endangered') to VU ('vulnerable'). A total of 46 native species of breeding birds were downgraded to a lower 'endangerment' category during the short-term period of the last few decades, while only 37 species had to be upgraded into more threatened categories (Südbeck *et al.*, 2007). This means that in the period from 1985 until the present day, there are now more Red List bird species in the less threatened categories than there are in the more threatened categories.

It is instructive to compare the minimally positive 'recovery trend' of breeding birds with the changes in butterfly populations in Germany, since in the latter case the short-term balance looks very different indeed (see Section 5.2.2). While in the long-term statistics the percentage of butterflies which have become more scarce (61%) is already higher than the percentage of birds that have become more scarce (44%), the prospects look even worse for butterflies in the short term (within the past 20–30 years). The number of butterflies which are becoming rare still remains very high; only a very small percentage of the butterfly species can show an increase in their numbers (Table 4.1) (Reinhardt and Bolz, 2011). In contrast to the birds, there seems to be absolutely no 'glimmer of hope'; most of these butterflies will in fact vanish.

The slightly positive trend in the recovery of some bird species can be misleading, because it paints too positive a picture about the true and frightening future of the species in Central Europe. Birds, like mammals, are apparently more resistant to environmental changes than, for example, insects. This is due to various reasons, which include mobility and a reduced susceptibility to temperatures due to their endothermic status. Butterflies are much more vulnerable than birds in Central Europe. The collapse of their populations has been much more pronounced, a fact which illustrates the recent destruction of many habitats more clearly (Thomas *et al.*, 2004).

Improved habitat conditions recognised for some breeding birds in the last few decades only apply to the majority of forest birds. The increases in the numbers of other bird species have nothing to do with improved habitats. Since the forest birds in Central Europe (at least the residents of dense forests) are in any case not at risk (and this was also the case 25 years ago), we can hardly say that the conservation efforts have been successful. The populations of the generalists which do not rely on special biotopes have mostly increased. These include the copse and forest bird species, for example, almost all warblers, the nuthatch and nearly all species of woodpeckers and several owl species. These birds benefit from the transformation of Central Europe from open country to copse and woodland. Even the population increase in such a spectacular species as the black stork is due to the increase in forested landscapes. Some species such as coots and ducks that benefit from eutrophication have also become more common, but most of these are species that have not been on the Red List from the outset.

In recent decades, species that have become more common have particularly benefited from direct protective measures (i.e. not from biotope improvements). These include the sea eagle, the peregrine falcon and other birds of prey, the cormorant and the eagle owl, because shooting them is meanwhile forbidden by law, the gulls, because their island breeding grounds are protected and the collared flycatcher in Bavaria, thanks to strategically placed nesting boxes. Only a few former Red List species such as the crane and the green sandpiper have actually benefited from actual improvements in their biotopes – rewetting of swamp woods had the desired effect. The remaining species continue to become less numerous.

It is the task of species protection and nature conservation to preserve endangered species by improving their habitats; and endangered species are specialists, not generalists. If we continue to have only generalists in Central Europe, this will deplete the former wealth of our landscape. Ornithologists who would associate their nature experiences in Germany with woodlarks, wheatears and various pipit, shrike and bunting species would not be content with the remaining feathered generalists in a Germany in which scrubland and wooded areas predominated.

Nature could no longer provide the exquisite species, only the 'standards' would be on offer. Friends of species richness in Central Europe must prepare for a desolate future.

5.1.2
The Red-Listed Birds on the Decline Are Almost Exclusively Open-Country Species

Why did the sharp population declines of many species of breeding birds occur in the middle of the last century? This is not an easy question; and the answer is weighted by opinions that range from the popular to the populist and cover all the hypotheses that pertain to pollution, industrialisation, global heating and contamination.

In Germany, 60% (32 species) of the 54 red-listed species that are on the decline and are listed in 'endangerment' Categories 1 and 2 inhabit open countryside. The rest are mostly flat-water breeding birds. Only four species (7%) of Categories 1 and 2 breed in large, dense forests; these are the capercaillie, hazel grouse, three-toed woodpecker and the white-backed woodpecker. These figures speak for themselves. If the goal of nature conservation is to promote endangered and vulnerable species, then the habitats that must be maintained and even created in Germany are above all sandy areas with poor grass growth, areas that have sparse scrub and wooded growth and the swampy flat-water regions. The protection of woods and forests is actually secondary in importance from the perspective of the species conservation organisations, but they usually place it 'first on the list', simply because it appeals to the general public.

If we were to examine the species which benefit most from the state of the Central European landscape today, the undoubted winners would be the forest birds, such as woodpeckers, owls, the black stork and others of that ilk. If forest-dwelling birds have been recently placed on 'endangered' lists, these are the species that require wooded zones with open corridors and clearcuts and areas where timber felling is carried out. It is these most endangered forest species that are most unlikely to benefit from the goals of the German national park concept (see Section 6.2). Neither the grey-headed woodpecker, the hazel grouse ('Endangerment' Category 2) nor the European nightjar ('Endangerment' Category 3) would benefit from a forest which is 'in a condition which has for the most part either not been affected or hardly been affected by human activity' and in which 'an undisturbed flow of nature's processes are safeguarded in their natural dynamics' (http://web.foerderverein-nationalpark-senne.de/?page_id=8). The hazel grouse and the grey-headed woodpecker need open forest glades with rich ground vegetation, while the nightjar needs clearcuts in the forest (Schumacher, 2012; Klaus, 2012).

The losers in today's Central European landscape include breeding birds in areas with sparse ground vegetation, such as open heaths and sandy, rocky and stony areas, because these areas are in a state of continuous growth or are even afforested in a targeted manner. Agricultural areas such as pastures and fields are almost the only spaces nowadays which are kept 'open'; but fertilisation has caused crop growth in these areas today to become much denser than it was in earlier times. In Central Europe today, there are hardly any more fields and pastureland with sparse growth and earthy or sandy areas. Such areas cannot be restored by 'organic' farming – because it is sandy areas with sparse growth in fields and mudflats with no vegetation in the meadows that are needed. Familiar field and meadow species that were very common in many regions 50 years ago, such as the grey partridge, lapwing, hoopoe and wryneck are being 'suffocated' in vegetation. Ground hunters like all species of shrikes and larks and most species of buntings, the European roller, tree pipits, tawny pipits and wheatear can no longer find food because the vegetation in their earlier habitats is now too dense.

Fifty years ago, all the meadow wader species suffered drastic population declines and these trends have only been moderately arrested by the rewetting programmes of the last 20 years. These waders, the 'Big Five', include the curlew, the black-tailed godwit, the snipe, the redshank and the lapwing and two very rare species, the ruff and the dunlin, which were brooding in the northern German interior until well into the 1950s of the last century, but almost became extinct in recent times. The population decline of the meadow waders is an outstanding example of some of the conservation measures' lack of success. In recent decades, many grassland areas have been ecologically rebuilt in locations where the meadow waders once lived. Rewetting, however, proved to be insufficient. At several places, the black-tailed godwits and snipes only returned in moderate numbers. Today's populations cannot be compared with those that inhabited these areas 60 years ago. The vegetation is now much denser than it was a few decades ago. The main reason behind the disappearance of these species is not the lack of moisture, but denser ground vegetation (Barkow, 2010). Nowadays, neither restored areas nor arable land and grassland have enough open earthy and muddy surfaces with very sparse vegetation.

Three impressive examples of breeding birds which have since disappeared illustrate how the open countryside in Germany still looked in the last century: the little bustard, the short-toed eagle and the stone curlew. Anyone who is familiar with the breeding habitats of these species can reconstruct how the landscape looked in Germany (and in the rest of Central Europe) less than a 100 years ago, with conditions offering suitable breeding biotopes for these three breeding birds.

The **little bustard** requires large, open areas of stony ground with low-growth ground vegetation as a breeding area. The last bird died out in 1924 in Thuringia, Germany. Today, Central European ornithologists associate the little bustard with the Crau region of the Rhone Delta in southern France. The Crau is a vast stony steppe with sparse scrub vegetation and no woods. Up until a 100 years ago, this kind of landscape existed in Thuringia and Saxony-Anhalt, but now forests cover the land.

The year 1969 saw the disappearance of the last **short-toed eagl**e from Brandenburg, Germany. In the first half of the twentieth century, it was still common throughout the Mark Brandenburg region. Here, it lived primarily on the wide open heaths that were home to the reptiles with which it fed its young.

Dense forests are not its territory. In the Rhineland-Palatinate, the Snake Eagle inhabited the Rhine and Moselle slopes, which were still largely free of vegetation at that time. Too many bushes and too much scrub growth would have prevented it from finding its prey. Since the French government passed laws protecting the short-toed eagle, it has once again become a common breeding bird in the south of France – but the biotopes which it can find in the Cevennes no longer exist in Germany.

The **stone curlew** vanished from the vicinity of Leipzig in Saxony as recently as 1987 (Nicolai, 2012). This bird inhabits large, gravelled, sandy areas which are near water and have only sparse ground vegetation. Until the beginning of the twentieth century, the stone curlew was a breeding bird in several German states, but then its population plummeted dramatically. The stone curlew was still breeding on Lake Neusiedl on the Austrian–Hungarian border until the 1960s. At that time, herds of cattle still grazed on the lake shores and their appetites and hooves kept the grass and vegetation down to a minimum. The stone curlew has since disappeared from this region along with the cattle (Dvorak, Ranner and Berg, 2012).

In addition to the disappearance of this spectacular species, there are a number of German breeding birds, such as the hoopoe, the woodchat shrike and the ortolan bunting, which are now either greatly diminished in number or have become extinct; remarkably, these birds are heat-loving (thermophilic) species and with the current global warming trend, the experts naturally predicted that they would flourish, but quite the opposite has occurred (Huntley *et al.*, 2007). Apparently, the climate warming of the atmosphere and the shading of the ground by thick wet grass and bushes are two counteracting processes, making it too cold on the ground for some of today's thermophilic species (Segerer, 2012) (Plates 1, 3 and 8).

The hoopoe and the woodchat shrike are still common breeding birds throughout the entire Mediterranean region. They were also common in Germany in recent centuries. The obvious conclusion here is that our Central European regions of that era must have had some characteristics that made them more Mediterranean-like than they are today, in spite of global warming. Global warming – present throughout Central Europe for decades now – has not brought back the hoopoe and the woodchat shrike.

The ortolan bunting is also a thermophilic bird. In the 1950s, the ortolan bunting population in north-western Germany reached its highest-ever level; there were around 5000 pairs in Lower Saxony alone. From 1955 onwards, there began a rapid population decline and this resulted in the virtual extinction of the bird in the region. The species completely disappeared from many large regions of north-western Germany between 1960 and 1970. The Dutch breeding population of the ortolan bunting vanished in 1992 and the birds disappeared from Belgium 2 years later. In Germany, the ortolan bunting breeds mostly in cereal crops and these must have a maximum height of 6–8 in. at the start of the birds' breeding season. The stalks must also be far enough apart that the birds have enough room to flit between them and find their food (Lanz, 2009). These were the low-yield fields that existed when the human population still suffered from famine (Plate 9). Today, however, we cannot recreate such fields. The come-back of the ortolan bunting is not a question of uncultivated margins around fields or 'lark windows' within the fields, nor is it a question of an insecticide and herbicide-free, biological agriculture; it is a question of too much atmospheric nitrogen which fertilises the entire areas.

5.2 Changes in the Butterfly Populations in Central Europe

We are only aware of a global threat to very few species of butterflies.

There are very many European endemics among the butterflies which live in higher mountain ranges.

By far the largest number of butterfly species in Germany has become rarer, both in the long term and the short term.

The disappearance of butterflies is significantly higher than the decline of birds in both the long term and short term.

The dramatic decline in butterflies puts a question mark against our widespread habit of using the birds' statistics as a yardstick for evaluating the loss of biodiversity.

The loss of habitat for butterflies negates the positive consequences of global warming.

Many butterflies in Central Europe need grasslands or light-rich woods, not pristine forests.

5.2.1 The Proportions of Germany and Europe in the Global Distribution Areas of Butterflies

Many butterfly species are so conspicuous that they catch the eye as aesthetic phenomena. However, much more research has been carried out on birds than on butterflies. As of 2014, the degree of endangerment to only 627 species of lepidopterans has so far been evaluated (http://cmsdocs.s3.amazonaws.com/summarystats/2014_2_Summary_StatsPage_Documents/2014_2_RL_Stats_Table4a.pdf), but roughly 160 000 species of lepidopterans have been described worldwide (Zhang, 2011). This is only 0.4% of all known species of lepidopterans. This means that for most species, we simply cannot compare the degree of species endangerment in the national and regional 'endangerment' lists to that of the global lists. Evaluating the responsibility borne by, for example, Germany for the preservation of its lepidopteran species is not possible in a global context, because the global degree of endangerment is not known for most of the lepidoptera species.

Even if we restricted our search to the diurnal (day-flying) species that have been thoroughly investigated among the lepidopterans, we would only know the global degree of endangerment for a few of these. The two well-known Central European *Papilionidae* species, the swallowtail (*Papilio machaon*) and the scarce swallowtail

(*Iphiclides podalirius*), for example, are not included in the IUCN Red List. Their global endangerment status is not known.

In Germany, the Danube clouded yellow (*Colias myrmidone*) is now extinct (Reinhardt and Bolz, 2011). During the past 60 years it abandoned most of its habitats in Bavaria and the last remaining population near Regensburg became extinct in 2000. There is reliable information about its distribution in Slovakia, Poland, Austria, Slovenia, Romania and Ukraine (http://www.iucnredlist.org/) (Kudrna *et al.*, 2011). For the EU member states (EU27), it is classified as 'Critically Endangered' (CR); for all of Europe, its classification is only 'Endangered' (EN) (Van Swaay *et al.*, 2010). Since the global threat to this species is not known, we cannot assess how much responsibility the Eastern European countries bear for the preservation of this spectacular butterfly.

For endemic European butterfly species, however, the deficit in non-European data is not an obstacle for the assessment of the national and European responsibility, because in the case of endemic species, the European and the global distribution are identical. Of the 482 European butterfly species, 142 species, almost a third of all European butterflies, are found only in Europe and nowhere else in the world. Most endemic European species live in the higher mountain ranges of the Pyrenees, the Alps and the Balkans and in the Mediterranean area. The high number of endemic species is surprising, because they must all have been able to survive the climate changes during the Pleistocene era. In contrast to many other animal species, they cannot have been extinguished by the ice ages, so they must have had the opportunity to avoid the cold periods in southern regions or lower altitudes and then have migrated back into their native country during the interglacials.

Endemism among the most Central European species is a definite rarity, because the recurring ice ages (and probably also the postglacial destruction by humans) wiped out the highly specialised species in Central Europe. The many European endemics among the butterflies are therefore an exception. This is illustrated by a comparison with the birds, of which there are very few endemic species in Europe, the red kite being one example of such species. Even those bird species that have their core occurrences in Europe (without being endemic to Europe) are rare; examples of these are the wood pigeon and the middle spotted woodpecker (http://www.iucnredlist.org/). The core populations of most birds occurring in Europe are located outside Europe.

However, this is not true of many butterflies. The reason why so many butterflies (in contrast to other groups of animals) survived the ice ages becomes clear when we realise that most endemic butterflies in Europe live in high mountainous areas or in the Mediterranean region (Van Swaay *et al.*, 2010). This gave them a greater chance of survival at the edge of the ice zones. The lowland butterflies were presumably not as fortunate. In contrast to the mountain species, the endemic Central European butterfly species which had adapted to the lowland forests of the interglacials probably could not survive the ice age and became extinct. This is why there are only a few European endemics among the Central European lowland butterflies as opposed to the mountain species. Examples for endemic

European lowland butterflies are the blues *Pseudophilotes baton* and *Polyommatus coridon* and the satyrines *Hipparchia semele* and *Hipparchia fagi.* Central Europe is only a peripheral zone of the core population area for most lowland butterfly species.

Since the degree of endangerment has been documented for almost all the butterflies in Europe, the European nations can focus their national responsibilities on European occurrences only and not global occurrences. Europe has 482 butterfly species (Van Swaay *et al.*, 2010). Of these, approximately 8% are considered to be threatened and have been added to the 'Critically Endangered', 'Endangered' or 'Vulnerable' categories. (http://ec.europa.eu/environment/nature/conservation/species/redlist/butterflies/status.htm) (Reinhardt and Bolz, 2011). Like other European nations, Germany can also focus its national responsibility for the preservation of certain butterfly species on the proportion of the populations that lives in Europe. If species are endemic to Europe, Germany's responsibility for the conservation of a species could even be assessed at the global level. However, this is not applicable because very few species of the European endemics occur in Germany and these few species find the limits (and by no means the main occurrences) of their distribution here. Not one single European endemic species has its core area of distribution in Germany.

5.2.2
The Drastic Decline of Butterflies in Germany and Central Europe

Germany has 189 butterfly species (Reinhardt and Bolz, 2011), none of which is endemic to that country. The geranium bronze (*Cacyreus marshalli*) does not actually belong to the native German fauna, because it was introduced from South Africa, probably with imported potted plants. It is the only Neozoan among German butterflies. This reduces the number of butterflies occurring in Germany to 188. Two irregularly occurring immigrants from the Mediterranean area are included among these 188 species and these are the blues *Lampides boeticus* and *Leptotes pirithous.* Both of these would have to be classified as Status II species based on the criteria underlying the German Red List of birds. The scarce tortoiseshell (*Nymphalis xanthomelas*) possibly belongs in this category status as well; the available data are, however, insufficient. If we regard only the above-mentioned two Mediterranean blues as occasional immigrants, this reduces the number of butterfly species that regularly occur in Germany to 186 (Status I species within the meaning of the Birds List) (Table 4.1).

Of these species, five species have become extinct in the past 100 years in Germany: the last extinctions were the false grayling (*Arethusana arethusa*) which disappeared in 1977 and the Danube clouded yellow (*C. myrmidone*) which vanished in 2000. This reduces the number of regularly occurring native butterflies in Germany to 181. This figure includes two new immigrants, the pale clouded yellow (*Colias erate*) and the southern small white (*Pieris manii*), which arrived from Eastern/south-eastern Europe only a few years ago. The degree of endangerment can of course not be evaluated for these two new species. Of the 181 recent and native, regularly occurring butterfly species, 7 species could not be evaluated and this means that data on the population situation, trends and risk factors are available for 174 butterfly species occurring in Germany.

In Germany, 77 butterfly species (i.e. 41% of all the species occurring in Germany for the last 100 years) are either endangered or extinct (Reinhardt and Bolz, 2011). The 22 very rare species in Category R are not included here, because no endangerment criteria are available that would justify classification into Categories 1–3. Very rare Category R species are, however, endangered simply because of their rarity, although no decline in their populations can be identified, either in the long term or in the short term. If we add the Category R species, the number of endangered species would be further increased.

In evaluating the future of species, population trends are more important than the actual number of endangered species. The population trend tells us how many species have become rarer and how many have become more common. If we look at the butterfly population trends in Germany, 113 species (61% of the butterfly species) have become rarer or extinct over the last 100 years and no apparent recovery can be determined in the last two decades. In the short-term period of the past 20–30 years, 81 butterfly species (45%) continued to become rarer, so the trend is continuous and appears to be never-ending.

It is worthwhile to compare these dramatic butterfly statistics with the decline in German breeding birds. This comparison gives us a picture that should make every conservationist sit up and pay attention. It tests our widespread habit of using the birds' statistics as a yardstick for evaluating the loss of biodiversity. While 'only' 79 of the 255 bird species (31%) that breed in Germany (Status I) are endangered (Südbeck *et al.*, 2007), 77 of 186 butterfly species (41%) are endangered (Reinhardt and Bolz, 2011).

The greater degree of endangerment to the butterflies becomes even more pronounced when we do not compare the species numbers, but the population trends of the birds with those of the butterflies. While a considerable number of bird species exhibit a positive trend both in the long term and short term, only a very small number of butterfly species have actually become more common. Over the past 100–150 years, 32% of the breeding bird species have a positive trend, while only 2% of the butterfly species have increased in number in Germany (Table 4.1). In the shorter period of the last two decades, the increase in the number of breeding birds amounts to 31%, while only 18% of bird species have become rarer. In stark contrast, no apparent recovery has been observed in German butterflies in the last two decades. Only 5% (10 species) of butterflies are more common today than they were two decades ago. The species which have become more common include the speckled wood (*Pararge aegeria*), the comma (*Polygonia c-album*), the large tortoiseshell (*Nymphalis polychloros*), the swallowtail (*Papilio machaon*), the brown argus (*Aricia agestis*), the short-tailed blue (*Cupido argiades*) and a few others. This is offset by 45% (81 species) of butterfly species that have become rarer during the last two decades in Germany.

The decline of the butterflies in Central Europe evinces an entirely different dimension than the decline of the breeding birds. This difference between the birds and the butterflies must have underlying reasons. Since the condition of habitats is the decisive factor for species frequency of occurrence, it follows that the dramatic landscape changes to which Central Europe has been subjected in the last 100 years, such as scrub encroachment and afforestation have harmed the butterflies significantly more than the birds. It is again remarkable, just as in the case of the birds, that the species of the open landscape are displaying a more negative population trend than that of the forest species.

From the outset, many more Central European butterfly species than bird species are dependent on sparsely wooded habitats; they live in open landscapes and often prefer sparsely vegetated soils (Blab *et al.*, 1984) (Plates 3 and 4), so it is self-evident why Central Europe's afforestation has a greater impact on butterflies than birds. There is no shortage of woods in Germany today, but there is a dearth of areas with sparse vegetation, such as dry meadows and open areas with sandy or stony soils. Only these areas provide sufficient heat when exposed to sunlight, especially if they are on hillsides which are exposed to the sun. The reforested southern slope of a mountain or even of a small hill is lost territory for most butterfly species. It could be an ideal habitat for butterflies, however, if only we had not planted the trees.

Since birds are much more well-known and popular than butterflies, the gaze of the nature and species protection world is focussed much more strongly on birds than on butterflies. This creates a false picture of the true state and the true causes of Central European species loss. Some conservation efforts even go in the wrong direction (Filz *et al.*, 2013).

A decline in individuals precedes the decline of the species. This is why the endangerment to a species is often not recognised early enough, since it can survive for a long time in some small niche before it becomes extinct. Species decline (at least in larger geographical areas) is thus hardly evident in a reduction in the number of species, but in the number of their individuals. Quantitative statements require long-term studies and these were carried out in the United Kingdom over two periods between 1970 and 1999 (Thomas *et al.*, 2004).

With the help of numerous amateur observers, Jeremy Thomas's team observed the butterflies on almost 3000 10-km-by-10 km areas from 1970. Fifty eight species were recorded between 1970 and 1982 and again between 1995 and 1999. The populations of more than two-thirds of these species (71%) declined significantly during the periods in question. It was found yet again that butterflies (and many other insects) are more affected by the destruction of open areas than other organisms. A total of 201 bird species were counted using the same methods throughout almost the same periods and in the same locations, but the birds only evinced a 50% decline in individuals. These studies proved yet again that birds and other vertebrates are not a good yardstick for the Central European species decline of recent decades.

Vertebrates are more adaptable than insects and are therefore less affected by habitat destruction. However, there are also other reasons why species decline is not as noticeable in birds as it is in butterflies and other insects. The loss of butterfly species has not been offset by new additions, neither through the immigration of new species nor by Neozoa. Among the butterflies, there are only a few species which have newly immigrated in the long term or in the short term; and no exotics (which could be considered as substitutes for lost species) have invaded the German countryside from parks or zoos and settled. Among the butterflies in Central Europe, there is only one example of an introduced species, the geranium bronze, *C. marshalli.* It was imported with potted plants which have been shipped from South Africa to Europe since 1990. In contrast, the loss of bird species in Central Europe has been offset by new, immigrated species, such as the Caspian gull, the collared dove, the melodious warbler, the common rosefinch and the serin. There are also numerous introduced species among birds (Neozoa), such as some parrot, duck and geese species. These compensate for the loss of species, at least in the statistics, but whether or not this compensation has any qualitative value is another question.

There is yet another reason why the birds in Central Europe have declined less than the butterflies in recent decades. Nowadays, birds are not shot as often as they were in previous centuries in Central Europe; and this means that cormorants, many species of heron, numerous birds of prey, nearly all species of duck and geese and all the crow species (Corvidae) are now more common in Central Europe than they ever have been in the last 1000 years. In contrast, butterflies and other insects were only captured and killed to a minor extent in the last few centuries. Those passionate entomologists who in the past decades caught butterflies with their nets to expand their collections only had a minimal local impact. Butterflies have never been as widely hunted as birds of prey. All the butterflies that have become rare in Central Europe in recent decades have not become rare because they have been hunted. The laws passed in the 1970s and 1980s against the capture and killing of insects in Central Europe have not helped butterfly populations to recover. Not one single butterfly has become more common in Germany because it is no longer hunted.

In recent decades, the loss of butterflies in Central Europe and in the United Kingdom has been dramatic. The Netherlands has lost a quarter of its butterfly species (Van Swaay *et al.*, 2010). Flanders (northern Belgium) lost 30% of its butterfly species in the twentieth century (Maes and Van Dyck, 2001) and the United Kingdom has lost almost 10% of its butterfly species (Warren, 1993a,1993b). These figures, however, only represent the percentage losses of the number of species that previously existed in these countries. The loss of 'mass' (the number of individuals) attains and even exceeds 50%.

5.2.3
The Positive Consequences of Global Warming for the Butterflies Are Negated by the Negative Effects of Habitat Destruction

From the point of view of environmental protection, current global warming is a major threat; but it does have a positive effect on most butterflies. For them, global warming is a win situation. Most butterfly species are thermophilic; they need warm temperatures, because this enables the caterpillars to develop fast enough and the butterflies to emerge early enough to propagate and continue their development cycle successfully. This has even led to the expansion of the population areas of some butterfly species.

When we realise that almost all butterfly species should have benefited from global warming, it is at first very difficult to understand why we have seen (and are seeing) such a drastic loss of butterflies in recent decades, particularly in Central and north-western Europe. In the case of Britain and north-western Europe, the temperature is the main reason why many butterfly species are limited in their more northerly distribution areas (Warren *et al.*, 2001), so it follows that most butterfly species should have benefited from the rising temperatures generated by the climate change of recent decades and should have spread further northwards.

These positive predictions, however, are more or less negated by the consequences of habitat destruction. Contemporaneous habitat loss has a greater impact than the increase in global warming. Micro-habitats on earthy, sandy and rocky surfaces that heat up strongly are a thing of the past in many areas; these areas have had to give way to dense, damp grassy growth, to scrub and bush encroachment and to afforestation. As a result, warm micro-habitats for insects have all but disappeared (Plates 2–4, 7 and 8). The general climate warming and habitat losses are mutually cancelling forces; but the latter factor largely predominates, so that in the end the positive effects of global warming are suppressed by the negative consequences of habitat loss. A striking example of this is the marbled fritillary (*Brenthis daphne*), which has largely disappeared from southern and south-eastern Germany in recent decades. At the same time, however, it immigrates into West Germany from the south-west (Kudrna *et al.*, 2011).

It was predicted that the 46 selected butterfly species occurring in Britain would react positively to global warming. Some did indeed move northwards a little, but three-quarters of the 46 species are in decline (Warren *et al.*, 2001). As expected, there is an important difference between the habitat generalists and the specialists. About half of the generalist butterflies have spread out in the last 30 years. These include the comma (*Polygonia c-album*) and the speckled wood (*Pararge aegeria*); but the other half of the habitat generalists and almost all habitat specialists have become rarer and have mostly retreated to small refuges. Britain has (almost) lost the majority of its butterflies in the last 30 years. What has remained are species-poor habitats dominated by a few mobile and widespread generalists (Warren *et al.*, 2001).

Things look alike, at least from the north-west down to the west of Germany. Anyone who sees lots of red admirals, small tortoiseshells, peacocks and commas flitting around the imported summer lilac in his garden in late summer must not forget that this is simply a poor remnant of a once much richer butterfly fauna. At least in north-western Europe, the few generalists will be the future of the species, but the rich abundance of specialists will be lost.

5.2.4 Many Butterflies in Central Europe Need Grasslands or Light-Rich Forests That Are Not Pristine

Since butterflies were collected all over in earlier decades and these collections can be viewed in museums today, local butterfly populations can be traced back for more than 100 years. When we compare earlier populations with those of today, our attention is drawn to the dramatic population declines in the late 1950s and 1960s of the last century from the north-west to the west of Germany. Older butterfly collectors can still remember seeing up to four Fritillary species of the genus *Argynnis*, up to five Fritillary species of the genus *Melitaea* and more than 10 Blue and Copper species of the Lycaenidae family in just 1 day in early summer on the dry grassland and in the marshy valley meadows of the 1950s. The heath fritillary (*Melitaea athalia*) and the small pearl-bordered fritillary (*Boloria selene*) were still common until the 1970s in the Eifel mountains in the west of the state of North Rhine-Westphalia. Today, both species have almost disappeared from this area.

Data comparisons in the west of the state of North Rhine-Westphalia near the cities of Düsseldorf (Lenz and Schulten, 2005), Münster/Coesfeld (Augustin, 2003), Krefeld (Bäumler, 2010) and Wuppertal (Laussmann *et al.*, 2010) show that in the last 60 years the number of butterfly species has declined in these locations by more than half. In the Düsseldorf area alone, 64 butterfly species existed around 1900; today only 27 species grace the landscape. This is a species loss of 58% (Lenz and Schulten, 2005).

The sometimes massive occurrence of many species of butterflies in the nineteenth century and first half of the twentieth century was focussed on sandy surfaces, rough pasture and dry and marshy meadows (Bourn and Thomas, 2002). It is the grasslands, not forests, that are the most important habitat for butterflies in Europe (WallisDeVries, Poschlod and Willems, 2002). In addition to the marshy meadow and moor specialists such as the fritillaries *Boloria aquilonaris* and *B. selene* and the moorland clouded yellow *Colias palaeno*, it is above all the inhabitants of the poor grasslands that have been in serious to profound decline in Germany for several decades now, for example, the hermit *Chazara briseis*, the idas blue *Plebeius idas* and the common branded skipper *Hesperia comma* (Binot-Hafke *et al.*, 2012).

The problem for German nature conservation is that it wants to retain and restore the original natural conditions. However, the original habitats in Europe (at least in Central Europe) are the forests (Küster, 2008) and these natural to semi-natural forests do not provide suitable habitats for most butterflies. About half of all butterfly species in Europe are residents of the grasslands, and more than half of these species are on the 'endangerment' categories of the European Red Lists (Van Swaay *et al.*, 2010).

However, even the other half of the butterflies of Europe (those which live in forests) are predominantly non-residents of closed forests; they need light-rich, thinned-out forests with beams of sunlight and flower-rich forest paths over which the tree canopy has not (yet) closed (Ulrich, 2002). The beech forests in particular, those jewels in the eyes of many conservationists (http://www.weltnaturerbe-buchenwaelder.de/) have very few butterflies (Fritzlar and Westhus, 2004). Central Europe's butterflies need light-rich forest which show signs of timber operations and are crisscrossed by clearcutting, road systems and power line routes, creating sufficient, open, sunny places with flowery border structures and earthy surfaces. Light-rich forests have become rare in Central Europe, because wind damage and fires are contained and human intervention in the form of forest pasture and undergrowth utilisation has declined.

In Germany, it is not only the grassland butterflies that are threatened; just over half of all the forest butterflies are also endangered and these numbers are on the increase (Ulrich and Caspari, 2007). The typical, endangered species of light-rich forest fauna include the fritillaries *Argynnis adippe* and *Boloria euphrosyne*, the white admirals *Limenitis camilla* and *Limenitis reducta* and the ilex hairstreak *Satyrium ilicis*. These species do not need pristine or semi-natural forests; they require human-designed, open and light-rich forest structures and forest edge formations. Most forest butterfly populations are threatened by the gradual change in our forests, which are getting ever darker, with canopies that are gradually closing. The decrease in light-rich forest species is making itself increasingly felt in Germany. The ilex hairstreak (*S. ilicis*), for example, was still widespread in Baden-Württemberg in 1991 (in 23 ordnance survey map quadrants). Ten years later, in 2001, only one population was known to exist in the Rheinaue region south of Freiburg (Ebert and Rennwald, 1991). In Saarland, only five sightings of this formerly widespread species have been recorded since 1990 (Ulrich and Caspari, 2007).

The desire of the German nature conservation associations to create national forest parks ignores the demand for conservation of the Central European butterfly fauna; because according to the German Federal Nature Conservation Act, national forest parks aim to maintain nature (for the most part) in a condition that is hardly or not at all affected by human beings, so that the natural momentum of an undisturbed flow of natural processes is ensured. A national forest park is desirable for many excellent reasons as far as nature conservation objectives are concerned; but it is not a suitable habitat for the conservation of the most endangered butterflies in Central Europe (see Section 6.2).

References

Augustin, A. (2003) Die Tagfalter des Kreises Coesfeld und der angrenzenden Davertbereiche (Lep., Rhopalocera et Hesperiidae) – Bestandsaufnahme von 1998 bis 2003. *Melanargia*, **15**, 85–158.

Barkow, A. (2010) Probleme der Wiesenvögel – Ist die Uferschnepfe noch zu retten? *Der Falke Taschenkalender*, 197–204.

Bäumler, A. (2010) Beiträge zur Makrolepidopterenfauna der Stadt Krefeld – Teil 1: Tag- und Dickkopffalter. *Melanargia*, **22**, 106–112.

Binot-Hafke, M., Belzer, S., Becker, N., Gruttke, H., Haupt, H., Hofbauer, N., Ludwig, G., Matzke-Hajek, G., and Stauch, M. (2012) *Rote Liste gefährdeter Tiere, Pflanzen und Pilze Deutschlands – Band*

3: Wirbellose Tiere (Teil 1), Bundesamt für Naturschutz, Bonn-Bad Godesberg, pp. 1–716.

Blab, J., Nowak, E., Trautmann, W., and Sukopp, H. (1984) *Rote Liste der gefährdeten Tiere und Pflanzen in der Bundesrepublik Deutschland*, 4th edn, Kilda-Verlag, Greven.

Bourn, N.A.D. and Thomas, J.A. (2002) The challenge of conserving grassland insects at the margins of their range in Europe. *Biological Conservation*, **104**, 285–292.

Dvorak, M., Ranner, A., and Berg, H.-M. (2012) *Atlas der Brutvögel Österreichs*, Styria, Graz.

Ebert, G. and Rennwald, E. (1991) *Die Schmetterlinge Baden-Württembergs: Tagfalter II*, Eugen Ulmer, Stuttgart.

Filz, K.J., Engler, J.O., Stoffels, J., Weitzel, M., and Schmitt, T. (2013) Missing the target? A critical view on butterfly conservation efforts on calcareous grasslands in south-western Germany. *Biodiversity and Conservation*, **22**, 2223–2241.

Fritzlar, F. and Westhus, W. (2004) Regionale Anwendung des Verantwortlichkeitskonzepts für globalen Artenschutz am Beispiel Thüringens, in *Naturschutz Und Biologische Vielfalt*, Bundesamt für Naturschutz, Bonn, pp. 255–265.

Huntley, B., Green, R.E., Collingham, Y.C., and Willis, S.G. (2007) *A Climatic Atlas of European Breeding Birds*, Durham University, Barcelona.

Klaus, S. (2012) Lückenbewohner im Natur- und Wirtschaftswald: Auerhuhn und Haselhuhn. *Der Falke*, **59**, 26–29.

Kudrna, O., Harpke, A., Lux, K., Pennerstorfer, J., Schweiger, O., Settele, J., and Wiemers, M. (2011) *Distribution Atlas of Butterflies in Europe*, Gesellschaft für Schmetterlingsschutz e.V, Halle.

Küster, H. (2008) *Geschichte des Waldes. Von der Urzeit bis zur Gegenwart*, 2nd edn, C.H. Beck, München.

Lanz, U. (2009) Vage Erfolgsaussichten im Ortolanschutz. *Der Falke*, **56**, 426–431.

Laussmann, T., Radtke, A., Wiemert, T., and Dahl, A. (2010) 150 Jahre Schmetterlingsbeobachtungen im Raum Wuppertal – langfristige Veränderungen in der Lepidopterenfauna der Region. *Jahresberichte des Naturwissenschaftlichen Vereins Wuppertal*, **61**, 31–100.

Lenz, N. and Schulten, D. (2005) Tagfalter (Lep., *Hesperioidea* et *Papilionoidea*) im Gebiet der Landeshauptstadt Düsseldorf um 1900 und um 2000 – ein Beispiel für alarmierende Artenverarmung im 20. Jahrhundert. *Melanargia*, **17**, 19–29.

Maes, D. and Van Dyck, H. (2001) Butterfly diversity loss in Flanders (north Belgium): Europe's worst case scenario? *Biological Conservation*, **99**, 263–276.

Nicolai, B. (2012) *Atlas Der Brutvögel Ostdeutschlands*, Gustav Fischer, Jena.

Reinhardt, R. and Bolz, R. (2011) Rote Liste und Gesamtliste der Tagfalter (Rhopalocera) (Lepidoptera: Papilionoidea et Hesperioidea) Deutschlands. *Naturschutz und Biologische Vielfalt – Bundesamt für Naturschutz*, **70**, 167–194.

Schumacher, H. (2012) Lieber licht als dicht: Ziegenmelker, Wendehals und Grauspecht. *Der Falke*, **59**, 21–25.

Segerer, A.H. (2012) Die physikalisch-geochemischen Grundlagen des planetaren Klimas und die Auswirkungen auf die öffentliche Diskussion – potenzielle Fallstricke für Ökofaunisten. *Nachrichtenblatt bayerischer Entomologen*, **61**, 32–45.

Südbeck, P., Bauer, H.-G., Boschert, M., Boye, P., and Knief, W. (2007) The Red List of breeding birds of Germany, 4th edition. *Berichte zum Vogelschutz*, **44**, 23–81.

Thomas, J.A., Telfer, M.G., Roy, D.B., Preston, C.D., Greenwood, J.J., Asher, J., Fox, R., Clarke, R.T., and Lawton, J.H. (2004) Comparative losses of British butterflies, birds, and plants and the global extinction crisis. *Science*, **303**, 1879–1881.

Ulrich, R. (2002) Vom Naturschutz vergessen: Die Lichtwaldarten. – Naturschutz im Saarland. *Naturschutz im Saarland*, **3**, 22–25.

Ulrich, R. and Caspari, S. (2007) Die Lichtwaldfalter im Saarland: erstes Modellprojekt im Warndt. Abh. *Delattinia*, **33**, 23–68.

Van Swaay, C., Cuttelod, A., Collins, S., Maes, D., Lopez Munguira, M., Sasic, M., Settele, J., Verovnik, R., Verstrael, T., Warren, M., Wiemers, M., and Wynhof, I. (2010) *European Red List of Butterflies*, Publications Office of the European Union, Luxembourg.

WallisDeVries, M.F., Poschlod, P., and Willems, J.H. (2002) Challenges for the conservation of calcareous grasslands in northwestern Europe: integrating the requirements of flora and fauna. *Biological Conservation*, **104**, 265–273.

Warren, M.S. (1993a) A review of butterfly conservation in central southern Britain: I. Protection, evaluation and extinction on prime sites. *Biological Conservation*, **64**, 25–35.

Warren, M.S. (1993b) A review of butterfly conservation in central southern Britain: II. Site management and habitat selection of key species. *Biological Conservation*, **64**, 37–49.

Warren, M.S., Hill, J.K., Thomas, J.A., Asher, J., Fox, R., Huntley, B., Roy, D.B., Telfer, M.G., Jeffcoate, S., Harding, P., Jefcoate, G., Willis, S.G., Greatorex-Davies, J.N., Moss, D., and Thomas, C.D. (2001) Rapid responses of British butterflies to opposing forces of climate and habitat change. *Nature*, **414**, 65–69.

Zhang, Z.-Q. (2011) Animal biodiversity: an outline of higher-level classification and survey of taxonomic richness. *Zootaxa*, **3148**, 1–237.

6
The Special Situation of Species in Central Europe

6.1
Central Europe's Landscape Is Not 'Nature'

The differences in the fauna and flora between the warm and cold periods of the Pleistocene were more pronounced in Central Europe than in most other parts of the world.

The alternation between warm and cold periods of the Quaternary in Central Europe led to a reoccurrence of species extinction and to remigration from the core population areas of the regions of the East and the Mediterranean area.

During the Pleistocene, Central Europe was always only a temporary, geographically marginal habitat for interglacial and ice-age fauna.

Europe has only few endemic species.

The thinning-out of forests by large animals and humans.

It is not the woods, but the anthropogenically created grasslands that are among the most species-rich habitats in Europe.

Former species richness in Central Europe as a result of the exploitation of nature.

The vegetation density of today's landscape seems to be the main factor for the loss of species in Central Europe.

The species conservation practices of the tropical countries cannot be used in Europe.

What lies behind the nature conservation movement's mission statements regarding a 'natural fauna' in Central Europe?

6.1.1
Alternating Forest and Steppe in Central Europe led to Waves of Extinctions and the Reimmigration of Species

Central Europe was more markedly characterised by the alternation of warm and cold periods than other parts of the northern hemisphere (von Koenigswald, 2004). The geological epoch of the Tertiary period lasted for 60 million years. In contrast, the Quaternary with almost 3 million years (the period in which we live today) is a relatively short period (Table 6.1). While the mammals and the birds evolved a high number of species during the Tertiary in Europe after

Species Conservation in Managed Habitats: The Myth of a Pristine Nature, First Edition. Werner Kunz.

Table 6.1 Duration of periods of quaternary in Central Europe.

Quaternary in Central Europe							
Geological period	**Pleistocene**		**Holocene**				
Archaeological period		**Palaeolithic**	**Mesolithic**	**Neolithic**	**Bronze age**	**Iron age**	**Historical time (written history)**
Begin of period in years before present (BP)	2.6 millions	0.8 millions	12 000	7 500	4 200	2 800	5 000

the collapse of the reptilian fauna, the Quaternary must be regarded as the age of extinction of many species, at least in Central Europe. Repeated Ice Ages decimated biodiversity.

The Quaternary (the geological present) includes the Pleistocene and the Holocene periods. The Pleistocene is the period which has seen up to nine cold and warm periods in Central Europe, depending on the geographic region. The Holocene is the last interglacial period and the one in which we live today. The constant alternation between cold and warm periods significantly changed the natural landscapes of Central Europe during the Quaternary. During the warm periods, Central Europe was covered by forest landscapes, and steppe landscapes covered the land during the Ice Ages. A radical change in the climate occurred every 50 000–100 000 years on (rough) average. At the beginning of each cold period, the forest animals lost their natural resources in Central Europe, followed by the plains animals. Each climate change was accompanied by a wave of extinctions, and immigration from neighbouring regions took place after the climate which prevailed at that time had stabilised.

It is because of this constant alternation between forest and steppe landscapes that we cannot decide what 'natural' fauna actually is in Central Europe. Forest animals or steppe animals? It is certainly short-sighted to simply look back a few thousand years and refer to the postglacial forest fauna as the natural fauna of Central Europe.

Strictly speaking, Central Europe has almost no natural fauna; this is because most species have been repeatedly destroyed by the disappearance of the forests at the beginning of the Ice Ages and by the disappearance of the steppes when the warm periods began. The result was that only a few endemic species survived the Pleistocene in Central Europe. Even today Europe differs from many other parts of the world in that only a few endemic species occur (von Koenigswald, 2002; Gruttke, 2004). Central Europe is a region of immigration, characterised by repeated waves of new colonisation. During the warm periods, the species of the Mediterranean region and of the south-eastern region immigrated into Central Europe, while the cold periods saw species immigrations from the Eastern European and Siberian steppes and from the northern tundra. Central Europe was continually being repopulated by species which had their core populations outside the region; and this still characterises the fauna and the flora of Central Europe today. There are only a very few typical Central European species. Many species are those of the Mediterranean region, the East and the North.

The differences in the fauna and flora between the warm and cold periods were more pronounced in Central Europe than in most other parts of the world (von Koenigswald, 2004). The temperature difference between the warm periods and Ice Ages played a role here, but other climatic differences also contributed. During the warm periods, Central Europe was characterised by an Atlantic, humid, rainy climate, while the cold periods were under the influence of a dry and sunny continental climate during which Central Europe was seamlessly connected with Eastern Europe and Siberia. These climate variations had a significant impact on the flora and fauna of Central Europe. It was probably the forest species which coped particularly poorly with the climate changes in Central Europe, suffering more than the forest species in Asia and North America.

The species poverty of Central Europe's forests differs from other parts of the world. During the Pleistocene, the rich Tertiary tree flora of Central Europe became increasingly species-poorer from warm period to warm period. Horse chestnut, redwood, *Magnolia*, hemlock spruce, chestnut and other original native trees died out during the Ice Ages in Central Europe and were not able to re-establish themselves during the interglacials (Küster, 2008). Other temperate zones of the world, such as North America, south-west Asia (e.g. the forests on the southern coast of the Black Sea) and eastern Asia retained their biodiversity during the Pleistocene, because they had no predetermined southern geographical boundary, and this opened the gates to postglacial immigrations from south to north. The species-rich flora from the Tertiary (and probably also the fauna, but this cannot be adequately proven, since pollen analysis is not possible) could therefore better survive the Ice Age era in America and in western and eastern Asia than it could in Europe.

We should not think of the Central European landscapes during the cold periods as being permanently covered with ice. The glaciers of the Alps and of Scandinavia have repeatedly developed and expanded throughout parts of Central Europe, but the ice only advanced for a few thousand years each time. Central Europe was free of ice the rest of the time, but it was characterised by permafrost, which hindered tree growth and encouraged the spread of the steppe landscape that was connected with the steppes of Eastern Europe and Siberia. This glacial steppe landscape (the 'Mammoth Steppe') was very continental in character and characterised by high pressure weather and strong sunlight.

The mammoth steppe is often compared with the Arctic tundra due to its permafrost. However, the sunlight in the southern latitudes of Central Europe was much more intense than in the tundra, with the result that plant growth was much stronger. The Ice Age mammoth steppes may therefore only be compared to the Arctic tundra in a limited way; a real counterpart of the mammoth steppe landscape no longer exists today and many of the characteristic species of mammoth steppe have consequently disappeared.

The almost 10-fold alternations between Ice Ages and warm periods in Central Europe during the Pleistocene in the last 2–3 million years have repeatedly led to the region being inhabited by steppe dwellers, and then again by forest dwellers. Each climatic cycle lasted for between 10 000 and 100 000 years. The species composition of the fauna and flora was thus continuously replaced at regular intervals. This process is often described by use of the terms *immigration* and *emigration*, but this projects a false picture. The truth is that only minimal animal emigrations occurred; they died out, to be replaced by new animals immigrating into the region. This peculiarity of the fauna and flora of Central Europe is a key element in understanding today's species composition – and this issue must be thoroughly investigated before any nature conservation and species protection measures can be taken. We cannot compare the nature and species protection measures needed in Central Europe to the measures required to save tropical species (see Sections 1.3 and 3.4). This especially applies to the priority (wrongly) accorded to the protection of forests in Germany.

For many species, the change in climate and vegetation created by the alternations between warm and cold periods meant that a continuation of their previous lives was no longer possible. Let us for a moment consider the emigration of the species into the neighbouring regions. These would have been the warm regions of the Mediterranean area at the beginning of the Central European cold periods, or the cold regions of Eastern Europe and Siberia at the beginning of the Central European interglacial periods. These regions, however, were not uninhabited, so they were not open and available for the refugees from the cold or the heat warmth. Animal refugees are definitely not made welcome in regions that are already occupied.

The alternation of cold and warm periods was a gradual process. It did not affect the individuals of a generation that might have had the urge to migrate because of cold or the lack of food. The true impact was first felt by their descendants, whose numbers decreased when the ecological conditions for specific animal species worsened due to the climate change. Climate change took place in surges. After a climatic deterioration, an improvement often occurred and the temporary decline in the numbers of individuals could then be offset. If, however, the climatic deterioration lasted for longer periods of time, entire Central European populations died out, because they usually could not escape to neighbouring regions due to strong competition. So there were probably few emigrations from Central Europe during the alternating cold and warm periods (von Koenigswald, 2004). Instead of emigrations, there was a decline in the birth rate, while adult animals were able to survive. The predominant factor in the gradual extinction process of the species was neither the death of individuals nor their emigration, but their low reproduction rates.

The waves of extinctions probably resulted in most Central European endemic species dying out; the lack of fossils, however, means that we have almost no information about the species that lived here in Central Europe between 1 and 2 million years ago. We only know the fossils of the larger animals. Non-endemic species that had a wide distribution area only died out from region to region in Central Europe; they survived, however, in their core population areas outside Central Europe. The thermophilic species had their core occurrences in the Mediterranean region and in North Africa, where they were protected against the cold from the north by the mountain ranges of the Pyrenees, Alps and Carpathians that ran in a west-to-east direction. The species which adapted to the cold climate survived in continental Eastern Europe and Siberia.

Following each wave of extinctions, however, a biotope for new flora and fauna opened up. The disappearance of the steppes resulted in a wave of animal immigration into Central Europe from the forest areas of the Mediterranean region. The birth rates of the species living in the Mediterranean region resulted in overpopulation and subsequent immigration to Central Europe. Conversely, the steppe animals migrated from the Eastern regions to Central Europe when a warm period was replaced by a cold period. The mammoth steppes then pushed into Central Europe, and the steppe animals migrated from continental Eastern Europe and Siberia to Central Europe. For an animal, immigrating into an area with hardly any other competing animals is easy; but emigrating into a populated area is often not possible.

Central Europe is peculiar in the sense that the intensive alternation between wet and dry-cold climate conditions prevented the region from ever becoming a long-term core population area, neither for the forest species nor for the steppe species; instead, Central Europe was always only a temporary and peripherally located distribution area for both (von Koenigswald, 2002). The core population areas of these species lay in the Mediterranean region, in Eastern Europe and in Siberia; and this situation still has not changed today.

Knowledge of this fact is vital if we really wish to understand the special situation of Central European wildlife and its protection measures. Hardly a landscape or vegetation exists that we could call typically Central European, so it follows that there are only a few species that could be called typically Central European. During the Pleistocene, Central Europe was constantly being populated by the species of the Mediterranean area and the Eastern regions, and this situation was exacerbated by the anthropogenic forest destruction of the last millennia, because man kept the landscape open and this drove the animals to immigrate from the Mediterranean region and the steppe plains. Only during recent decades has Central Europe lost its open-land character; its landscape became different from Mediterranean and Eastern landscapes; and with that a major proportion of its species has disappeared.

6.1.2
The Disappearance of Large Animals at the End of the Pleistocene

The current interglacial, the Holocene, is radically different from the previous interglacial periods of the Pleistocene. Most large mammals are now extinct (probably along with many other animals, the fossil remains of which have not been found). Both the forest landscapes of the interglacials and the steppes of the Ice Ages were populated by large animals, most of which died out at the end of the Pleistocene, although they survived all the earlier climate changes. There was something different about the last 100 000 years compared with the 2 million years before that.

The remarkable thing is that both the large mammals of the warm periods and those of the Ice Ages lived for millions of years until the end of the last interglacial period or until the end of the last Ice Age, only to die out during the geologically short period of the last 100 000 years (von Koenigswald, 2002). At the end of the last interglacial, the European forests lost the European forest elephant (*Elephas antiquus*) and the forest rhinoceros (*Stephanorhinus kirchbergensis*); only the European water buffalo (*Bubalus murrensis*) survived until the early Holocene. The large mammals of the Ice Age steppes, namely the woolly mammoth (*Mammuthus primigenius*), the woolly rhinoceros (*Coelodonta antiquitatis*), the steppe bison (*Bison priscus*) and the giant elk (*Megaloceros giganteus*) lived longer; but they only survived to the end of the last Ice Age. Some predators of the Pleistocene, such as the cave lion (*Panthera leo spelaea*) and the cave bear (*Ursus spelaeus*) became extinct at the end of the last Ice Age, although they were not wholly dependent on climate and vegetation and they lived in Central Europe during both the warm and the cold periods.

What was so different about the end of the last interglacial and glacial periods compared with all the previous climate and vegetation changes and why were these processes a global phenomenon? It is often assumed that man eradicated the animals of that period. One argument for this hypothesis is that the extinction of large mammals took place in the same geological period during which *Homo sapiens* emigrated from Africa and invaded most parts of Eurasia. That was at the end of the Palaeolithic. Only Neanderthals and *Homo erectus* previously lived in these regions, but apparently there were not so many of them and they possessed less hunting intelligence.

During more than 200 000 years of living together with *H. sapiens*, it is possible that the large mammals in Africa developed co-evolutionary instincts for flight, learning to escape from the humans after a time. That may have saved these mammals from extinction. In contrast, man's appearance during the Pleistocene surprised the large mammals in Europe and Asia; they did not recognise the danger and did not develop flight behaviour when humans approached them. Who knows, perhaps man was able to slaughter them as if they were stationary mountains of meat, a fate similar to that of many endemic bird species on remote ocean islands in historical times.

However, this hypothesis is contradicted by von Koenigswald (2004) who argues only a few Ice Age hunters existed, with the result that man could only have had the opportunity of killing the last animals of a species when the population of that species had already been more or less decimated through other causes. He also points out that the core populations of the large Ice Age animals on the mammoth steppes were located in Siberia, and there were even fewer humans there than in Central Europe at the end of the Pleistocene.

It remains a mystery why the richly varied fauna of the Pleistocene, or at least the many large mammal species, died out in the last 100 000 years. If climate change and not man was the main cause, then it must have occurred in two climatically very different core regions, namely in Siberia and in the Mediterranean region; and such changes did not occur in the previous 2 million years during the entire Pleistocene period.

During the interglacials, was the natural landscape in Central Europe not closed woodland, but more like semi-open, park-like landscape meadows because of the large herbivorous mammals? There are also conflicting views about this. Thanks to fossil pollen mainly preserved in bogs and lake sediments, we can nowadays reconstruct the vegetation of primeval landscapes. There *were* open areas during the heydays of the forest, but pollen analysis does not give us any reliable information about how large these open areas really were. There is no compelling evidence to suggest how frequently the large animals in the interglacial forest landscapes really occurred and how many and how large were the areas they possibly kept open. Fossil records cannot tell us anything about game population densities. Even skeletal remains in Stone Age settlements only provide scanty evidence about the frequency of species occurrence.

The large mammals were grazers, so there must have been pastureland during the interglacials. Large open spaces probably existed only after storms with high winds, forest fires and in the flooded areas of the riparian zones (floodplains). The idea that Central Europe was open parkland with large grasslands, shrubs and forest islands during the interglacial warm periods is probably not true. However, it does seem certain that the huge herbivorous fauna did visibly thin out the forests. A dense, dark forest (which would hence forward become prevalent in Central Europe if all silvicultural encroachments had somehow been halted) probably did not exist during the interglacial periods.

The large mammals restricted the growth of the forests during the interglacials; but in the Holocene, prehistoric and historic man prevented the growth of dense forests through deforestation, clear cutting, agriculture and animal husbandry. After the prehistoric disappearance of the elephant and the rhino, the eradication of the aurochs, and the decimation of the European bison (*Bison bonasus*) and moose in historical times, man thinned out the postglacial forests.

The evidence hidden in the many pollen profiles and the fossil traces left initially by Stone Age hunters, then by early farmers and cattle breeders, makes it quite clear that postglacial forests were (at all times) thinned out more intensively than they have been very recently and will be in the future. When reforestation occurred after the Ice Age, dense, closed-canopied forests did exist in Central Europe, but not to the same extent as the forest growth that has gradually appeared in Germany during recent decades. The current objectives of the German Forest National Park authorities in Germany to create something 'natural/primeval' by allowing the forests to develop naturally by themselves, and to avoid silvicultural encroachments where possible are targeting the creation of a forest structure that has probably never existed in Germany.

6.1.3
Past Exploitation of Nature Created the Open Landscapes and the Species Richness of Central Europe

It was not only the Ice Ages that heavily decimated the forest species in Central Europe. In the postglacial period it was man who intervened in the re-emergence of the forests. With the decline of the steppes, man tried to hinder the development of forests, because the forest took away his main source of food. Humans originally came from the savannahs of East Africa and they feel at home in the open countryside, where they have a wide view to spot prey and detect approaching danger in time. Postglacial man apparently tried to hinder the emergence of forests (Remmert, 1988). Gradual reforestation occurred parallel to the onset of his sedentary lifestyle, and Mesolithic and Neolithic man used forest fires and clearcuts to keep the landscape open, because he needed the cleared land for his settlements and farmland.

These open areas became habitable for rabbits, hares and larks. Some steppe animals (such as the great bustard, see Section 6.5.3.3) were probably able to make a comeback relatively quickly in the postglacial period, or perhaps they always stayed close to the human settlements from the outset. Man kept the old Ice Age landscape clear to a limited degree. Mossy areas, heaths and shrub landscapes could develop again because some seeds still remained in the ground and the immigration routes for the plant life of the north and east were open. It seems quite possible that after the Ice Age, prehistoric man actually prevented primeval, closed forest cover from developing in many accessible locations from the outset. Forests certainly grew back after the fires and clearcuts. However, the forest that grew back was not primeval; the trees which gained the upper hand were those which benefited from human destruction (Küster, 2008). When we speak of natural ecosystems, we mean those which are not affected by humans; and such ecosystems have hardly existed in many regions since the last Ice Age in Central Europe (Remmert, 1988) (see Sections 3.1.1 and 7.1.2).

As regards the animals and plants, Central Europe still bears the legacy of a steppe country. Many Central European species are steppe dwellers, particularly those which are vulnerable today. These are immigrants from the open lands of the north and the east, such as black grouse from the treeline region, golden plovers from the shrub-free Nordic Fell plateaus and skylarks, grey partridges and bustards from the steppes and semi-steppes of western Asia. Many bird species in Germany are open-land species, which are not adapted to living in undisturbed, primeval forest (Section 5.1.2). These species include the normal field species such as the grey partridge and the skylark, plus swallows, shrikes, wagtails, pipits, chats, buntings and many more. Most butterfly species and many other insects are inhabitants of the open country. The grasslands of Central and south-eastern Europe are among the most species-rich environments in the world. If we consider some selected small areas, European semi-arid grasslands may even have more animal and plant species than the tropical forests (Connell, 1978; Dengler, 2012), but most semi-arid grasslands are not natural habitats in Central Europe.

This is why we cannot compare species protection in Central Europe with that of the tropical forests, because the tropical forest species which are endangered today have adapted to the forest as a stable habitat and often occur there endemically. These species are threatened today by the current destruction of tropical rainforests. In Central Europe on the other hand, almost no species are threatened by the destruction of forests. They are threatened by the destruction of other habitats.

Nowadays, we should neither find it surprising nor strange that large open spaces such as airports and military training grounds quickly develop into 'species havens', because they imitate the original glacial steppe landscapes. Despite postglacial forestation, human destruction of nature has never totally eradicated these habitats. The last millennium saw almost all of Central Europe turn into a steppe landscape and the forests pushed back to small areas; and this has restored the glacial primeval landscape. Open land has now been steadily disappearing for several decades now, after surviving for thousands of years. Central Europe is endangered as a steppe region and as such is losing many of its species. To halt this process, we must act against nature. Central Europe's open areas with their species richness were created in the postglacial era by human intervention – and only man's intervention can save these areas. The human desire for primeval states and naturalness in the Central European landscape far outweighs the species' silent cry for help. The open lands of Central Europe do not meet the current climate and soil conditions, so they are not 'natural'. They must be protected from nature and must not be left to themselves.

Agricultural man has shaped Central Europe since the Neolithic period. Slash and burn, sod cutting and forest pasture were the predominant forms of agro-economy. After the third or fourth year of cultivation, the soils were mostly exhausted and were left to themselves, serving as poor pasture for between 15 and 20 years. The sparse greensward was eaten quickly; only the plants that were rejected by the animals remained. The areas which were infertile now became wastelands of grass, heather and gorse bushes, very often after only a short time. To fertilise the remaining arable land, the grass turf layer of this fallow land was peeled, dried, burned and spread on the fields together with the autumn seeds (Paffen, 1940; Pyne and Goldammer, 1997; Rösch *et al.*, 2001, 2002). In time, this created the typical heathland. Heather and the undemanding matgrass grew and flourished. The heather was cut regularly. After some years of this treatment, a peat-like ground cover formed and the peaty soil was cut with a shovel, the sods being used as fuel for man's homes and bedding for his cattle.

This form of agro-economy led to the degradation of the soils in Central Europe. It was exactly like the agro-economic methods in many of today's developing countries, methods that are greatly criticised. These agro-economic methods, however, led to an increase in the number of species in Central Europe during the last millennia. What we complain about in developing countries today is that they are eradicating their species by destroying what is almost primary nature; but the same process is believed to have taken place thousands of years ago in Central Europe. In Central Europe today, the destruction of primary nature no longer endangers species.

During the past centuries, the plundering of the countryside did not harm the majority of species – on the contrary, it even formed the foundation for the species richness of the past. This fact, one that is so hard for some people to accept, only seems so paradoxical because it contradicts an almost ideologically accepted view, which was propagated in promotionally effective manner even as late as the 1980s of the last century. This view states that a sustainable, sparing use of resources would also simultaneously benefit the species; but this is definitely not the case for many Central European species, or at least not for many endangered bird and butterfly species.

The destruction of the forests in Central Europe opened the gates to thousands of immigrants from the northern, eastern and southern regions for many millennia. Open, nutrient poor soils and sparse vegetation formed the habitats for open-land species that have their core populations on the edge of Europe and as far distant as western Asia. In the ruined nature of Central Europe, the immigrants found a home which some had already populated during the Ice Ages on the steppes of Central Europe. In the form of remnant populations or returning emigrants, they succeeded in repopulating the anthropogenically created wastelands and fields. The anthropogenically deforested areas in Central Europe were similar to the Nordic Fell plateaus of Scandinavia, the moorlands of Northern and north-eastern Europe and the steppes of the Near East. This allowed many open land species to immigrate or reimmigrate to Central Europe, such as the black grouse, golden plover, bustard, grey partridge and brown hare. These immigrants are described as being synanthropic. The species that were not dependent on large steppe areas, but rather on small plots of open land (e.g. most lark, pipit and shrike species) also benefited during the past millennia from the destructive agricultural practices, because they need open areas to spot and hunt their prey (Coudrain, 2011; Schumacher, 2012).

During the past centuries, the plundering of the countryside did not harm the majority of species – on the contrary, it even created the biodiversity of the past. Central Europe reached its highest level of species richness at a time when agriculture and forestry were anything but ecological and sustainable (Häpke, 1990; Plachter, 1997). Ruthless exploitation of nature was part and parcel of the Central European agricultural methods of earlier centuries and millennia. Such methods completely contradicted the general requirements (which are today legally required) for proper and correct agriculture and forestry and for the sustainable use of resources.

Most of the Central European butterflies rely on open ground, because there are too few warm sunny spots on lush, overgrown meadows, in bushy thickets and in the forest (Ulrich, 2002). Almost all of the current reports published in Germany about the disappearance of butterflies point to this landscape change (Cungs, 1991; Augustin, 2003; Lenz and Schulten, 2005; Laussmann, Radtke and Wiemert, 2005; Laussmann *et al.*, 2010; Bäumler, 2010; Retzlaff and Robrecht, 2011; Filz *et al.*, 2013). The occasional massive occurrence of many species of butterflies in the nineteenth century and in the first half of the twentieth century was focused on sandy areas, dry slopes and marshy meadows (Bourn and Thomas, 2002).

Of course there are still large open spaces in Central Europe. These are the agricultural farmland and pasture areas, the use of which ensures that they do not become overgrown with scrub and woodland. However, the agricultural fields and pastures have become unusable for most of the open-land animals such as skylarks, grey partridges and brown hares, simply because modern cultivation and harvesting methods with farm machinery leave nothing else for the animals. Nowadays, fields and meadows are exploited down to the last square yard, weeds are gone, and crop residues hardly remain on the ground. To make matters worse, the grain fields themselves have changed; the vegetation is much too dense. There are no more sparsely vegetated fields of grain or other agricultural crops. The areas used for agriculture today are densely vegetated and therefore almost uninhabitable for most birds, butterflies and other species which formerly lived on the grain fields. Today, agricultural areas are 'deader' than any industrial area (Plate 4).

When we speak of the disappearance of open landscapes, we do not mean the agricultural land, but rather the remaining open land adjacent to the agricultural land. This is not preserved as open land, however; it is afforested and developed into woodland and this is why most Central European endangered animals are increasingly being squeezed into the few open areas spared by both agriculture and afforestation; but even these remaining open areas are disappearing, because vegetation is becoming denser. They become overgrown because most of the slope faces, hilltops and stream valleys are no longer grazed and the shrubbery and the matgrasses are no longer required for stables, or making fires in the home. Excessive nitrogen levels in the atmosphere are an additional factor (see Section 6.3). The increasing over-fertilisation and the lack of tree and ground vegetation removal are the main reasons why the remaining areas (those left over by agriculture and forestry) become rapidly overgrown with grass and shrubbery. Such dense and highly overgrown areas with no sparse vegetation did not exist in Central Europe in earlier centuries; but today uncontrolled growth has returned. These areas are no longer habitable for many Red List species.

The only sparsely vegetated open spaces that still remain today are industrial and traffic areas and those around towns and villages (Plate 4). This is one of the reasons why many species abandon nature and retreat into the open areas of the more 'machined' surfaces. Many species, such as the grey partridge, lapwing and skylark find refuge in the cities, on the wastelands of factories, beside airport runways, on and around railway embankments, on the areas adjoining motorways, on surface mining excavation sites and in military training areas. The call of nature conservation organisations for a return to 'primeval nature' totally ignores the urgent needs of species like this.

Even the cities today are not spared from tree planting and an increase in shady vegetation. As has happened in the countryside, a subtle structural change, hardly noticed by the public, has taken place in the cities of Central Europe during the past 30 years. The urban landscape is becoming more forest-like, resulting in an increase in shade levels. Open, sun-exposed areas are also becoming rarer in the cities (Mitschke, 2009). Open spaces and vacant building plots have largely disappeared, kitchen gardens with fruit trees and vegetable beds with sandy, open ground have become rare and the planting of trees in green belts and gardens is on the increase. Many homeowners in residential areas turn their gardens into small woods. Based on the increase in the volume of wood, cities are a direct reflection of the entire landscape change in Central Europe today.

A corresponding change can be seen in the breeding avifauna in German cities. The house sparrow, starling, kestrel, greenfinch and black redstart were among the most specialised bird species in the urban environment. The populations of all five species are evincing a comprehensive and ongoing decline in Central Europe. Several species that were originally forest breeders have benefited from the gradual structural change in our cities and are immigrating into them in increasing numbers. The long-tailed tit, short-toed treecreeper, blackcap, goldcrest, marsh tit, coal tit and great spotted woodpecker are increasingly replacing our typical city dwellers such as sparrows, black redstarts and starlings (Mitschke, 2014).

Yet again, we can see that this 'compensation' by new immigrants has no counterpart in the world of butterflies. Most butterfly species in Central Europe are originally residents of sunlit waste ground, ranging from sparse vegetation to sandy soil. If such areas disappear from our cities, no immigrant shadow species will arrive from the deciduous and coniferous forests to 'replace' them (as in the case of the birds), because there are not enough butterfly species like this in Central Europe.

Overall, the number of birds in cities is on the increase. It is the specialised bird species that are disappearing, increasingly displaced by the generalists. In the German state of North Rhine-Westphalia, the 10 most common breeding bird species now amount to 60% of the total breeding bird population (Weiss and Schulze-Hagen, 2014). The number of species is decreasing, although the total number of individual birds is not on the decrease.

6.1.4
What Is 'Natural' Fauna and Flora in Central Europe? – The Implications for Nature Conservation

Europe differs from most regions of the rest of the world. It has relatively few species, which, however, are found in high population densities (Baillie, Hilton-Taylor and Stuart, 2004). This is partly due to the geographically high latitude location; a general rule states that the number of species is highest near the equator, but these numbers decrease the more we move to the north (Godfray and Lawton, 2001). However, it is also likely that Europe lost many of its original species a long time ago because of the Ice Ages and the encroachments of postglacial man (Palm, 1957; von Koenigswald, 2002). Species such as the mammoth, woolly rhinoceros and the aurochs have left behind easily found fossils or even historical documents in, for example, cave paintings, and these provide us with evidence of past biodiversity loss. The many other species, evidence of which neither exists in the form of fossils, cave drawings nor ancient writings, have disappeared without a trace. What is now happening in the developing countries, especially in tropical areas, probably took place in Europe thousands of years ago.

Central Europe hardly has any more sensitive endemic species with close ties to original habitats. Most of the remaining species are the rugged ones and the immigrants from the northern tundra, the Asian steppes and the open areas of the Mediterranean. Central Europe today is home to a range of species, which is probably (to a large extent) not really original, but instead consists of interglacial and postglacial immigrants or of those species which entered the region in the wake of past deforestation. However, it is these same flora and fauna that have characterised Central Europe for millennia. They are not the original, primeval forest flora and fauna, which have been under threat in Central Europe for several decades now and which currently grace the Red List. The absence of an original flora and fauna is probably the reason why Europe today has only a relatively small number of globally threatened species (Baillie, Hilton-Taylor and Stuart, 2004). The real species decimation is taking place today in other parts of the world, not in Europe. That is already behind us.

In total contrast to global species extinction, today's Central European species extinction has nothing to do with the arrival and settlement of humans in new areas. It has more to do with our high standard of living, simply because Central Europe's biodiversity was based on centuries of poverty for the populace and the exploitation of nature caused by that poverty. The exploitation of natural resources has made Central Europe a steppe country, and most of the species that are threatened today in the region are open-land species. These include almost all buntings, larks, pipits, shrikes and most butterflies.

Seen in this light, the popular call of the nature conservation associations for pristine forests appears to be questionable. The guiding principles of nature conservation often refer to the preservation of a 'natural fauna' that once existed here. What do they mean by this 'natural fauna' in Central Europe? Do they mean the steppes of the glacials or the forests of the interglacials? Central Europe is home to both steppe biocoenoses and forest biocoenoses. Both of these habitats and biocoenoses characterised Central Europe for 2 million years. During the last millennia, humans increasingly thinned out the forests, so Central Europe became both woodland and a steppe country.

The Pleistocene migrations of animals from the eastern and southern regions were complex territorial movements that regularly resulted in population-genetic changes. Species which migrate back to an area are not the same as those which retreat to it. If climate warming causes new, more northerly areas to be taken over, an entire species does not usually migrate to the north. Those that migrate are often only the peripheral populations of the previous core population areas of the species. The immigrants of these peripheral populations then create new populations that may differ genetically from the previous populations. The populations of the species' core area often remain unaffected by these changes.

Conversely, if living conditions worsen through dropping temperatures, and some affected individuals try to migrate, these then enter refuges that are already occupied by other populations. This results in competitions with the resident individuals of the same species or of other species, which in turn can lead to genetic changes because of newly created selection pressures.

So what is the 'natural species population' in Central Europe, one to which we could refer if it were our objective here and now? One ecologically similar situation would be the flora and fauna of the last interglacial, because a climate prevailed at the time that would be comparable to the present climate. The last interglacial had a 'nature' that was not yet influenced by man; but the current interglacial period (the Holocene) is in many ways incomparable with the last interglacial. The large mammals, such as the forest elephant, forest rhinoceros, hyena and lion, are no longer with us and they impacted on the species composition and the vegetation of the last interglacial. If we refer to a *natural fauna and flora* from today's perspective, the biodiversity loss at the end of the Late Pleistocene must be taken into account. If we created a nature area with no human encroachment in Central Europe today, and we left it to grow naturally, a new type of nature would be created, one that possibly has never been seen before in Central Europe.

Given the uniqueness of Central Europe, it follows that the many nature conservation goals and recommendations for species conservation that have achieved global importance for most countries do not apply to Europe. It is vital that we in Central Europe realise that we cannot afford to leave nature to its own resources. We would allow a nature to develop through which we would lose many species. Rapidly emerging dense grass, bush and tree growth would only provide an optimal habitat for part of the Central European flora and fauna; many species would become extinct.

Today, we live in the age of species extinction (see Sections 5.2.2 and 8.2.3). However, the causes of species extinction in Central Europe differ from those of the tropical rain forests. Forest species are endangered in the rainforests, but in Central Europe these species are in no danger. Species extinction in the rainforests is the result of the destruction of almost-primary nature, whereas in Central Europe most species extinction is the result of nature reclaiming areas through afforestation, areas that were open and free of forests for thousands of years. The agricultural land (agricultural fields, meadows and pastures) is lost to the species today, and the open areas that still remain in Central Europe today are no longer what they were for thousands of years, namely sparsely vegetated areas with thinly spread, low vegetation growth and enclosed areas consisting of bare stone, earth, sand and mud. Nature in Central Europe today has to a large extent no more areas with sparse vegetation.

The destruction of nature cannot be blamed as being a major contributor to the loss of the Central European species because the loss of species has only become really noticeable in the last 100 years or so in Central Europe; but nature in the region was actually destroyed much earlier. To a much greater extent, the blame for the destruction of nature lies with the conversion of Central Europe from sparsely vegetated open land to a land of dense green vegetation, scrub and thick forests.

While the developing countries strive to keep nature as original as possible in order to prevent the extinction of species, almost the opposite conservation rules apply in Central Europe. Here, our objective is to protect endangered species from too much nature. The endangered species in Central Europe are the open-land species, including many of the agricultural areas. These agricultural areas were meagre and barren and were exploited for millennia; they cannot be compared with agricultural land today, where artificial fertilisation and advanced farm machinery have made prosperous production areas out of meadows and fields that are almost species-free. Central Europe probably reached its highest level of species richness around 1850 (Blab *et al.*, 1984; Häpke, 1990).

This is why Central Europe, with its species conservation issues, faces a very different task than the other continents, where the rescue of endemic species is to a large extent the top priority, especially in the rain forest areas; but in Central Europe the priority is the rescue of species that have immigrated from other countries into the secondary habitats of Central Europe. In non-European countries, the destruction of forests decimates a great number of species, but the opposite is true of Central Europe. Here, the worst species killers are thickly forested areas and excessive afforestation.

This places a large question mark over one ideologically motivated principle of nature conservation. At least in Central Europe, the protection of a landscape cannot be justified by 'naturalness', because the Central European nature is 'second-hand'. The motive to protect the tropical rainforests from destruction differs from the motive to protect an endangered landscape in Central Europe, because the identity of the Central European landscape has been shaped by man.

It is undisputed that the current, large-scale destruction of the forests in South America, the Congo Basin and South East Asia destroys those species dependent on habitats that are in an (almost) original state. Rescuing these species can only be achieved by preserving the originality of nature, at least in some limited regions; but this is not true of Central Europe. The Central European species that are threatened today are residents of habitats created by the clearing of forests centuries and millennia ago. The majority of the endangered species living in Central Europe today do not live in forested areas, but in open landscapes that cannot be regarded as the original, warm period biotopes of Europe.

Nature conservation in Central Europe must decide which arguments exist for the protection of nature, and must continue to think hard about which methods it can adopt to protect nature. The protection of the rainforests is much easier to substantiate than the protection of species in Central Europe because there (and only there), the main priority is the preservation of nativeness. The method through which the rainforests can be protected is simple in theory, though difficult in practice: leave Mother Nature alone to take her of herself.

These principles cannot be applied to Central Europe. We could hardly create virgin, original nature in Central Europe, even if this were desirable. And if we leave nature to itself, that is, if we avoid human intervention and changes, we run the risk of creating landscapes and habitats that may never have existed in post-glacial Central Europe.

Even if it were possible to create a Central European pristine forest area, one that approximated a habitat in which no human encroachment had ever existed, then the important question arises as to which animals would live in this primeval forest. It could be that a primeval forest like this would not have very many animals, because the wildlife in the warm period after the last glacial was not able to adjust sufficiently to such a habitat. And let us not forget that the large animals would be missing, those that would hinder the growth of young trees and create clearings. The forest elephant and the aurochs are extinct; the former lived in Central Europe during the interglacial warm periods, while the latter lived in the present millennium. The European bison would possibly not allow itself to be settled in sufficient population numbers, quite apart from the fact that this would be a human encroachment, one which would contradict the concept of 'unspoiled state', since the artificially released European bison population would require management, and thus would involve artificial human intervention in unspoiled nature; in other words, nature would not be left to itself.

There is also the fact that you can hardly leave forest fires to rage as wildfires in Central Europe at the present time; but that would have to be ensured if we were to leave nature to its own devices. We also must consider the precipitation of nitrogen from the atmosphere; it is 20 times higher in today's Europe than it was 60 years ago (see Section 6.3). This (expected) fertilisation of the forests, even in those areas outlying the agriculture areas, would probably make the forests denser, and the closed canopy of trees would darken the sun-starved forest floor; in other words, we would create forest structures that never existed here in the past.

6.2
Can Endangered Species in Central Europe Be Saved through Forest National Parks?

The National Park concept relies on the natural regenerative capacity of an ecosystem and promises the creation of original nature.

However, the creation of forest national parks in Germany will not restore original nature; it will allow the current forest ecosystem to return to a wild state.

The forest national parks help only a few of the red-listed species threatened by extinction today.

The Teutoburg Forest and the Eifel in Germany are not Brazil.

The hazel grouse is a bird that lives in dwarf shrub growth and benefits from anthropogenic encroachments in the forest.

Almost all kinds of forest butterflies in Central Europe are species that live in light-rich forests; they depend on forestry management operations.

The National Park concept aims to establish large-scale habitats that will be largely left to themselves and almost never tended. The guidelines of the Federal Nature Conservation Act in Germany (Bundesnaturschutzgesetz) dictate that national parks should be 'predominantly in a state that is either not at all or only very slightly influenced by human encroachment', a state which ensures 'an uninterrupted flow of natural processes in their natural dynamics' (German Federal Nature Conservation Act BNatSchG, §§23–27). 'Real wilderness' is the objective in the national parks; nature should be allowed to develop as 'nature' as much as possible without human intervention. Flocks of sheep, for example, would have no place in a national park (König, 2014).

National parks go hand-in-glove with the basic ideology of German nature conservation, because 'original', 'untouched' and 'humans out' are all priority one in German nature conservation. It would undoubtedly be fascinating to experience a nature with no human encroachment. It is also of great scientific interest to await the development of a nature left exposed to the natural dynamics of its biocoenoses. However, these objectives must not be equated with the objective of increasing species richness – and the German National Park ideology does not differentiate between commitment to the 'wilderness aspect' and the quite dissimilar objective of halting species decline (Riechelmann, 2013). There is a lack of reasoning behind why national parks are seen as 'something good'. Instead, national parks are rather a 'commitment to pristine nature' and thus primarily based on ideology (Tornede and Pieper, 2012).

It is often claimed that national parks are necessary to halt the loss of species richness; but this statement lacks sufficient substantiation. One thing is certain, however: the goal of 'pristine wilderness' and the objective of a high level of species richness in Central Europe cannot be brought together under one roof. The expectation that we can save a significant portion of the endangered Red List species in Germany by creating national parks seems to apply only to the three national parks of the German Wadden Sea and the two national parks on the Baltic coast because apart from the high mountain regions, it is almost exclusively Central Europe's ocean and coastal ecosystems that have preserved the character of naturalness in the past millennia. The forest landscapes of the Central European interior have not been in a natural state in recent millennia, indeed, they have hardly ever been in a near-natural state. Under today's conditions of increased eutrophication throughout the entire landscape of Central Europe, the objective of original forest wilderness untouched by humans would potentially create habitats that have never existed in this form in the postglacial era; and the question is whether or not this can save the species on the Red List?

In Germany, there are 16 national parks, 15 biosphere reserves and roughly 35 nature parks (http://de.wikipedia.org/wiki/Nationalparks_in_Deutschland). In the past 20 years, nature conservation areas have been expanded greatly in Germany, according to the Federal Statistics Office. The areas of the national parks have increased by 48%, nature reserves by 174%, and biosphere reserves were almost tripled (Riechelmann, 2013). These numbers make very effective advertising for politicians and they are also used for popular support. It is a fact that protected areas promote nature. However, the claim that these areas also serve the protection of species must still be put to the test.

The species often projected by nature conservation organisation as their flagships, namely the black stork, the eagle owl and the wild cat (König, 2014) have indeed increased in number in recent decades, but that is not due to the establishment of national parks. The black stork has flourished throughout Central Europe during the past half century (parallel with the reforestation process), but these population increases have not only occurred in the national parks (Janssen, Hormann and Rohde, 2004). Likewise, the reason why eagle owls have become more common in recent decades has nothing to do with national parks – it is because we have stopped hunting them.

If the main goal behind regaining the species richness of earlier centuries is to rescue the Red List species and increase the numbers of species and individuals, it will not be achieved through the current national park concept of 'nature without human encroachment'. In Central Europe, this objective is much more likely to be achieved by conserving the historically developed, cultural landscape; and, in total contrast to the national park concept, conservation like this requires especially intensive human intervention in both vegetation and landscape. A national park, however, is focused on unspoiled nature – and therefore not on the protection of many endangered species in Central Europe. In a cultural landscape that has been processed for millennia (and still is being processed), the majority of plants and animals living there have necessarily adapted to the artificial conditions, and many are dependent on them. If forest management and usage disappear and nature unleashes its own momentum, the vegetation will run wild throughout the landscape, causing many species to lose their habitats.

For a long time now, Germany has had no more pristine forests; the country has almost exclusively consisted of cultivated land and secondary forests. So what actually happens when forests and open spaces are allowed to run wild? This is the intention in national parks. No one can know the outcome. The promise made by the National Park promotional associations that the national park 'pristine forests' would become species-rich simply cannot be substantiated. The objectives of existing and planned German national parks to ensure an uninterrupted flow of natural processes in their natural dynamics are diametrically opposed to the protection of our Red List species. Among the 68 birds on the current Red List of German breeding birds, there are very few species that would benefit from the creation of national forest parks; and for most of the remaining species, the National Park ideals are counterproductive.

More than two thirds of all the butterfly species in Germany are on the Red List, at least in most states (http://www.schmetterlinge-deutschlands.de/index.htm). The collapse of most butterfly species during the last 60 years has continued; and today the situation is even more drastic. To date, no evidence exists that the establishment of national parks has helped any butterfly species at all. The Eifel National Park, the 14th German National Park founded in 2004, has not brought back *Melitaea athalia*, the heath fritillary, a butterfly which was common in the Eifel Mountains everywhere 50 years ago. And this also applies to all the other butterflies the populations of which have declined in recent decades. Neither the fritillaries *Argynnis adippe*, *Boloria euphrosyne*, *Boloria selene* and *Euphydryas aurinia*, nor the Blues *Polyommatus bellargus*, *Polyommatus dorylas* and *Pseudophilotes baton* have benefited from the foundation of the German Eifel National Park; these butterflies have partially – or in some cases almost completely – disappeared from the Eifel region during the last 40 years, despite the foundation of a National Park. Some endangered butterfly species in several regions have instead been saved on mining areas, industrial sites, ammunition depots and military training areas.

The experience of recent decades nullifies the idea that the establishment of a national park may be the right way to save the butterflies.

The German nature conservation organisations support the creation of national parks with tens and even hundreds of millions (Anonymous, 2012). There is almost no discussion about how similar sums would be better spent on rescuing many Red List species by maintaining training areas abandoned by the military, or preserving mining areas when they fall victim to 'renaturation'.

The National Park concept is born of a longing for the pristine and the unspoiled, created by distaste for the increasing industrialisation and mechanisation of our life processes. National parks with their landscapes that are untouched by human hand are perceived as being 'fortresses against industrialisation' (Schwägerl, 2003). We should be able to experience 'real wilderness' in national parks. The pursuit and fulfilment of these aspirations has an intrinsic value and should not be denigrated here. However, the conservation organisations have reinterpreted this human demand for virgin nature as an objective bid to preserve species richness – and this is not based on any objective reasoning. Central Europe's flora and fauna has very little to do with originality and original nature. The Teutoburg Forest and the Eifel are not Brazil. The efforts to save the rainforests in Brazil involve the preservation of original, often endemic fauna. We have no such fauna in Central Europe; instead we have the postglacial immigrants and the hemerophiles which gathered after the thinning out or eradication of the forests.

The Šumava (Bohemian) Forest/Bavarian Forest was the first national park to be founded on German and Czech soil. The German area was first declared a national park in 1970 (Bibelriether, 1997), followed by the Šumava in 1991 (Klaus, 2012b). The stated aim of this national park is to create a 'nature that is totally left to its own resources', the counterpart of a 'restrained, well-managed' nature, 'completely in the spirit of the modern conservation concept' (Bibelriether, 1997). Whether or not this is really the 'modern conservation concept' is under scrutiny. It is certainly not a modern concept for species conservation.

The Šumava/Bavarian Forest National Park is the most famous of its kind in Germany. It embodies the concept of leaving nature to its own devices, thus progressively protecting the park from any human intervention and enabling the original forest to regenerate itself (Bibelriether, 1997). The National Park powers-that-be do not define exactly what this 'original forest' might be, nor is the issue addressed regarding whether or not a forest can actually ever regenerate in a Central Europe that has been anthropogenically landscaped for millennia.

The National Park Guide 'Naturland Germany' (Bibelriether, 1997) states that 'Man needs to be patient. He must learn to rely on the natural regenerative capacity of an ecosystem. He has to abandon the idea that he must always "intervene". Doing absolutely nothing, that is, giving the forces of nature a free hand is difficult for us'.

Two facts are ignored here:

- First, almost all Central European forests have been destroyed in past prehistoric and historic times, either by clearing or by fire. The forests which then

regrew were no longer the same as the original forests. The y were characterised by several emergent species (e.g. the beech), so they must be referred to as *secondary forests*. The structure of the Central European forests is not only the result of climatic conditions and soil conditions on site, it is also a testimony to the historical events that have taken place in the past (Küster, 2008).
- Second, the Central European flora and fauna has not lived in forests for thousands of years, but it has adapted to anthropogenically thinned-out forests and open meadows after the last ice age. Even if it were possible to successfully regenerate original, virgin forests in Central Europe someday, hardly any species on the Red List would be saved. A nature conservation success would then be applauded, but it would be highly unlikely if it even partially stemmed the current tide of species extinction.

One example here is the hazel grouse, the 'symbolic species of Šumava'; it has disappeared from most Central European forests outside of the Alps. The hazel grouse loves open woodlands with rich ground vegetation. It feeds on flowers and the seeds of perennials and shrubs. In late summer and autumn, its staple diet is mainly berries. Its distribution is therefore the open taiga of Russia and eastern Scandinavia. The hazel grouse is not a bird of virgin and dense forests. Although it is referred to as the *symbolic species of the Šumava*, it has benefited from the results of traditional forest management methods such as short and medium-rotation coppicing (Niederwald and Mittelwald). Such forms of woodland usage have either declined sharply in many locations in Central Europe or completely disappeared.

The human 'war' against softwoods proved to be extremely detrimental as did the substantial suppression of pioneer forests through afforestation (Flade and Sudfeldt, 2008; Klaus, 2012a). In sharp contrast to the National Park objectives of creating original forests, the Hazel Grouse needs light-rich forests, the canopies of which must not be too tightly closed, enabling the penetration of enough light to help a dwarf shrub layer grow on the forest floor. The berries of dwarf shrubs constitute an important part of the Hazel Grouse's diet.

In the Šumava, forestry and agricultural land usage ended when the Sudeten Germans were expelled after the Second World War. A 'natural' succession occurred in thinned-out forest areas as a result: a whole range of different deciduous trees grew in the clearings. These were trees that would not have grown without previous human forest encroachment (or all the trees on specific forest plots could not have reached a certain age at the same time). In the 1960s, these half-grown forests reached an age described by hazel grouse research specialist Dr Siegfried Klaus as being 'suitable for the hazel grouse' (Klaus, 2007). Hedges also grew on the tops of the man-made fragments of bedrock, and this enabled the grouse to spread out in all directions. The result was a significant increase in the hazel grouse population, a direct result of earlier anthropogenic interventions in the 'primeval forest'. This example shows us that it is not the 'pristine' forest which is a guarantee for healthy hazel grouse populations, but very specific interventions in nature followed by 'natural' succession.

Butterflies rely much more than birds on human intervention in the uncontrolled proliferation of the forest. Almost all forest butterflies in Central Europe are species which thrive in light-rich forests (Ulrich, 2002). They colonise warm, open clearings in the forest and can only live where the sunlight falls on the forest floor. They need clearings, forest paths with their edges, forest glades and clearcuts or power lines that cut through the woods. Their very existence is threatened by the creation of forests in a near-natural state.

In Saarland in Germany, 52% of the total of 25 forest butterflies is already on the Red List. Some of the species which were widespread in the 1970s in Saarland were the poplar admiral (*Limenitis populi*), the scarce copper (*Lycaena virgaureae*), the ilex hairstreak (*Satyrium ilicis*), the high brown fritillary (*Argynnis adippe*) and the pearl-bordered fritillary (*Boloria euphrosyne*); their populations have either been drastically reduced, or are in immediate danger of extinction (Ulrich, 2002). The negative trend is nationwide for most of these species. The ilex hairstreak (*S. ilicis*), for example, was still widespread in Baden-Württemberg in 1991 (in 23 ordnance survey map quadrants). Ten years later, in 2001, only one population was known to exist in the Rheinaue region south of Freiburg (Ebert and Rennwald, 1991b). In Saarland, only five sightings of this formerly widespread species have been recorded since 1990.

There are now more high forests in Germany than there have been for over 1000 years; and these forest interiors are becoming denser and thus darker inside. Less and less light is falling on the forest floor, and this is even happening on forest pathways and roads. Current aerial photographs no longer show forest paths, because they are obscured by the closed canopies of the trees. This never used to be the case in the past.

How can we help the species that live in light-rich forests? The answer is that any targeted help for these species must involve the following human interventions in the forest:

- Introduce light to the forests and 'air' them (e.g. by trimming trees).
- Create clearings (made by, e.g. power lines), widen roads and crossways, open up brook valleys.
- Let sunlight reach the ground, especially on areas exposed to the south, and along roadsides and slopes.

The suggestion that we should counteract the current loss of species through the establishment of national parks in Germany is based on an illusion. Even the current eutrophication of our landscape through nitrogen fertilisers delivered from the air to almost every square metre of ground will ensure that the creation of untouched forest structures simply cannot restore that which existed for thousands of years in Central Europe. There is a potential danger that we will end up with forest structures which have little to do with the restoration of original nature. The current forest ecosystem will return to a wild state. This could threaten and not help the plant and animal biocoenoses that have adapted to disruptive encroachments into the forests for thousands of years.

6.3
Nitrogen Is Suffocating Species Richness in Central Europe

In the 1950s, intensive use of nitrogen fertilisers in Central Europe initiated a disaster for species richness.

Nitrogen seems to be more at fault for the decline of insect populations than the insecticides themselves.

Specialists suspect that neonicotinoids endanger birds; but other reasons also lie behind the decline of some bird species.

Extensive human utilisation of the soils in Central, Southern and south-eastern Europe lasted for centuries – and it was one of the reasons for the high level of biodiversity. Some of the man-made grasslands in south-eastern Europe are still preserved today; and they are among the most species-rich habitats on earth. If we compare small areas of these grasslands to the tropical rainforests, the species richness of the grasslands even exceeds that of the rainforest in the case of some species groups (Dengler, 2012).

It is the sparsely vegetated areas of Europe that are now just as endangered as the tropical rainforests. At present, a paradoxical situation prevails in which many species of Central Europe are jeopardised by the cessation of land use, but tropical forest species are threatened by the onset of land utilisation. The species in decline today are not those of Europe's forests. It is one of the great misperceptions in the public consciousness that the devastating destruction of the tropical forests is placed on an equal footing with an alleged threat to European forests and the alleged benefits of afforestation and tree plantations in Central Europe.

Only a few red-listed species in Central Europe are forest species. It is not the species of an intact and untouched nature that are threatened, but all of the species that have adapted to extreme habitats, such as those of the sand and gravel plains, mudflats and escarpments. In the past, habitats like these were created by floods, hurricanes and major fires. Then came man; he destroyed nature and exploited the soil, and that created the extreme habitats. The settlement areas and the arable land had to be kept forest-free, and wood gradually became a coveted commodity for house and road building, stockade fences, fuel and later for the smelting of ore. Where the forests still stood, they were thinned out by cattle browsing and wood collectors. Arable land was created where the forests were cleared, but it suffered from nitrogen deficiency after a few years and such areas consequently had sparse vegetation or were temporarily unused as fallow land. Larks, buntings and shrikes found homes on such land. Riverbanks consisted of mud, gravel and sandbanks and their landward areas were often limited to escarpments. It was here that stone curlews, terns, common sandpipers, little ringed plovers and other waders made their homes (Pfeuffer, 2011; Stickroth, 2012).

Nitrogen in agriculture was in short supply for thousands of years. In large parts of Central Europe, nitrogen-rich soil was exhausted after just a few years of cultivation, and areas like this then became unsuitable for growing cereals (Lethmate, 2005). Fertilisers in organic form had to be used on such fields, and this meant that manure, slurry, liquid manure, green waste and mulch were taken to the fields and spread there. The organic fertilisers were not enough, however, and the fields only produced meagre growth of cereals and crops, making them ideal habitats for many open land species.

A big turning point came only 100 years ago, when Fritz Haber (1886–1934) and Carl Bosch (1874–1940) succeeded in synthesising ammonia in industrial dimensions. Nitrogen was suddenly available in sufficient quantities for the artificial fertilisation of agricultural land. At first, the high price of the new fertiliser discouraged farmers who wanted to use it over large areas of the landscape. Most of the poor rural population could not afford the new product in sufficient quantities to fertilise their fields. Comprehensive artificial fertilisation only began on most fields in Central Europe in the 1950s and that initiated the disaster for Central European species richness. In those days, if the population had been sensitised against nitrogen fertilisers as it has been today against genetic engineering, protests and demonstrations against the destruction of our biodiversity would have taken place.

The radical change that was initiated by artificial fertilisation in the Central European landscape was hardly perceived by the population, and that in itself is amazing. The open landscape in Central Europe, one that had existed for millennia simply disappeared in less than a century. Nowadays, the former image of the landscape can be easily compared with that of the present by means of the many castles that stand on hills. These can be easily recognised today, because the structures themselves have not changed for centuries. However, the cliffs and bare slopes on which they stood 100–200 years ago are now covered by dense woods. In almost all cases, a comparison shows that the landscape has changed so much that today it seems as if we are looking at a totally new country.

Around 100–200 years ago, the landscape was still characterised by an almost treeless and shrubless vista (Tauch, 1974), whereas today it is dominated by shrubs and woodland. Almost all of the sparsely vegetated earth and sand surfaces are overgrown today. The last 1000 years never saw the dense vegetation we have today, nor were the mountain and hill slopes completely overgrown with bushes as they are today. In contrast to earlier times, almost all waysides, embankments, roadsides and stream and ditch edges are today covered with dense layers of weeds and herbs.

The high standard of industrialisation in Central Europe and intensive farming in confined spaces generate an enormous excess of nitrogen. For several decades now, Central Europe has suffered from extensive eutrophication, and to a greater extent than many other global continents and regions. The nitrogen emissions from internal combustion engines are in themselves bad enough; but agricultural fertilisation allows similar amounts of nitrogen to escape into the atmosphere. Today's Central European agriculture would seem to be a much greater species killer than industry (Reichholf, 2005). Public awareness, however, seems more willing to condemn the industrial world than to lay the blame at the farmer's feet.

The nitrogen that escapes into the atmosphere from many locations rains down again elsewhere, so all of Central Europe's areas today are consequently fertilised with nitrogen. In earlier centuries, between 0.4 and 0.8 kg of nitrogen rained down on an acre of Central European land every year; that figure has now risen to more than 16 kg per acre (Lethmate, 2005); this means that even areas remote from any agricultural land are regularly fertilised from the air; not even nature reserve areas are spared. Open spaces with sparse vegetation have become rare habitats, because at the same time grazing of the land has also greatly declined and the removal of vegetation through firewood harvesting, turf cutting and the raking of fallen foliage layers no longer occurs. Such open areas are far more important for the protection of species than forests (Plate 4).

Nitrogen is not directly toxic for species, it eradicates them indirectly; and nitrogen-rich soil does actually boost the growth of a few plants such as nettles and the majority of sweet grasses (Poaceae). These few species do get a competitive advantage from this, but it is an advantage which is denied to the major proportion of the species in the habitat. Three-quarters of all Central European plant species suffer in the competition against the 'nitrogen-species' and are literally overwhelmed by them. As a passive poison, nitrogen is much more dangerous than the active plant and insect poisons. According to some authors, artificial fertilisation has eliminated more insect species than all the range of insecticides used by man (Reichholf, 2005).

The few plant species that benefit from nitrogen grow taller and denser; they also darken the ground, which then becomes cool and moist. This is why the numbers of ground-dwelling insects have sharply declined (Litzbarski and Litzbarski, 1996). Contrary to the expectation that the populations of most heat-loving butterflies would increase as a result of the global warming of recent decades, the exact opposite is the case (Warren *et al.*, 2001): the dense vegetation has initiated a 'cooling' process at ground level. In recent decades, it is the heat-loving species of butterflies that have disappeared (Plates 2–4). Thanks to the unshaded sunlight in Germany 150 years ago during the 'Little Ice Age', they apparently found the ground warmer than it is today (Segerer, 2012). For butterfly populations, average temperatures are not as important as the maximal heating up of local ground surfaces, even if this process occurs for only a few hours a day. This explains the presence of so many butterflies in the upper montane and alpine regions, although average temperatures there are relatively low. The open surfaces, and the inclined surfaces that receive vertical sunlight in the higher elevations near to and above the tree line are heated up intensively for a few hours of the day, and that is sufficient for many butterfly species. Surface areas overgrown with dense vegetation, however, are not insect-friendly.

DDT-based insecticides were taken off the market during the course of the past few decades (at least in Europe and the United States), and a new generation of insecticides has since been held responsible for the decline of Central European species: the neonicotinoids. Neonicotinoids bind to the nicotinic acetylcholine receptors in the central nervous system of invertebrates, but have significantly lower affinity to the corresponding receptors of vertebrates. This is why neonicotinoids are regarded as being neurotoxins that are lethal to insects, but non-harmful to vertebrates (Hallmann *et al.*, 2014; Goulson, 2014). Crop and garden plants have been treated with these insecticides all over the world since the mid-1990s of the last century. Some of the seeds are already 'stained' with neonicotinoids before sowing, and the insecticides migrate into the plant from the stem right up to the blossom and the pollen. Neonicotinoids are water-soluble. They have already been detected in the catchment area of Holland's agricultural areas (Tennekes, 2010). This could also affect underwater insect larvae which are not meant to be targeted by the pesticides (Hallmann *et al.*, 2014).

It is hypothesised that insect-eating birds are also affected by the neonicotinoids, since population declines of certain bird species in Holland have been observed in areas that are near to the agricultural land where the neonicotinoids have been used (Tennekes, 2010). This correlation between the areas where certain bird populations were determined to be on the decline and the land on which the neonicotinoids were used is extremely troubling. Correlation, however, is not proof of a causal link. A critical review of published data (Tennekes, 2010) makes it clear that while wheatear populations are on the decrease, the stonechats are on the increase; and while the green woodpecker's numbers are increasing, the wrynecks are disappearing. Since the wheatear and stonechat have similar nutritional requirements and the wryneck and the green woodpecker are both ant-eating species, it is hardly credible that the insecticide can be blamed for the decline of the wheatears and the wrynecks, because the stonechat and the green woodpecker would also be affected. The more likely conclusion is that the toxins are not to blame.

During recent decades, wheatear and wryneck populations have declined sharply throughout the whole of the north-western region of Central Europe, and not only in the areas affected by neonicotinoids. The wheatear and the wryneck require ground with extremely sparse vegetation to find food and areas like this have disappeared in the last few decades. The stonechat and the green woodpecker, on the other hand, can find their food on more overgrown areas. wheatear numbers are drastically decreasing because they require vegetation-free areas near rock piles or areas with stacked, freshly cut turf mounds; stonechats, on the other hand, can even see insects on grassy areas from their higher perching positions, if the vegetation is not too dense. The same applies to wrynecks and green woodpeckers. The wryneck has to actually see the ants running around in an open area to catch them, whereas the green woodpecker can hack them from the loose earth. To prevent the wheatear and the wryneck from dying out in Central Europe, we need more stony and sandy surfaces, as in earlier times. It appears to be unlikely that insecticides have caused the decline of the wheatear and the wryneck; but it is more likely that the disappearance of these species is being hastened by the nitrogen-based fertilisers that have caused the dense grass growth.

Very few Central European butterflies can live on land with dense grass growth and slopes overgrown with scrub (Ebert and Rennwald, 1991a). They need open ground areas with (at least intermittent) strong sunlight, otherwise they cannot warm up enough, and can find neither suitable nectar plants, nor warm sites for laying eggs and developing caterpillars. Well-fertilised, green grass surfaces are lethal for many insects.

Although birds are not usually directly dependent on soil heat, several species of birds are also threatened by dense grass growth, because their food consists of insects which they must see on the ground in order to catch them. These include the hunters which sit on high perches, such as the shrike and the common redstart, plus almost all of the bunting and lark species. They require foraging areas with only partial ground vegetation, otherwise they cannot see the insects moving on the ground. These birds can only prey on insects in sparsely vegetated areas (Martinez, 2010).

The example of what happened to the wryneck has shown us that conservation measures can go terribly wrong if they ignore the preservation or creation of open ground areas. In past centuries, wrynecks in Central Europe preferred to live in orchard meadows (Streuobstwiesen), which were once heavily grazed, and the rough pastures between the fruit trees, often mowed, were used for the cultivation of fruit and for the harvesting of hay at the same time. To protect the orchard meadows (mainly in south-west Germany), the nature conservation bodies decreed that these meadows should be mowed later to avoid the destruction of wildlife through early mowing: this was put into place by means of the landscape management directives and the contractual nature conservation programme (Vertragsnaturschutzprogramm). This action, however, which was seen as a *protective* measure, did not benefit the wryneck, it actually *drove it away* (Hübner, Rauh and Will, 2013). The wryneck is a ground hunter that feeds on ants and long grass hides ants. This bird feels more at home on disused railway tracks, marshalling yards and industrial areas than in well-kept parkland that only appears attractive to humans.

For centuries, agriculture was the driving force behind species richness in Central Europe. Now everything has changed and the exact opposite holds true. Fertilisation has made deserts of today's agricultural land, a land in which the skylark sings no more, no lapwings breed and no more brown hares can be seen. We need unfertilised land and rough-surfaced 'grouted' areas. It is therefore not surprising that military areas, brownfield sites and surface mining excavations have become the last refuges for many red-listed species. As areas of species richness, they have taken on a role that should be the domain of the nature conservation areas (see Sections 3.5.3 and 6.4.3).

6.4
The Rescue of Many Red List Species in Central Europe Requires Technical Interventions in Nature

'Habitat-Heterogeneity Hypothesis': uneven surface structures harbour more species than even surfaces.

The concept that the species richness found on agricultural land in the past can be restored by means of 'biological' farming methods, toxin-free cultivation and seeds that have not been genetically engineered does not address the real causes behind our current loss of species richness.

Consumer protection is not wildlife conservation.

An extensive 'biological' agriculture will save only a few agricultural species. Modern agriculture must be competitive on the international market; we cannot relegate the farmer to museum activities.

The species-rich fields and meadows of past centuries cannot be reconciled with agricultural methods that are oriented on profit.

It is impossible to scale down our current high-yield agriculture to low-yield agricultural production in order to save the Red List species.

The meagre croplands of past eras were associated with famine and extreme poverty throughout the rural population, so any resurgence of these agricultural methods would not be acceptable to modern farmers.

The lands earmarked for food production would appear to be lost for species conservation.

The future of the species does not lie in agricultural reform, but in the parallel existence of agriculturally used land and artificial areas which from the outset are not intended to yield profit-making crops, but to protect the species: land sparing versus land sharing.

The fact that many of the threatened species need more dirt, morass and sand than the cultivation of healthy vegetables and fresh green growth does not correspond with the ideals of the great majority of the human population.

The rescue of endangered species through 'wildlife engineering' and the creation of habitats 'on the drawing board', which of course are not wildlife sanctuaries.

At least in technologically advanced and densely populated countries, there will be no other alternative than to create artificial habitats or special areas that have been technically altered in order to save and protect threatened species.

The use of excavators and wood mulchers for the conservation of insect habitats.

The 'restoration to nature' of mining areas after the termination of coal mining destroys species.

6.4.1
Untouched Habitats, Compared to Habitats with Disruptive Encroachments

Even areas are usually more species-impoverished than comparable surface structures in the same habitats, which, however, have been made uneven by natural disasters, damage caused by game or human interventions. Heterogeneity in habitats stimulates biodiversity because most animals and plants do not colonise exactly the same habitat. This would lead to strong competitive struggles. 'Disturbing' influences in nature create heterogeneity, thereby giving most plants and animals the opportunity to avoid one other. If dense forest is broken by corridors or if small areas of ground are uneven and the vegetation has been partially removed, this creates niches (at least to a moderate degree) that are used by different species. The result of this is that a homogeneous habitat is species-impoverished, while a heterogeneous habitat is more species-rich. A wealth of variously structured habitats, even mini-habitats in a small area, provides suitable habitats for more species than a uniformly structured biotope (Topp, 2011).

This is described as a 'habitat-heterogeneity hypothesis' (Krebs, 1972). According to this theory, species numbers should increase on deciduous forest floors when a planar forest floor is replaced by an undulating structure – and this is indeed the case. The 'habitat-heterogeneity hypothesis' was verified using the species richness of beetles (*inter alia*) in the leaf litter of forests in Germany (Topp, 2011). Beetle population densities are higher in depressions on the forest floor; and additional species that are not found on flat and even forest floors were also found in these indentations.

In the Central European forests, structures like this were caused more by anthropogenic encroachments than by natural processes. In the forests of the German Central Uplands in historical times, mining activity was prevalent in very many locations and evidence of this can still be seen today in the uneven surface structure of these forests. Forests on historical mining areas have more beetle species and beetle individuals than forests on natural land (Topp, 2011). In previous centuries, mining transformed the originally flat floors of the forests into areas with small-scale sinks and elevated knolls (at that time, however, these areas were not actually forests). Today's hilly surface structures and sinkholes are the recognisable remains of the former test pits and the excavated rock dumps. These traces of mining date back to the early historical periods of the Celts and Romans.

However, man also changed the structure of forest floors in more recent historical times. In nineteenth-century Holland and in the Lower Rhine region, flax pits (Flachskuhlen) were excavated for cloth manufacture, and many craters were created by bomb blasts during the Second World War. This created valuable mini-habitats in which a high biodiversity was able to develop; but this species richness would not have arisen if humans had not created such habitats. In today's forests with former mining areas and historical flax preparation, there are more species and individuals than there are in forests with no such human intervention (Topp, 2011).

During the nineteenth century, the production of fabrics from flax was widespread in several regions in northwestern Central Europe. In order to produce spinnable yarns from the flax stalks, flax pits (Flachskuhlen) were excavated and filled with water. The harvested stalks then fermented for several weeks to extract the fibres from the stalks. The fermentation process caused the fibres to detach themselves from the woody core of the stalks, so the fibres could then be used for the production of yarns that were spun into linen fabrics.

Flachskuhlen like this produced a strong stench during fermentation and the dirty degradation products were then simply channelled into the streams. The process of cloth production in the nineteenth century is one more example of the enormous amount of environmental pollution that has taken place in our past; in Central Europe during this pollutant-rich era, however, species richness was much greater than today. During the 1960s to 1980s of the last century, the degree of environmental pollution (especially water pollution) was much higher in the German Democratic Republic (GDR) than in the former West Germany. Despite this, many endangered species have survived in the GDR (including the aquatic beavers) while the same species disappeared during these decades in West Germany.

The 'Habitat-Heterogeneity Hypothesis' (Krebs, 1972) can be applied to both the structures of tiny sections of a biotope and to global dimensions. In a forest or in a meadow an uneven ground surface is richer in species than a flat ground surface (Tews *et al.*, 2004), and even a lake shore with different kinds of sands and gravels is richer in species than a shore with a uniform sand grain size (Stein, 1984). However, differences in the habitat heterogeneity also exist on much larger areas. For example, the greater species richness of bird life in the tropics in comparison with that of the temperate latitudes can also be explained (among other reasons) by the fact that the individual areas in tropical forests are more heterogeneous than they are in the homogeneous and more even boreal forest floors (MacArthur, Recher and Cody, 1966; MacArthur, 1969).

Modern habitat maintenance measures are also confronted by this problem. Meadows that have been traditionally grazed by pigs are today often only mown, in order to preserve the habitat structure. This has led to the decline or even extinction of numerous species that were dependent on the foraging activities of the pigs (Schreiber *et al.*, 2013). This means that to preserve species richness, it is not enough to simply keep meadow vegetation short. Only the reintroduction of old breeds of domestic animals and the imitation of historical agricultural practices could restore the original biodiversity of meadows again. But nature conservation goals like this are unrealistic, because we cannot turn convert today's mechanised agriculture into a museum exhibition. The argument put forward by the nature conservation organisations that they can restore past biodiversity on agricultural land using 'biological' cultivation methods, toxin-free farming and genetic engineering-free seeds ignores the real causes of the current biodiversity loss. Policies like this do have their value as measures for consumer protection, but not as measures to restore earlier species richness.

If the nature conservation movement (which is organised into associations) would single-mindedly pursue increased biodiversity as its first objective, it would have to welcome, continually demand and constantly promote the human encroachments upon many unspoilt habitats. This would include churning of the soil, which can be performed just as efficiently by motocross riders and tanks as by pigs, and it would also include the removal of the top layer of vegetation, which in former centuries was used for litter in cowsheds. This encouraged the development of heaths, something which could also be done today by forest milling, and open-cast mining, which removes the 'natural' upper soil layers, bringing up deeper-lying soils that are not so polluted with nitrogen. All these measures promote the abundance of species, although they are in effect interventions that destroy nature.

6.4.2
The Animals of the Agricultural Land Are Lost: Land Sparing versus Land Sharing

The restoration of poorer agricultural land must be given the highest priority in species protection – but this goal is unrealistic. Modern agriculture must be competitive in the international market. We cannot ask the farmer to carry out activities that belong in a museum. The centuries-long species richness of the agricultural land in Central Europe has probably been lost forever; and bio-farming methods will not bring it back.

Repeated attempts have been made to rescue the endangered agricultural species by reforming modern agriculture. There are efforts to practise agriculture more extensively and to reduce insect control measures (Tscharntke *et al.*, 2012). These efforts are seen as being 'biological' and nature-friendly, but they are actually consumer protection measures that serve the interests of 'organic food'; they must not be confused with the rescue of the agricultural species that are truly threatened. 'Organic farming' meets the needs of the population for healthier eating; but it will hardly curb the loss of species. The few species that profit from organic farming are mainly those that have indeed declined sharply in numbers, but the existence of which is not actually endangered in Central Europe. Measures conceived for a more extensive agriculture are of course very welcome; skylarks will once again be heard in the spring and summer and tortoiseshells, peacocks, admirals, painted ladies and cabbage whites will return; but the species richness of earlier centuries will never return.

The specialists among the agricultural and meadowland species, including most butterfly species, the whinchat, the ortolan bunting, the black-tailed godwit and the common snipe will hardly benefit from a biological and more extensively driven agriculture. The specialists rely on sparsely vegetated areas in the fields or on swampy land in meadows and pastures (Plate 9). common snipes need meadows with mud and sour grasses and sedges, the way meadows looked in previous centuries. In the past, meadows were the embodiment of the wasteland. Most meadows in Central Europe were wetlands, consisting of 'Filz und Moos und einigem Gehügel' ('tomentum and moss on hilly ground'), as described in a report from 1802 (von Hazzi; cited in (Schreiber *et al.*, 2013)). The common snipe used to breed on such meadows. Attempting to restore lost habitats like this through ecologically driven agriculture would be an unrealistic illusion. The species-rich fields and meadows of past centuries cannot be reconciled with agricultural methods that are oriented on profit.

There would seem to be only poor prospects of success for any project that aims to restore the endangered species of the agricultural landscape through 'natural' agriculture and contractual conservation management agreements. Even the grey partridge is unlikely to benefit from this, and the former agricultural species that are even more reliant on sandy open spaces need more than flower strips at the edges of fields and skylark plots. Species such as the ortolan bunting need sparsely vegetated fields with bare soil areas and low stalk growth (Lanz, 2009) and this can only be realised with considerable crop yield losses, something which no farmer would even consider (Plate 9). Many of today's endangered agricultural species depend on the fallow lands of the former three-field system of agriculture, weed-infested fields, sandy heathlands and on the clover and lucerne fields that were previously created for nitrogen enrichment. In addition to the ortolan bunting, these species include the woodlark, the great bustard and all four Central European species of shrike and in the butterfly world, the clouded yellow (*Colias hyale*), the bath white (*Pontia daplidice*), the wall brown (*Lasiommata megera*), the large wall brown (*Lasiommata maera*), the hermit (*Chazara briseis*), the grayling (*Hipparchia semele*) and many blues (Plate 5).

Reserving small areas on agricultural land as marginal field strips and lark plots is successful in a small way. The skylark is a beneficiary of these measures, but the agricultural species threatened by extinction in Central Europe do not profit by them. The flower strips at the edges of fields, however, pose a very different threat. Colourful flower beds are perceived by the population as being intact 'nature' and give the impression that 'everything is all right now'. Yes, they look good; but they are not the habitat for red-listed species. The corn bunting does not need any blooming flower strips; it needs soil, poor vegetation and a little neglect (Plate 1).

Today's machining of agricultural land and the perfection and cleanliness of the harvesting methods leaves hardly any crop residue for animals on the farmland areas. The 'weeds' that formerly grew profusely on the fields are absent, and the fields themselves, on which the absence of fertilisation meant that cornstalks grew sparsely and in patches, no longer exist. It is impossible to scale down our current high-yield agriculture to low-yield agricultural production in order to save the Red List species. The trend towards organic farming is of little use in counteracting the loss of species in Central Europe. Modern agriculture has to be economically competitive. The progressive mechanisation of agriculture cannot be stopped, since competition in the global market is a must. Reforming to an 'ecological' agricultural economy would have to be much too radical if it were to have any effect at all on the really endangered hemerophile species.

As long as the fields are not interspersed with large sandy or muddy areas where the growth of cornstalks is at best sparse, the lapwing, ortolan bunting and corn bunting will not return in sufficient numbers to agricultural areas. The meagre croplands of earlier times were associated with a rural population plagued by famine and great poverty. No modern farmer would accept a resurgence of such agriculture (Plate 9). The idea that we could return to the less productive agricultural practices of past centuries and sustain its cultivation and harvesting methods in the present is an unrealistic illusion. Given the rapid growth of the world's population and the high number of starving people in the world, can we really envision a backward agricultural reform as an appropriate means for preventing species extinction? The lands earmarked for food production would appear to be lost for species conservation.

The future of species does not lie in agricultural reform, but in the parallel existence of agriculturally used land and artificially created and managed areas which from the outset are not intended to yield profit-making crops, but to protect the species. The only way to save most of the highly endangered agricultural species would seem to be the removal of agriculturally used areas from the species protection concept and to accept that these areas are dead for the species. Then, we must create separate areas for species protection. If the rich nations of Central Europe continue to generate huge profits from agricultural products, then these nations should also be able to afford special areas for species protection.

This solution has been drafted on several occasions and proposed to the politicians (Phalan *et al.*, 2011). It is known as *land sparing* instead of *land sharing*.

This solution, however, is not widely accepted. This is not only due to the fact that the establishment and maintenance of special areas costs money; a different factor seems to be more critical here: it would be difficult to convince the population of the sense of such special areas. An agrarian reform in the sense of 'land sharing' meets the population's need for a healthy lifestyle. Political aspirations like this are acceptable to a wide majority, because it is generally expected that measures that promote 'organic food' and thus human health will benefit the animals at the same time. But in many cases, this is simply not true. The fact that many of the threatened species need more dirt, morass and sand than the cultivation of healthy vegetables and fresh green growth does not correspond with the ideals of the great majority of the human population. The establishment of special habitats for agricultural species will therefore be an act that has almost no chance of being realised, because most people have a totally different concept of 'nature protection'.

6.4.3
A Future for the Protection of Species: Technically Manipulated Habitats

In the Central Europe of today, the protection of habitats alone (nature conservation) is no longer enough to stem the loss of species. We are too technically sophisticated, have a too high level of prosperity and are very overpopulated. We cannot turn back the wheel of history. Where certain species are almost at the 'Extinct' mark, the habitats for these species need to be designed by technical means – literally on the drawing board – and recreated. There are meanwhile examples that show how species that are endangered or even those that are threatened with extinction could be saved through the artificial creation of habitats (Kunz, 2013).

The creation of species conservation areas (which are of course not nature reserves) is called *Wildlife Engineering*. Large-scale projects for 'Wildlife Engineering' exist in the United Kingdom and in the Netherlands, where more of a pragmatic attitude prevails; the German Nature Conservation movement, on the other hand appends rather ideologically motivated ideas, aimed at 'untouched nature'.

In Lakenheath Fen near Cambridge in East Anglia, an 840-acre wetland area with extensive reedbeds was technically created by the British Royal Society for the Protection of Birds (RSPB) on a former carrot field bought between 1995 and 1997, and great bitterns and cranes settled here (www.rspb.org.uk/reserves/guide/l/lakenheathfen/). The great bittern almost became extinct in Britain in the 1990s. In 1997, a total of only 11 'booming' males was counted (the contact call of the male bird is like a deep, sighing foghorn or bull-like boom). Based on the knowledge of exactly how the habitat of the great bittern had to look, the ground relief was designed on higher ground, the water supply was regulated and major new reedbeds were created. Electric fences protect the areas from the predations of the fox. These measures have succeeded in increasing the number of booming male great bitterns from 11 to 120 between 1997 and 2013 (Schäffer and Flade, 2013). This is undoubtedly a species conservation measure that has almost nothing to do with nature conservation.

Similar breeding grounds were created in the Ouse Washes (also in East Anglia) for meadow waders (www.rspb.org.uk/reserves/guide/o/ousewashes/). One hundred and nine acres of arable land was purchased and the ground relief remodelled by creating sinks. The edges of this area were partially sealed by a foil, and the water level had to be regulated by pumping targeted amounts. The area was sown with a grass mixture. The nests with eggs and young birds were protected by electric fences and the targeted shooting of individual birds of prey and corvids. The development of the area actually took the expected course. Today this technically designed area has become an important breeding area for lapwings, common snipes and godwits.

Similar projects have been carried out in the Netherlands. Open land species such as the woodlark and nightjars were saved from vanishing from the Maas Dunes on the Dutch–German border by removing the upper soil and humus layers on designated areas, thus recreating barren heaths that had been destroyed by afforestation in recent decades (http://www.np-demeinweg.nl/documents/home.xml?lang=de).

At least in technologically advanced and densely populated countries, there will be no other alternative than to create artificial habitats or special areas that have been technically altered in order to save and protect threatened species. There are more examples of this than the many in Europe. Even in the United States, it has been shown that only protecting habitats is not sufficient when it comes to rescuing very specialised species from extinction. In certain circumstances, machines and fire must be used for landscaping. One example of this is Kirtland's warbler (*Dendroica kirtlandii*), a breeding bird of the United States (Probst *et al.*, 2003). The breeding habitat for this species is stands of young jack pine's (*Pinus banksiana*), which may not be more than 20 years old. This is why today's breeding range is restricted to a small district on Michigan's Lower Peninsula. In the 1950s, singing males numbered around 500, but this number had shrunk to roughly 200 by 1971. The number of young pine trees was doubled between 1987 and 1990 through controlled fire and the elimination of older trees and this tripled the number of warblers in the 10 years from 1990 to 2000. A goal like this could not have been attained by a nature conservation movement totally oriented on preservation and the avoidance of 'disruptive interventions'.

In Germany, measures to help increase the number of individuals in endangered species through the creation of artificial habitats are only picking up steam very slowly. An ideology oriented on nature conservation dominates the opinions of the conservation 'establishment', which firmly believes that the creation of artificial habitats would turn nature into a zoo. But there are species which cannot be saved by any other means.

One simple example of how to save a species by means of an artificial habitat is the wheatear. In Germany, the wheatear is a species threatened with extinction (Red List Category 1), although the bird is very habitat-tolerant. It needs large barren areas with cairns in which it builds its nests. It does not even always have to be cairns, however; the wheatear also inhabits peat-stripped moors where sods are piled up on ground that has little or no vegetation; and the external structures of these piled-up sods look somewhat like cairns. One thing that the wheatear cannot tolerate is what is threatening all of Central Europe today: and that is the loss of barren ground and pieces of rock through all-pervading vegetation. The wheatear is currently very much in decline, because peat cutting in north-western Germany has gone out of fashion. It would be a simple matter to resettle the wheatear by using machines to first remove the bushes and grass, and then erect cairns in a sufficiently large area. Something like this should really be possible everywhere, and it seems that no other measures are needed to reintroduce the wheatear (Plate 6).

A task that is much more complicated is the artificial planting of reed and sedge areas in zones with low water levels. However, such measures will be essential for the protection of species in the future. One hundred years ago, the aquatic warbler was prevalent in northern Germany. It needs wet meadows with many sedge beds. Habitats like this have almost completely disappeared in Germany, and the aquatic warbler has been reduced to a total of 6–7 pairs in Germany. There is no possibility of recreating a biotope like this by adopting different forms of agricultural farming or by leaving nature to itself. Only the establishment of artificial habitats could now save the aquatic warbler from completely disappearing in Germany.

It is much easier to preserve endangered species of insects by manipulating habitats, because insects are smaller than birds and can achieve a higher diversity of species even in smaller areas, so designing only a few acres of land may be enough to meet the requirements of certain insect species. Endangered insects can be resettled on small areas by using technical equipment with minimum effort. Here, we should only familiarise ourselves with the concept of a future species conservation through habitat management – and finally distance ourselves from conservation values such as preservation, untouched nature and the promotion of wilderness.

In this way, the rare dusky large blue (*Phengaris nausithous*) has been resettled at several locations in Germany, thanks to planned habitat measures (Stevens *et al.*, 2008). During its first stages of development, the caterpillar of this butterfly feeds on the great burnet plant (*Sanguisorba officinalis*). In its later stages of development, the caterpillar needs the European fire ants (*Myrmica rubra*) which feed it in their nests below the earth. The required population density of this ant species must exist in the area which the dusky large blue inhabits; otherwise the artificial reintroduction of this butterfly would not be possible. If the habitat is too sparsely populated by the host ant, there is a danger that the caterpillar in its correct stage of development will not be carried into the ants' nest quickly enough, and other species of ants would then attack and devour it. However, in spite of this complex life cycle, artificial habitat design has succeeded in resettling this butterfly species in the areas where it used to occur (Stevens *et al.*, 2008).

The question is whether small-scale, isolated settlements can remain genetically stable in the long run. Small isolated populations lose their allelic diversity over time through genetic drift and this endangers their long-term survival. Many species (but not all) need to exchange genes with neighbouring populations in order to survive (Habel *et al.*, 2009). Many detailed studies on the mobility of certain insect species have been carried out (and the mobility is often very limited)

(Ehrlich, 1965), but these tests only acquire data over selected limited periods. It cannot therefore be ruled out that species which live the majority of their lives in a small area are sometimes prone to large migratory movements, covering long distances to genetically connect remote populations with one another.

Some Hymenoptera species are highly mobile, even if they do live in a small area; they can fly great distances. There is a 74-acre former ammunition depot located in the 'Friesheimer Busch'. This area lies in the south-western zone of Cologne in North Rhine-Westphalia (Cölln and Jakubzik, 2010b). It consists of monotonous rows of square plots that formerly contained ammunition stores. These plots are surrounded by earthworks which were intended to provide protection from explosions. This erstwhile ammunition depot looks like a system of uniform rectangles which has been designed by architects, and it certainly cannot be described as 'nature'. The plots are surrounded by intensively farmed areas which are therefore dead as far as species are concerned. Two hundred and seventy eight different species of stinging wasps (Aculeata) were discovered around the protective walls of this former munitions depot. That is almost 40% of all the species that occur in North Rhine-Westphalia, and a third of these 278 species are on the North Rhine-Westphalia Red List (Cölln and Jakubzik, 2010a).

Another example from Luxembourg shows just how successful the use of excavators and forestry milling can be. A post-mining landscape in a Luxembourg ore basin, with south-facing rock walls and an extremely broad diversification of species was at risk of becoming overgrown by grass and shrubs, thus losing its value as an insect habitat. Excavator work eliminated the vegetation, restoring the state the landscape was in when the ore was excavated (Cungs, 1991). Only 150 m long and 6 m high, this (now bush-free) cliff of pronounced xerothermic character with a seam of species and flower-rich pioneer vegetation located directly in front of the cliff resulted in the settlement of 117 bee and wasp species (Cölln and Jakubzik, 2010a). Such habitats, created from the drawing board (so to speak), cannot, however, be left to themselves. They must be periodically reworked.

However, these experiences with Aculeata may not be generalised because Aculeata are distinguished by specific features that enable them to quickly achieve a high number of individuals in small areas. Individual Aculeata can quickly colonise new areas, because they can establish founder populations there due to the parthenogenetic origin of the male individuals, without the need for more sexual partners on site (Bull, 1983).

These examples underscore the fact that human intervention in the landscape can create habitat islands for biodiversity; and these are islands that lie in a landscape that would otherwise be dead for biodiversity. The key is to provide or manufacture the habitat structure that is needed for the nutrition and reproduction of the species concerned. Scenic beauty is not one of the measures necessary for the rescue of species threatened with extinction. Just as in the past centuries and millennia, the biodiversity of Central Europe was not created by human interventions that targeted healthy ecosystems or scenic beauty.

Essential tools of species conservation will be the future agricultural and forestry machinery (Cölln and Jakubzik, 2010a), which can encroach much more intensively upon Mother Nature than the volunteers from the local entomological associations who remove bushes manually. Many insect habitats could be saved by working with an excavator or a forest rotary hoe, habitats which have become overgrown with bushes in the last few decades and have thus lost their value as an insect habitat.

The International Union for Conservation of Nature (IUCN) identifies 36 animal species and 25 plant species that have become extinct in the wild and only exist in captivity, but for which there is a realistic chance of living in the wild again if the appropriate habitats in nature were restored by technical means (Baillie, Hilton-Taylor and Stuart, 2004). Many of the amphibians which are at present endangered or living only in captivity could be saved in this way.

However, the future of species conservation through the creation of special habitats for the purpose of preserving species that are threatened with extinction is not simply a financial problem. Interventions like this must prevail against a widespread ideology that equates species conservation with nature conservation (particularly in Germany); and the same ideology warns us that the establishment of artificial habitats for the conservation of certain species will turn us away from the wilderness and make nature a zoo; but the persons behind this ideology do not seem to consider that we have long been on this path since we started hanging out nest boxes and installing winter feeding for animals and birds.

We must take a new attitude towards that which is described as the destruction of nature. In any case, there is hardly any untouched nature in Central Europe. Many species that live in Central Europe, have (for millennia) adapted to habitats that have been manipulated by man. In Germany, a large proportion of the nature conservation areas are man-made. If man had not ruined nature in the past, then many nature reserves would not exist today. What is described as 'nature' and which we strive to protect as nature reserves often consists of former peat workings and fish-breeding ponds, and they would not exist at all if man had not destroyed the original, virgin nature. The dry grasslands of the Central German mountain landscapes, so rich in orchids and butterflies, go back to the intensive goat and sheep grazing of earlier centuries. This was undoubtedly a significant destruction of nature; today these areas are 'nature reserves'.

The large open-cast mining excavations for obtaining lignite in Central Europe, and the gravel pits are a significant demolition of nature. But these excavations have created 'species paradises'. Numerous endangered species have resettled on the nitrogen-poor layers of earth taken from the depths of the mine; species that were destroyed elsewhere by nitrogen fertilisation and forestry (see Sections 2.3, 5.1.2 and 5.2.4). When the major nature protection associations complain about the destruction of nature by lignite mining (which they do, and perhaps to excess), no reference is made to the promotion of rare species that is actually accomplished by these interventions in nature. A frank admission that nature conservation associations in general assign a higher priority to nature protection than to species protection is totally lacking.

The law requires that the open-cast excavation sites should be 'restored to a natural state' after coal or gravel extraction has finished. The general understanding of this is that the previous state is again restored. It is agricultural land that is largely restored, and that has often been legally defined before the start of any excavations. This side of 'restoration to a natural state' is difficult to attack with arguments. The destruction of biodiversity through agriculture must be accepted (see above). However, the measures for 'restoration to a natural state' which are executed outside the agricultural land have to be attacked. Wells are filled, steep slopes are flattened and escarpments are levelled; trees and shrubs are planted. The underlying motivation here (besides prescribed safety measures) is the 'beautification' of the landscape. The majority of the population sees barren areas, stony plains and steep depressions as ugly, but gently sloping areas with trees and trails as perceived as being beautiful. In a landscape like this, you can relax, go for a 'proper' walk and take the dog for a walk. But the animals that are worth preserving have disappeared thanks to these 'beautification' measures.

The terms *nature* and *species richness* are quite clearly being confused with one another. The so-called *restoration to a natural state* of the mining areas is more a species-destroyer than a species-promoting measure, especially when the restoration is linked to afforestation (Kunz, 2004) (Plate 7). The large lignite excavation areas in Central Europe have offered numerous species a replacement landscape, because the former habitats where these species lived have been destroyed by eutrophication. This applies to several species of birds, including the corn bunting, the tree pipit and the woodlark (Plates 1 and 8). This applies to many species of butterflies and many orchids. The so-called restoration of nature on the excavation areas after mining has finished destroys these species' habitats; habitats that were only short-lived thanks to the destruction of nature through brown coal mining. The importance of restoration measures (*Renaturierung* or *Rekultivierung*) is misunderstood again and again. On the restored surfaces, there in first instance is a species richness which is larger than on comparable surfaces which are not the product of restoration. This phenomenon is 'sold' to the public as a credit of the restoration measures. In truth, however, not the restoration, but the earlier destruction of nature by open-cast mining has created the valuable habitats, since nitrogen-poor soils were transported to the surface. The restored habitats will lose their value in a few years already, since nitrogen will accumulate soon (Kunz, 2004).

6.5
The Golden Plover, the Godwit and the Great Bustard – hemerophiles in Central Europe

The golden plover, a resident of the large heaths in the northern region of Central Europe.

The golden plover, beneficiaries of human intervention in nature: Habitat optimisation through bog drainage and peat cutting.

The ideal conception of the golden plover as a bird of the original and pristine upland moors is misleading; in northern Central Europe the golden plover actually inhabited unnatural secondary biotopes created by destructive human intervention.

The original breeding habitats of the black-tailed godwit in Central and Western Europe no longer exist today. It is only in recent centuries that the black-tailed godwit has colonised areas worked by humans.

The black-tailed godwit is bound to agriculturally machined surfaces in the European interior, but agriculture should not look the way it looks today.

The great bustard: treated as an agricultural pest for centuries, today this bird would have become extinct in Central Europe without human assistance.

In Central Europe, the great bustard lives only on farmed land.

The occurrence of the great bustard in historic and prehistoric times in Central Europe.

What does the future hold for the great bustard in Central Europe?

6.5.1
The Golden Plover, a Bird of the Sheep Pastures and the Partly Drained High Bogs in the Northern Region of Central Europe

The golden plover is a bird that breeds on open areas on damp heaths and tundras; its core populations are in northern Eurasia. In the north-western region of Central Europe in earlier centuries, the golden plover found man-made habitats which were similar to the wet heathlands of northern and north-eastern Europe. This is why it was able to settle here for centuries, although none of these habitats in north-western Central Europe were natural or near-natural.

Until the beginning of the twentieth century in Germany, the golden plover populated a networked breeding area, which comprised the region Hohes Venn (high fens) near the Belgian border, Western Münsterland (the county of Borken) and all of south-western Lower Saxony (almost all the moors between the rivers Ems, Hunte and Weser) (Heckenroth and Laske, 1997) to the Lüneburger Heide (heathland in the county of Soltau) (Schmidt, 2001). Today less than 10 pairs remain in Lower Saxony; the golden plover has all but disappeared as a breeding bird in Germany.

There are various answers as to why the golden plover has been brought to the edge of extinction. Some see the golden plover as an indicator of the originality and undestroyed purity of a landscape, namely the upland moors (or bogs), and they associate its disappearance with the destruction of these moors (Bibelriether, 1997). Others see the golden plover as a beneficiary of landscape destruction (van Noorden, 1997).

The core areas of occurrence of the golden plover are the wet heaths (Fjells) of the flat mountain slopes, the high plateaus above the coniferous forest timberline and the tundras in Fennoscandia and Northern Russia. It can be assumed that Central Europe was populated by the golden plover during the glacial periods that were outside of the actual glaciation zones. The habitat of the golden plover disappeared with the decline of glaciation and the resulting emergence of dense grasslands, bushes and forests. The only open spaces left in north-western Central Europe were the large upland moor expanses in the Netherlands, Belgium, northern Germany and Denmark.

However, the original postglacial upland moors were most probably unsuitable breeding grounds for the golden plover, since its movements over the ground must be unhampered while it searches for food. The predominantly wet sphagnum moss areas interspersed with dry Bulten (raised domes of peat and sphagnum mosses) cannot have made good breeding grounds for the golden plover. Only the moor landscapes altered by humans became habitats for the bird. During recent centuries, man changed the upland moors, optimising them as habitats for the golden plover. For thousands of years the upland moors of northern Central Europe were a pristine habitat. They remained untouched nature for a longer period than most forests. The large moors could only be accessed at their edges by humans, because they were too wet and had no locations for orientation, so anyone venturing into them could get lost or even sucked down to drown. The pristine upland moors were probably not populated by the golden plover for a long time; it is more likely that the heaths allowed this northern bird species to immigrate into Central Europe. Sheep grazing, the burning of vegetation, the removal of the upper layers of turf and other human interventions created large open areas with low-growth vegetation in the Netherlands, Belgium, Denmark and Northern Germany (van Noorden, 1997). These areas became suitable breeding habitats for the golden plover. It was only after this that the golden plover apparently settled in the moors.

Thousands of years ago, peat production began with the drainage of Central Europe's upland moors, and this meant that humans were able to colonise the outlying areas of the large moors. The drainage of the moors enabled drier heathland to replace the wet sphagnum moss surfaces, so the upland moors gradually became habitable for the golden plover. The golden plover locates its food visually. It runs around very quickly looking for insects, so it needs level surfaces. The bird cannot run between dense tussocks of grass and Bulten, so it was not present on the peaty moss surfaces of the growing upland moors. The ideal conception of the golden plover as a bird of the original and pristine upland moors (Bibelriether, 1997) does not seem to be correct. Decades ago, scientific publications already showed that the golden plover is a breeding bird of the short-grass heaths that were only created when the upland moors were drained (Hölscher, Müller and Petersen, 1959).

The image of the golden plover as a resident of the original, unspoilt moors, however, has been firmly ingrained into the public consciousness. Until the middle of last century, the upland moors of north-western Germany (even during drainage) were still wide, unpopulated areas where the view reached to the far horizons, and the golden plover became a symbol of unspoilt nature (Bibelriether, 1997). After the Second World War, the predictably rapid and final destruction of the last remaining moors occurred. This combination reinforced the illusion of the end of an unspoilt nature brought about by destructive human activities. The melancholy fluting calls of the golden plover in the total silence of the wide moorlands conveyed the feeling of a pristine and original landscape. However, these illusions had nothing in common with the actual situation, which was that the golden plover in northern Central Europe inhabited unnatural secondary biotopes, which had only been created by destructive human intervention.

While the still numerous broods of plovers in the moors of north-western Germany were attracting the attention of nature lovers during the 1950s, it was observed that the bird preferably nested near sites where the turf was about to be cut away. This was not seen as a sign that the suitable breeding habitats were thanks to the activities of the excavators; on the contrary, it was assumed that the excavation work was the reason behind the destruction of the golden plover. This assumption was not at completely wrong, because the excavators did ultimately destroy the golden plover. It has always been (and still is) exceedingly difficult to make it clear to the followers of our conservation ideology that the destruction of nature through human activities *can create habitats* for rare species, but on the other hand, that these *same activities can also destroy these habitats*. We have a seemingly paradoxical situation here. On the one hand, the turf excavators have created really suitable habitats for the golden plover, but on the other hand the advance of the large machines is threatening the existence of these habitats. Man threatens the habitats that he himself has created, and this applies more to Central Europe than to other parts of the world.

In previous centuries, the golden plover was a common breeding bird in the vast, treeless heaths that had been created in the Netherlands and the neighbouring regions of North Rhine-Westphalia in Germany (i.e. the county of Borken) by sheep grazing, the burning of vegetation and the removal of the upper layers of turf. Here, the golden plover found open areas with low-growth vegetation, areas which were mostly treeless all the way to the far horizons. These areas resembled the wet heaths or fells (Fjälls) of northern Scandinavia. In the Netherlands and in north-western Germany, these treeless regions attained their maximum areas around 1800, and the golden plover also attained its highest breeding population at this time. By 1900, these ideal habitats had vanished due to afforestation. Parallel to this, sheep farming declined in the non-forested areas, and in the absence of grazing the vegetation grew too high and too dense for the golden plover. The breeding population of the golden plover on the heathlands declined rapidly, dying out in western North Rhine-Westphalia in 1915 and in the Netherlands in 1937 (van Noorden, 1997). The golden plover thus disappeared as a heathland breeder in Central Europe. It only survived on the upland moors because heather-covered dry areas were created as a result of peat extraction. These areas replaced the lost sheep pastures and indeed were partially used as such. After the Second World War, there were more than 100 pairs of golden plovers on the north-western German upland moors in Emsland and in the Diepholzer Moorniederung (Diepholz Lower Moor area).

These areas were mostly stripped of their peat during the 1950s and 1960s of the last century. Only a few remnants of the great (but from now on dry) moors were rescued by being given the status of nature reserves, so the removal of peat was discontinued and the drainage ditches were filled. This led to the rewetting of limited areas, and the birch trees, which had meanwhile grown to a fair height, could now be drowned. However, these renaturation measures did not restore what had been there before. The original sphagnum upland moors were not restored, nor was the heathland with its low-growing vegetation, which had been the former successional biotope of upland moors after peat excavation started. Purple moor grass (*Molinia caerulea*) became predominant on the ecologically rebuilt open areas instead of peat moss and heather. These areas are no longer the breeding habitat of the golden plover, because the vegetation is simply too high and too dense.

The last golden plovers survive today on a few of the much-reduced former moor areas in western Lower Saxony. These are the dry areas near the peat excavation sites. Since the large peat excavation machines were an immediate threat for the nests and the young birds, the only way to save the breeding sites was to tell the peat excavation companies exactly where the nests were. The machines could then avoid them (Rolfes, 2006). The Niedersächsische Umweltministerium (Lower Saxony Ministry of Environment) donates €140 000 per year specifically for the protection of the golden plover. In 2005, there were only eight breeding pairs throughout Germany, but recent years have seen the bird populations recover somewhat (http://www.pressrelations.de/new/standard/result_main.cfm?r=277562&aktion=jour_pm).

The story of the golden plover in northern Central Europe is a perfect example of the dilemma in which Nature Conservation finds itself. The rescue of the golden plover has nothing to do with the preservation of original nature, nor could Nature Conservation achieve anything at all by simply protecting the last remaining habitats from the invasion of man. In the 1980s, the remaining moor areas in the Diepholzer Moorniederung were patrolled by walkers during the breeding season. These persons were employed to prevent visitors from entering the moors and thereby save the golden plovers – a totally ineffective move.

During the last century, the golden plover neither had to be protected from humans nor from the destruction of nature; and this in effect meant that the bird could not possibly lie within the focal range of the standard conservation practices. The golden plover owed its existence in Central Europe to the past destruction of nature and the emergence of transitory, successional landscapes. The upkeep of successional habitats can only be carried out by machines. Technical habitat design is, however, expensive and thus cannot be realised in most cases. A much more serious obstacle here is that the destruction of nature as a species conservation measure is not well received by the great majority of the population; and this is why the Nature Conservation organisations neither support nor promote it.

6.5.2
The Occurrence of the Black-Tailed Godwit in the Western and Central Europe of Today Has Nothing to Do with Pristine and Untouched Nature

Like the curlew, the common snipe, the ruff, the dunlin and most other meadow breeders in Central Europe, the black-tailed godwit is also threatened with extinction (Category 1 in the German Red List) (Südbeck *et al.*, 2007). Significant losses are being reported from every country in Central and Western Europe. In the Netherlands, where around 90% of Western European godwits breed, the population has declined by more than 50% in the last decade alone.

The reason for this is that the black-tailed godwits are a hemerophile species in Central and Western Europe, so their decline has not been caused by man destroying unspoilt nature; the black-tailed godwit is a clear example of how a bird binds itself to a human-designed habitat. The occurrence of the black-tailed godwit in today's Western and Central Europe has nothing to do with near-natural habitats. In the Central European interior, this species needs damp, farmed grassland. Nature cannot provide a habitat like this, because human landscaping created this habitat (Barkow, 2010).

The original breeding habitats of the black-tailed godwit in Central and Western Europe were extended flat meadows with vegetation that was not too high and not too dense, like fens and broad estuaries exposed to the tides. Such habitats no longer exist in today's Central Europe. It is only in recent centuries that the black-tailed godwit has colonised areas worked by humans. In Northern Germany, and especially in the Netherlands, the dykes resulted in the creation of breeding areas, grassy flat areas that were no longer flooded by the sea.

The original habitats of the black-tailed godwit have long since vanished in Central and Western Europe. This meant that to survive in the long term during the past 1–200 years, the black-tailed godwit could only inhabit meadow structures created by permanent human intervention. Agricultural work on the land (as carried out into the 1960s) created a suitable habitat structure for the black-tailed godwit. Godwits need moist ground into which they can 'poke' their beaks to catch food; but the ground may not be too densely covered with vegetation. The bird cannot even get a foot on the ground if the grass growth is too dense. It is also too cold for its chicks in wet grass, and they can no longer find enough insects and other arthropods. Only sparse grass coverage is easily penetrated by sunlight, creating a warmer soil in which more insects can live, and are also easier for the bird to see. The newly hatched young birds in particular feed exclusively on insects and other arthropods during their first few weeks of life (Barkow, 2010).

The black-tailed godwit requires a long breeding time from mid-April for 3 weeks, and then another 5 weeks to mid-June until the chicks are able to fly. Grass may not be mowed during this period. This means that the black-tailed godwit needs a habitat that agriculture simply cannot provide today because delayed mowing and the foregoing of fertilisation at the same time are not actions that are in line with an agriculture oriented on economic yield; and this is quite apart from the fact that grass today grows faster and denser than ever before due to the general eutrophication of our entire land. The preservation of the black-tailed godwit in this day and age is difficult to achieve. The bird is bound to agriculturally machined surfaces in the European interior, but agriculture should not look the way it looks today. It was only the way that agricultural work was carried out until the 1960s that created a suitable habitat structure for the black-tailed godwit.

The black-tailed Godwit's fate is shared by almost all grassland birds today. These include not only the 'Big Five', that is, the curlew, the black-tailed godwit, the redshank, the common snipe and the lapwing, but also small songbirds such as the whinchat. The problem is that the grassland birds cannot be saved by nature protection methods, because these species are dependent on the management of land by the farmer. Land management in its historical form promoted the meadow breeders, but modern agriculture has the opposite effect – it destroys almost all grassland birds. But there is no alternative, because if we leave nature to its own devices, we shall drive these birds away for good.

6.5.3
The Great Bustard in Central Europe: From Frequent Agricultural Pest to a Bird Threatened with Extinction

Past and Present Populations of the great bustard in Central Europe. Among the bird species that breed in Germany there are only two species that are classified as Red List species in their global distributions: the great bustard and the aquatic warbler. All other breeding birds in Germany are not classified as being globally vulnerable. This is why Germany has a special responsibility for the preservation of the great bustard, especially since a significant proportion of the world's great bustard population was breeding in Germany just 100 years ago.

The great bustard is classified as 'vulnerable' (VU) on the IUCN Red List. It breeds from Morocco, Spain, Central/south-eastern Europe, Russia and the Near East to Central Asia and China (Baillie, Hilton-Taylor and Stuart, 2004). Today's great bustard world population consists of approximately 40 000 individuals. Of these, half live in Spain (Lane, Alonso and Martin, 2001). Until 1930, Germany had nearly 7000 individuals, almost 18% of great bustards living in the world today.

Since the great bustard today can only live in Germany in habitats specifically manufactured and maintained for them, objections have been raised that the great bustard does not belong to Germany and its preservation would be comparable to the zoo concept. Some nature conservationists say that because the bustard can only be kept alive artificially here, it is not a true wild bird and its preservation would have nothing to do with nature protection. This objection is correct. The great bustard is a bird of the steppes; it cannot survive in natural and near-natural landscapes in Central Europe. However, nature conservation and species protection are two totally different objectives. Less than 100 years ago, Germany was still home to one-fifth of the current world's great bustard population, and the bird was probably a 'natural' breeding bird in Germany on the mammoth steppes during the ice ages.

In the postglacial warm period, the great bustard was a hemerophile species in Central Europe. It lived only in the vicinity of humans, in areas that had been altered by human activities. The fact that the bird is now threatened with extinction in Central Europe has nothing to do with the fact that man destroys nature. It is solely a result of the changing farming methods, which are no longer the same as they were in past millennia. Only the simulation of past farming practices can preserve the great bustard in Central Europe and *not* the protection of nature.

In previous centuries, the great bustard was prevalent in almost all European countries, such as France, England and Southern Sweden, that is, countries that are not immediately associated with steppe landscapes and which are far from the south-east European and Central Asian steppes. The last Central European populations of the great bustard can be found today in Germany, Austria, Hungary,

Czech Republic and Slovakia (http://www.grosstrappe.de/). They were more or less common and very widespread in these five countries a little more than half a century ago, but their numbers declined dramatically within just two to three decades, reaching an all-time low by the 1970s and 1980s. They would probably not exist in these countries today if comprehensive protection measures had not been taken from around 1990. The great bustard is a strict hemerophile which can live exclusively on agricultural land in Central and Western Europe. The bird avoids the near-natural steppes in Spain's interior, which are no longer used for agricultural purposes (Lane, Alonso and Martin, 2001).

The great bustard has been classified as 'threatened with extinction' (Category 1) in the Red List of breeding birds in Germany (http://www.nabu.de/tiereundpflanzen/voegel/wissenswertes/roteliste/10221.html). Until a 100 years ago, Germany had nearly 7000 great bustards in Saxony-Anhalt, Brandenburg, Mecklenburg-Western Pomerania and Saxony. Some birds still bred on the eastern border of Lower Saxony and in the Thuringian basin, and in 1935 a record of a brood was even documented in Baden-Württemberg. The animals declined in number from year to year, but the really dramatic disappearance of these birds only started after the war. In Mecklenburg-Western Pomerania around 1980, they vanished from the vicinity of Anklam and 1994 saw the last birds disappear in Saxony to the north of Leipzig. A few birds still held out in Saxony-Anhalt and Brandenburg.

The exact dates of the collapse of the great bustard populations are difficult to determine. great bustards are extremely territorial and sedentary. We should not see their 'extinction' as being caused by individuals dying or by the birds migrating. 'Extinction' in this case means that reproduction has come to an end. The young birds starved first, because there were not enough insects on the fields. However, several individual adult birds (which have a mainly vegetarian diet) remained in their ranges for many years, and this gave the false impression that the great bustard had not yet died out.

At the end of the last century, the final disappearance of the great bustard in Germany and the neighbouring countries could only be prevented by the initiation of labour-intensive protection programmes. Such programmes involved careful research into the bird's needs, and individuals were even reared by hand. Even today, many bustards are artificially reared in aviaries and subsequently set free at locations where the few survivors still exist in the wild.

Great bustards have a long laying period, from mid-April to July. This helps to counteract the multiple raids on their nests carried out by predators. The birds are hormonally controlled in such a way that they can always lay more eggs if the chicks or eggs are lost. The conservation programmes benefit from this characteristic which great bustards fortunately possess. The eggs can be collected at the beginning of laying period and are then hatched artificially in breeding establishments. The hens then lay another clutch of eggs and hatch their young. The chicks from artificially hatched eggs are reared in aviaries and then reintroduced in the wild to increase the numbers of wild birds.

In the 1990s of the last century in Germany, the great bustard survived in only three locations: in the 'Havelländische Luch', the 'Belziger Landschaftswiesen' in Brandenburg County and in the 'Fiener Bruch' in Saxony-Anhalt (http://www.grosstrappe.de/mainerleben_gebietskarten.html). Appropriate agricultural measures are now being taken to landscape larger areas as habitats suitable for great bustards and efforts are being made to fence in their breeding grounds against foxes. Following the population low point of the early 1990s, a total of 123 birds now live in these areas (current count carried out in February 2012). Without the landscaping and the 'last-minute' reintroduction programmes, the great bustard would now be extinct in Germany. The leading initiative for this rescue programme was taken by the Litzbarskis, a married couple who were awarded the Bundesverdienstkreuz (The Order of Merit of the Federal Republic of Germany) in October 2011 for their unprecedented success in saving the great bustard.

Similar projects are underway in other European countries; a reintroduction project has been launched on a former military site in England that covers 100 and 11 000-acres.

The Habitat Requirements of the great bustard. Today, bustards can only live in Central Europe in a habitat that is continuously created and recreated by humans. Habitats left to themselves become unsuitable for bustards in just a few years. Even in Spain, where there are natural open areas, the bustards stay on the areas that are ploughed and machined by humans (Hellmich, 1996; Lane, Alonso and Martin, 2001).

In the past, alfalfa and rape fields contributed much to the existence of the bustard's frequent breeding occurrences in Germany. The multi-year alfalfa fields in particular offer an all-year-round habitat with a rich supply of food (Dornbusch, 1996). As a result, the Landesumweltamt Brandenburg (Brandenburg Environmental Agency) has concluded contracts with farmers to grow certain crops on designated areas since 1990. These crops provide enough food for the great bustard and are also farmed extensively, enabling enough insects to live in the fields. Alfalfa and rape are mainly prescribed, but lupins, peas and clover are also cultivated. The cultivation of these crops is interrupted every few years, so that certain areas then lie fallow. This is a return to the historical three-field system of yesteryear, a system which had to give way to modern yield management and fertilisation. The Landesumweltamt pays annual compensation to farmers for these usage restrictions and cultivation requirements, because the economic yield from these fields is low.

Fallow land alone does not solve the problem. Land which regularly lies fallow quickly loses its attraction for the bustards. Where arable land and areas that have been uncultivated for years lie adjacent to one another, the great bustard quickly gravitates to the cultivated fields and ignores the fallow areas (Litzbarski *et al.*, 1996). Even in the Western Spanish ranges of the great bustard, it has been documented that old fallow fields quickly lose their appeal for the bird in the areas where agricultural use has been abandoned; but its populations increased in the areas where traditional agricultural methods were practised, which meant land lying fallow for only 1 or 2 years (Hellmich, 1996).

During the day, great bustards mainly prefer areas with low vegetation growth, especially the meadows that have been recently mowed or grazed. Areas where the vegetation grows in height become increasingly unattractive for the birds. The adjoining grain fields are mostly visited after the harvest (Eisenberg, 1996). Rape is the main winter food of the great bustards in Eastern Germany.

Alfalfa is also very important due to its abundance of insects (especially grasshoppers) and other arthropods. The hen feeds the chicks almost exclusively with arthropods for the first 8 days of their lives (Litzbarski and Litzbarski, 1996). An insufficient supply of arthropods was the main cause of the high great bustard chick mortality in eastern Germany in the 1970–1980s. This was also the main reason why bustard populations collapsed in Central Europe a few decades ago. The most important prerequisite for rescuing the last great bustard populations in eastern Germany is the re-establishment of sufficient numbers of arthropods. Only in this way can enough food be guaranteed for the great bustard chicks in late May and early June.

The reason for the decline of the insects in the 1970s and 1980s was the widespread eutrophication of the landscape through nitrogen fertilisation and the nitrogen content in the rainfall. In the 'Havelländisches Luch' bustard range, an annual nitrogen deposition of 20–22 kg N per acre was measured (Litzbarski *et al.*, 1996). So we have free nitrogen fertilisation raining down from the atmosphere (even today), and in an amount that is more than 20 times greater than in previous centuries. This fertilisation resulted in strong vegetation growth that did not exist years ago. In contrast to earlier times, most areas were no longer grazed; and for those who wished to design a bustard-friendly habitat, this created a very difficult hurdle to surmount. On the one hand, the first mowing of grassland sites had to be delayed long enough to prevent the bustard chicks from being killed by the mowing machines; but on the other hand, late mowing led to the vegetation becoming denser, which in turn kept the earth cool and damp, and that meant that insects could not develop sufficiently (Litzbarski *et al.*, 1996).

Yet again it was clear for all to see that the countrywide fertilisation of agricultural areas (first carried out only a few decades ago) is the main enemy of species richness. Other grassland birds, like the lapwing and the black-tailed godwit are also destroyed by this fertilisation. This is not about the straightforward killing of insects and birds by poisons – it concerns the suffocating effect caused by fast-growing grass. The rapidly growing dense vegetation forces early meadow mowing to take place at a time when eggs in nests and young birds not yet fully fledged are hidden in the meadow grass. And if this moves the nature protection organisations to support later mowing, then the young birds will starve because the dense grass reduces the number of insects or hides them from the chicks.

The preservation of traditional farming methods is only possible with substantial government funding. Such funding now exists in all great bustard breeding areas in the European Union. Supported by EU funds, societies for the protection of the great bustard have been established in Germany and in other countries, for example, the Pannonische Gesellschaft Großtrappenschutz (Pannonian Society for great bustard protection), established in 1995, in which members from Austria, Hungary, Slovakia and the Czech Republic work together to protect the bird (Kollar and Wurm, 1996).

Is the great bustard a Central European Bird? The occurrence of the great bustard in Central Europe sheds light on the particular situation threatening the fauna of Central Europe, a fauna beset by dangers which often have other causes than the threat to animals in many other parts of the world; and the preservation of these animals therefore requires other 'conservation' measures. More than in many other parts of the world, the fauna of Central Europe was characterised by the cold steppes of the ice ages and anthropogenic influences during the subsequent reforestation in the last postglacial warm period. In Central Europe, the cold steppes of the ice ages harboured fauna of the steppes, while the significantly shorter warm periods hosted forest fauna. Quaternary Central Europe was therefore steppe land for longer than it was forest land.

Nomadic people in Central Europe had to rely on this steppe fauna as their basic food source. Since the forests took away his hunting grounds, man apparently fought back against the increasing forest cover at the end of the ice age. Afforestation during the last interglacial period (in which we now live) has been hampered by man and this has benefited several steppe animals. The deforestation and fires caused by man probably enabled the steppe fauna to partially survive on open and semi-open areas in the postglacial period, or they were perhaps able to settle on deforested sites again after having retreated from them earlier.

Since it is a large bird, great bustard fossils exist and historical sources dating back to the thirteenth century have also mentioned it in Central Europe (Klafs, 1983), so we can partially reconstruct the state of the Central European landscape from the well-known habitat requirements of the great bustard, right up until the Paleolithic. Fossil discoveries show that the great bustard occurred in prehistoric postglacial times in Central Europe; it did not simply immigrate from the East during the widespread deforestation of the Middle Ages. The great bustard is not a new immigrant into the Central European fauna. In Central Europe, it was a bird of the cold Pleistocene steppes. In the postglacial era, it was a bird of crop areas in the Neolithic and Bronze Ages, one of the hemerophiles accompanying man through time.

The oldest evidence of great bustard populations in Central Europe dates back to the Paleolithic; this evidence is more than 15 000 years old, from the final phase of the last ice age. The bones of the glacial bustards were found together with the bones of Wild Horses, woolly rhinoceros and black grouse, confirming the presence of the bird in Central Europe in the then-prevailing ice age steppes (the 'mammoth steppes') (Klafs, 1983). In order to ascertain the wealth of animals on the steppes of Central Europe, we must bear in mind that the glacial steppes were different from the present-day northern tundra landscapes due to the prevailing strong sunlight and long vegetation periods.

The hazel woods that emerged with climate warming and the (later) mixed forests of oak and other trees made the Central European occurrence areas of the great bustard increasingly uninhabitable for the bird. However, parallel to the afforestation, Neolithic man was now adopting a sedentary way of life, mainly on the loess areas (Aeolian sediment) and on the ground and terminal moraine areas of Central Europe, and he kept afforestation under control by his slash and burn methods, by driving his cattle into the forests to browse and by his farming, thereby creating significant open spaces. Bones from northern Germany, Hungary, Ukraine and Yugoslavia prove that the great bustard accompanied the agricultural people from the Neolithic Age to the Bronze Age, in the role of a beneficiary. This means that there must have been considerable open areas in Central Europe 8000 to 3000 years ago, mainly in sandy areas that were easier for man to colonise. It is remarkable that in those far-off times, the great bustard lived in the same areas where it also occurred from the Middle Ages right up to the present.

The Brandenburg areas in Germany which the bird still inhabits today were densely populated by human beings since the Neolithic period and especially so at the end of the Bronze Age (approximately 3000 years ago). In the subsequent Roman and Celtic times, there is no fossil or historical evidence for the occurrence of the great bustard in Central Europe. Perhaps the birds retreated during that period, only returning with the great forest clearances during the Teutonic colonisation of the Eastern Elbe areas (Eschholz, 1996).

The great bustards immigrated from the East. It is unlikely that they came from Southwestern Europe (Spain). This has been shown by comparative DNA analyses of blood samples of bustards from Spain, Hungary, Slovakia, Germany and Russia (http://www.grosstrappe.de/). Before the start of the Middle Ages in many parts of Central Europe, the bustards found suitable habitats in the previously forested areas which had been cleared by man's fire and slash and burn methods.

However, the man-made open plains were only one of the conditions which the great bustard needed to stay close to human beings. Another important condition was the three-field system of agriculture used until the mid-nineteenth century. It involved an annual cycle between summer cereals, winter cereals and fallow land. One thousand and two hundred years ago, this economic system helped the great bustard to spread throughout Southern Germany, Denmark and Southern Sweden, and the birds probably became more common from that time on. The eighteenth century was the time of the most frequent occurrence for the Great Bustard, due to an excess supply of arable, fallow and pasture land in Central Europe. The Great Bustard is a very climate-tolerant bird. It thrived in the cool summers of England and Scotland, reaching optimal numbers in Germany in the eighteenth century during the very cold winters and cool summers of the 'Little Ice Age'. This fact is particularly noteworthy because the Great Bustard cannot rear its chicks without an abundance of insects – so it follows that in the Central Europe of the 'Little Ice Age', there must have been more insects than there are today.

Many areas of bustard occurrence in Western Pomerania and Brandenburg in Germany, which were populated by the Great Bustard (in the last 200 years before the collapse of the populations) and probably first developed into suitable breeding habitats for the bustards from 1800 on. Before 1800 they were fen areas, covered by trees and shrubs, which became habitable for the bustards only after the drainage of the swamp and fen landscapes carried out under Frederick the Great. We should not think of the pre-eighteenth century swamps of Northern and Northeastern Germany in the time before Frederick the Great as being meadowland or sedgeland, which are ideal breeding grounds for snipe and other waders. They were alder fens, and only the drainage projects during the reign of Frederick the Great transformed the fen woodland landscape into one of meadows and pastures. It was only this drastic, anthropogenic 'conquest of nature' (Blackbourn, 2007) (and not 'original' nature) that transformed the fens of Brandenburg and Pomerania into paradises for birds where the now-extinct Great Snipe, the almost extinct Aquatic Warbler and the Spotted Crake (a rarity today) were common breeding birds (Hesse, 1914).

The high propagation rate of Great Bustards in the fields of Prussia was based on the multi-field system with fallow land (Schulze-Hagen, 2005). Hunting the great bustard in the eighteenth century was a sport reserved for the nobles. Farmers were not allowed to shoot them. The birds were hated as agricultural pests because they fed on the crops of the fields (e.g. rape and pulses). Petitions were delivered to Frederick the Great, requesting that the King should reduce the 'high number' of great bustards, to reduce the damage to crops in the fields (Klafs, 1983). Schoolchildren were sent into the fields to collect great bustard eggs to help keep the bird's population down (Schulze-Hagen, 2005) (http://www.grosstrappe.de/).

In this day and age, it is hard to imagine that such a shy and sensitive bird (threatened with extinction in Central Europe today) was common just a few decades ago in Central Europe and a hemerophile which sought the proximity of human beings. There is hardly any other bird that exemplifies so very clearly what has happened during the last few decades. It was not until the 1960s of the last century that almost all habitats were so drastically altered by fertilisation, that the entire Central European landscape was transformed into a mutation that had never been seen for centuries, even millennia. And this transformation has almost nothing to do with the destruction of nature; on the contrary, it could be described as being a return to nature. The core objectives of the recent conservation movement to leave nature to its own resources, untouched and undisturbed, do not even touch the reality of Central European species richness of earlier centuries. The main causes behind the decline of the Great Bustard are neither traffic, human settlement, nor industrial and environmental pollution. The bird's decline is caused mainly by the fertiliser used in agriculture, a form of land management which, after 1950, was so very different to that of all the preceding centuries and millennia (http://www.grosstrappe.de/).

In historic and prehistoric times, the Great Bustard was part of the German countryside. It characterised the image of the landscape and the still-recognisable field names. It can also be found in German literature (not least books relating to hunting) and in paintings. The Great Bustard is an impressive bird, the heaviest flying bird in Eurasia. In the spring, the courting males unfold a gleaming white, splendid plumage that makes them visible for miles on open areas. Great Bustards were much sought-after game birds. Various place names bear testimony to this, such as 'Trappenschieß' in Alzey in Rheinhessen, where a bustard range was located between Kreuznach, Bingen, Mainz and Worms around 1500 (Palm, 1957).

It cannot be that the Great Bustard is classified as 'non-German' bird, simply because it has no 'natural' future in Germany today and can survive only thanks to the creation of artificial habitats. Modern agricultural methods leave no room for the Great Bustard in Germany. The bustards of Central Europe and England (and perhaps even in Spain) can probably never survive on their own again, they will always depend on human assistance. We must ask ourselves, however, is that really an argument against the Great Bustard being part of the Central European landscape? In today's technological and wealth-dominated Central Europe, many other animal and plant species are dependent on artificial habitats. To create habitats using the technology available today would be a price to pay for the achievement of human prosperity.

We cannot save many of the endangered species in Central Europe today with a nature conservation approach that is committed to an untouched nature, simply because that objective ignores a whole range of species that have been part of Central Europe for centuries and even millennia. Nature conservation organisations must finally make up their collective minds and start to entertain a vitally important concept: the technical design of the landscape. If Central Europe were a 'nature untouched by human hand', the Great Bustard would not belong here. It would be an 'alien species'. We cannot make a yardstick out of nature in order to arrive at a decision whether or not Great Bustards belong here. It was only the dramatic destruction of nature in its original form – something which already happened in Central Europe during the Neolithic age – that created habitats for the Great Bustard and for many other plants and animals.

References

Anonymous (2012) *Bürger begehren gegen Naturparks und Reservate auf*, Frankfurter Allgemeine Zeitung 06.29.

Augustin, A. (2003) Die Tagfalter des Kreises Coesfeld und der angrenzenden Davertbereiche (Lep., Rhopalocera et Hesperiidae). – Bestandsaufnahme von 1998 bis 2003. *Melanargia*, **15**, 85–158.

Baillie, J.E.M., Hilton-Taylor, C., and Stuart, S.N. (2004) *2004 IUCN Red List of Threatened Species – A Global Species Assessment*, International Union for Conservation of Nature and Natural Resources (IUCN), Gland and Cambridge.

Barkow, A. (2010) Probleme der Wiesenvögel – Ist die Uferschnepfe noch zu retten? Der Falke Taschenkalender, pp. 197–204.

Bäumler, A. (2010) Beiträge zur Makrolepidopterenfauna der Stadt Krefeld – Teil 1: Tag- und Dickkopffalter. *Melanargia*, **22**, 106–112.

Bibelriether, H. (1997) *Naturland Deutschland*, Franckh-Kosmos Verlag, Stuttgart.

Blab, J., Nowak, E., Trautmann, W., and Sukopp, H. (1984) *Rote Liste der gefährdeten Tiere und Pflanzen in der Bundesrepublik Deutschland*, 4th edn, Kilda-Verlag, Greven.

Blackbourn, D. (2007) *Die Eroberung Der Natur*, Deutsche Verlags-Anstalt (DVA), München.

Bourn, N.A.D. and Thomas, J.A. (2002) The challenge of conserving grassland insects at the margins of their range in Europe. *Biological Conservation*, **104**, 285–292.

Bull, J.J. (1983) *Evolution of Sex Determining Mechanims*, Cummings Publishing Company, Menlo Park, CA.

Cölln, K. and Jakubzik, A. (2010a) Nature-Building – Technische Unterstützung heimischer Biodiversität – dargestellt am Beispiel der Stechimmen. *Dendrocopos*, **37**, 57–76.

Cölln, K. and Jakubzik, A. (2010b) Technisch überformte Landschaft als Lebensraum für Stechimmen (*Hymenoptera Aculeata*). *Insecta*, **12**, 37–49.

Connell, J.H. (1978) Diversity in tropical rain forests and coral reefs. *Science*, **199**, 1302–1310.

Coudrain, V. (2011) Wendehals – Gefährdung durch zu wenig offenen Boden. *Der Falke*, **58**, 2.

Cungs, J. (1991) Beitrag zur Faunistik und Ökologie der Schmetterlinge im ehemaligen Erzabbaugebiet "Haardt" bei Düdelingen (Insecta, Lepidoptera). *Travaux Scientifiquesdu Musée National d'histoire Naturelle de Luxembourg*, **17**, 1–364.

Dengler, J. (2012) Europäische Trockenrasen schlagen tropische Regenwälder. *Biologie in unserer Zeit*, **42**, 148–149.

Dornbusch, M. (1996) Situation und Schutz der Großtrappe (*Otis t. tarda* L., 1758) in Sachsen-Anhalt. *Naturschutz und Landschaftspflege in Brandenburg*, **5**, 28–29.

Ebert, G. and Rennwald, E. (1991a) *Die Schmetterlinge Baden-Württembergs: Tagfalter I*, Eugen Ulmer, Stuttgart.

Ebert, G. and Rennwald, E. (1991b) *Die Schmetterlinge Baden-Württembergs: Tagfalter II*, Eugen Ulmer, Stuttgart.

Ehrlich, P.R. (1965) The population biology of the butterfly, *Euphydryas editha*. II. The structure of the Jasper ridge colony. *Evolution*, **19**, 327–336.

Eisenberg, A. (1996) Zur Raum- und Habitatnutzung handaufgezogener Großtrappen (*Otis t. tarda* L., 1758). *Naturschutz und Landschaftspflege in Brandenburg*, **5**, 70–95.

Eschholz, N. (1996) Großtrappen (*Otis t. tarda* L., 1758) in den Belziger Landschaftswiesen. *Naturschutz und Landschaftspflege in Brandenburg*, **5**, 37–40.

Filz, K.J., Engler, J.O., Stoffels, J., Weitzel, M., and Schmitt, T. (2013) Missing the target? A critical view on butterfly conservation efforts on calcareous grasslands in south-western Germany. *Biodiversity and Conservation*, **22**, 2223–2241.

Flade, M. and Sudfeldt, C. (2008) Vögel und Schutz der biologischen Vielfalt in Deutschland – Bilanz zur 9. Vertragsstaatenkonferenz der Konvention über die Biologische Vielfalt. *Der Falke*, **55**, 170–178.

Godfray, H.C.J. and Lawton, J.H. (2001) Scale and species numbers. *Trends in Ecology and Evolution*, **16**, 400–404.

Goulson, D. (2014) Ecology: pesticides linked to bird declines. *Nature*, **511**, 295–296.

Gruttke, H. (2004) *Naturschutz Und Biologische Vielfalt*, Bundesamt für Naturschutz, Bonn, pp. 7–23 (in German).

Habel, J.C., Finger, A., Meyer, M., Louy, D., Zachos, F., Assmann, T., and Schmitt, T. (2009) Unprecedented long-term genetic monomorphism in an endangered relict butterfly species. *Conservation Genetics*, **10**, 1659–1665.

Hallmann, C.A., Foppen, R.P.B., van Turnhout, C.A.M., de Kroon, H., and Jongejans, E. (2014) Declines in insectivorous birds are associated with high neonicotinoid concentrations. *Nature*, **511**, 341–343.

Häpke, U. (1990) Die Unwirtlichkeit des Naturschutzes – Böse Thesen. *Kommune*, **2**, 48–53.

Heckenroth, H. and Laske, V. (1997) Atlas der Brutvögel Niedersachsens. *Naturschutz und Landschaftspflege in Niedersachsen*, **37**, 1–329.

Hellmich, J. (1996) Gibt es eine Abhängigkeit der Großtrappenbestände in Cáceres (Extremadura, Spanien) von der traditionell betriebenen Landwirtschaft? *Naturschutz und Landschaftspflege in Brandenburg*, **5**, 54–58.

Hesse, E. (1914) Die Vögel der Havelländischen Luchgebiete. *Journal für Ornithologie*, **62**, 334–386.

Hölscher, R., Müller, G.B.K., and Petersen, B. (1959) Die Vogelwelt des Dümmer-Gebietes. *Biologische Abhandlungen*, **18–21**, 1–124.

Hübner, G., Rauh, M., and Will, D. (2013) Wendehals und Landschaftspflegepraxis – Erkenntnisse einer Lebensraumanalyse im westlichen Oberfranken (Wrynecks and landscape management – lessons from habitat analysis in western Upper Franconia). *Natur und Landschaft*, **79**, 118–123.

Janssen, G., Hormann, M., and Rohde, C. (2004) *Der Schwarzstorch*, Westarp Wissenschaften.

Klafs, G. (1983) Die historische Entwicklung des Bestandes der Großtrappe (Otis tarda L.) auf dem Territorium der DDR und in einigen Nachbargebieten. Proceedings of the 4th Symposium Sozialistischer Länder über die Grosstrappe in der DDR, pp. 11–16.

Klaus, S. (2007) A 33-year study of hazel grouse Bonasa bonasia in the Bohemian Forest, Sumava, Czech Republic: effects of weather on density in autumn. *Wildlife Biology*, **13** (Suppl. 1), 105–108.

Klaus, S. (2012a) Lückenbewohner im Natur- und Wirtschaftswald: Auerhuhn und Haselhuhn. *Der Falke*, **59**, 26–29.

Klaus, S. (2012b) Nationalpark Sumava: Herzloser Umgang mit Europas "wildem Herzen". *Der Falke*, **59**, 172–177.

von Koenigswald, W. (2002) *Lebendige Eiszeit – Klima und Tierwelt Im Wandel*, Wissenschaftliche Buchgesellschaft, Darmstadt.

von Koenigswald, W. (2004) Das Quartär: Klima und Tierwelt im Eiszeitalter Mitteleuropas. *Biologie in unserer Zeit*, **34**, 151–158.

Kollar, H.P. and Wurm, H. (1996) Zur Bestandssituation der Großtrappe (*Otis t. tarda* L., 1758) in Österreich. *Naturschutz und Landschaftspflege in Brandenburg*, **1/2**, 7–9.

König, B. (2014) 10 Jahre Nationalpark Eifel – Ein Meilenstein trotz durchwachsener Bilanz. *Naturschutz heute – NRW*, **2**, 3.

Krebs, C.J. (1972) *Ecology: The Experimental Analysis of Distribution and Abundance*, Harper & Row, New York.

Kunz, W. (2004) Der Braunkohletagebau als Ort der Wiederansiedlung seltener Tagfalter und anderer Organismen – Was wird durch Rekultivierung zerstört? (The brown coal open-cast mining as a place for resettlement of rare butterflies and other organisms: what will be destroyed by recultivation?). *Entomologie heute*, **16**, 245–255 (in German).

Kunz, W. (2013) Artenförderung durch technische Gestaltung der Habitate – Neue Wege für den Artenschutz (Species promotion by technical shaping of habitats – new ways in species protection). *Entomologie heute*, **25**, 161–192 (in German).

Küster, H. (2008) *Geschichte des Waldes. Von der Urzeit bis zur Gegenwart*, 2nd edn, C.H. Beck, München.

Lane, S.J., Alonso, J.C., and Martin, C.A. (2001) Habitat preferences of great bustard *Otis tarda* flocks in the arable steppes of central Spain: are potentially suitable areas unoccupied? *Journal of Applied Ecology*, **38**, 193–203.

Lanz, U. (2009) Vage Erfolgsaussichten im Ortolanschutz. *Der Falke*, **56**, 426–431.

Laussmann, T., Radtke, A., and Wiemert, T. (2005) Schmetterlinge beobachten im Raum Wuppertal. *Jahresberichte des Naturwissenschaftlichen Vereins Wuppertal*, **57/58**, 1–270.

Laussmann, T., Radtke, A., Wiemert, T., and Dahl, A. (2010) 150 Jahre Schmetterlingsbeobachtungen im Raum Wuppertal – langfristige Veränderungen in der Lepidopterenfauna der Region. *Jahresberichte des Naturwissenschaftlichen Vereins Wuppertal*, **61**, 31–100.

Lenz, N. and Schulten, D. (2005) Tagfalter (Lep., *Hesperioidea* et *Papilionoidea*) im Gebiet der Landeshauptstadt Düsseldorf um 1900 und um 2000 – ein Beispiel für alarmierende Artenverarmung im 20. Jahrhundert. *Melanargia*, **17**, 19–29.

Lethmate, J. (2005) Ein globales Eutrophierungsexperiment: Stickstoff-Regen. *Biologie in unserer Zeit*, **35**, 108–117.

Litzbarski, H., Block, B., Block, P., Holländer, K., Jaschke, W., Litzbarski, B. & Petrick, S. 1996. Untersuchungen zur Habitatstruktur und zum Nahrungsangebot an Brutplätzen der Großtrappen (*Otis t. tarda*, L.1758) in Spanien, Ungarn und Deutschland.

Naturschutz und Landschaftspflege in Brandenburg, **5**, 41–50.

Litzbarski, B. & Litzbarski, H. 1996. Einfluß von Habitatstruktur und Entomofauna auf die Kükenaufzucht bei der Großtrappe (*Otis t. tarda* L., 1758). *Naturschutz und Landschaftspflege in Brandenburg*, **5**, 59–64.

MacArthur, R. (1969) Patterns of communities in the tropics. *Biological Journal of the Linnean Society*, **1**, 19–30.

MacArthur, R., Recher, H., and Cody, M. (1966) On the relation between habitat selection and species diversity. *American Naturalist*, **100**, 319–325.

Martinez, N. (2010) Gartenrotschwanz – Wiesenstruktur entscheidet. *Der Falke*, **57**, 388.

Mitschke, A. (2009) Wo sind all die Haussperlinge geblieben? – 25 Jahre Stadtkorridorkartierung in Hamburg. *Hamburger avifaunistische Beiträge*, **36**, 147–196.

Mitschke, A. (2014) Schwanzmeisen statt Haussperlinge – vom Wandel der Vogelwelt in einer nordwestdeutschen Großstadt. 147. Jahresversammlung der DO-G – Tagungsband, p. 96.

van Noorden, B. (1997) Why did the Golden Plover Pluvialis apricaria disappear as a breeding bird from The Netherlands? *Limosa*, **70**, 89–96.

Paffen, K. (1940) *Heidevegetation und Ödlandwirtschaft in der Eifel*, Ludwig Röhrscheid, Bonn.

Palm, V. (1957) Ehemalige Reviere der Großtrappe (Otis tarda) in Deutschland. *Ornithologische Mitteilungen*, **9**, 203–207.

Pfeuffer, E. (2011) *Der Ungebändigte Lech – Eine Verlorene Landschaft in Bildern*, Wißner-Verlag, Augsburg.

Phalan, B., Onial, M., Balmford, A., and Green, R.E. (2011) Reconciling food production and biodiversity conservation: land sharing and land sparing compared. *Science*, **333**, 1289–1291.

Plachter, H. (1997) Naturschutz im Abseits? *Biologie in unserer Zeit*, **27**, 306–316.

Probst, J.R., Donner, D.M., Bocetti, C.I., and Sjogren, S. (2003) Population increase in Kirtland's Warbler and summer range expansion to Wisconsin and Michigan's Upper Peninsula, USA. *Oryx*, **37**, 365–373.

Pyne, S.J. and Goldammer, J.G. (1997) in *Sediment Records of Biomass Burning and Global Change* (eds J.S. Clark, H. Cachier, J.G. Goldammer, and B.J. Stocks), Springer-Verlag, Berlin, Heidelberg, New York, pp. 1–489.

Reichholf, J. (2005) *Die Zukunft Der Arten*, C.H. Beck, München.

Remmert, H. (1988) *Naturschutz. Ein Lesebuch nicht nur für Planer, Politiker und Polizisten, Publizisten und Juristen*, 1st edn, Springer-Verlag, Berlin, Heidelberg.

Retzlaff, H. and Robrecht, D. (2011) Die Veränderung der Fauna der tagaktiven Großschmetterlinge im Kreis Lippe im Lauf der vergangenen 140 Jahre. *Melanargia*, **24**, 193–268.

Riechelmann, C. (2013) *Was heißt hier schon Natur?*, Frankfurter Allgemeine Sonntagszeitung, p. 65.

Rolfes, W. (2006) Die letzten Goldregenpfeifer in Niedersächsischen Mooren. *Fotografie draußen*, **37**, 26.

Rösch, M., Ehrmann, O., Herrmann, L., Bogenrieder, A., Deil, U., Goldammer, J.G., Page, H., Hall, M., Schier, W., and Schulz, E. (2001) in *Albersdorfer Forschungen zur Archäologie und Umweltgeschichte* (ed R. Kelm), Boysens & Company, Heide, pp. 96–119.

Rösch, M., Ehrmann, O., Herrmann, L., Schulz, E., Bogenrieder, A., Goldammer, J.G., Hall, M., Page, H., and Schier, W. (2002) An experimental approach to Neolithic shifting cultivation. *Vegetation History and Archaeobotany*, **11**, 143–154.

Schäffer, N. and Flade, M. (2013) Elektrozaun, Prädatorenbekämpfung, Ablenkfütterung: Welchen Vogelschutz wollen wir? *Der Falke*, **60**, 396–403.

Schmidt, F.-U. (2001) Die Vogelwelt im Landkreis Soltau-Fallingbostel. *Naturkundliche Beiträge Soltau-Fallingbostel*, Heft, **7/8**, 1–358.

Schreiber, K.-F., Brauckmann, H.-J., Broll, G., Krebs, S., and Poschlod, P. (2013) *Artenreiches Grünland in der Kulturlandschaft – 35 Jahre Offenhaltungsversuche Baden-Württemberg*, Verlag Regionalkultur, Heidelberg.

Schulze-Hagen, K. (2005) Allmenden und ihr Vogelreichtum – Wandel von Landschaft, Landwirtschaft und Avifauna in den letzten 250 Jahren. *Charadrius*, **40**, 97–121.

Schumacher, H. (2012) Lieber licht als dicht: Ziegenmelker, Wendehals und Grauspecht. *Der Falke*, **59**, 21–25.

Schwägerl, C. (2003) *Jüterbog-West: In einer ausgedehnten brandenburgischen Kunstwüste wird der Naturschutz neu erfunden*, vol. **190**, Frankfurter Allgemeine Zeitung, p. 38.

Segerer, A.H. (2012) Die physikalisch-geochemischen Grundlagen des planetaren Klimas und die Auswirkungen auf die öffentliche Diskussion – potenzielle Fallstricke für Ökofaunisten. *Nachrichtenblatt bayerischer Entomologen*, **61**, 32–45.

Stein, W. (1984) Untersuchungen zur Mikrohabitatbindung von Laufkäfern des Hypolithions eines Seeufers (Col., Carabidae). *Zeitschrift für angewandte Entomologie*, **98**, 190–200.

Stevens, M., Braun, T., Schwan, H., Sorg, M., Große, V., Kaiser, M., and Kiel, E.-F. (2008) Die Rückkehr des Dunklen Wiesenknopf-Ameisenbläulings – Kooperationsprojekt hilft *Phengaris nausithous* im Rhein-Kreis Neuss wieder zu etablieren. *Natur in NRW*, **4**, 37–41.

Stickroth, H. (2012) Die versunkene Vogelwelt des ungebändigten Lechs. *Der Falke*, **59**, 332–338.

Südbeck, P., Bauer, H.-G., Boschert, M., Boye, P., and Knief, W. (2007) *The Red List of Breeding Birds of Germany*, 4th edn, vol. **44**, Berichte zum Vogelschutz, pp. 23–81.

Tauch, M. (1974) *Rheinische Landschaften – Gemälde und Aquarelle aus dem 19. und 20. Jahrhundert*, Gesellschaft für Buchdruckerei A.G, Neuss.

Tennekes, H. (2010) *The Systemic Insecticides: A Disaster in the Making*, Weevers Walburg Communicatie, Zutphen.

Tews, J., Brose, U., Grimm, V., Tielbröger, K., Wichmann, M.C., Schwager, M.C., and Jeltsch, F. (2004) Animal species diversity driven by habitat heterogeneity/diversity. The importance of keystone structures. *Journal of Biogeography*, **31**, 79–92.

Topp, W. (2011) Historische Bergbaufelder: Pingen sind Schlüsselstrukturen und Quellenstandorte für die streulebenden Käfer in Buchenwäldern (Historical Mining Fields: mining pits are keystone structures and sources for the litter dwelling beetles in beech forests). *Entomologie heute*, **23**, 123–143.

Tornede, D. and Pieper, B. (2012) Keine Frage – Der Nationalpark Teutoburger Wald wird kommen. *Naturschutz heute – NRW*, 4–6.

Tscharntke, T., Clough, Y., Wanger, T.C., Jackson, L., Motzke, I., Perfecto, I., Vandermeer, J., and Whitbread, A. (2012) Global food security, biodiversity conservation and the future of agricultural intensification. *Biological Conservation*, **151**, 53–59.

Ulrich, R. (2002) Vom Naturschutz vergessen: Die Lichtwaldarten. *Naturschutz im Saarland*, **3**, 22–25.

Warren, M.S., Hill, J.K., Thomas, J.A., Asher, J., Fox, R., Huntley, B., Roy, D.B., Telfer, M.G., Jeffcoate, S., Harding, P., Jefcoate, G., Willis, S.G., Greatorex-Davies, J.N., Moss, D., and Thomas, C.D. (2001) Rapid responses of British butterflies to opposing forces of climate and habitat change. *Nature*, **414**, 65–69.

Weiss, J. and Schulze-Hagen, K. (2014) Viele Menschen – viele Vögel? Ornithologie und Vogelschutz in Nordrhein-Westfalen. *Charadrius*, **50**, 3–22.

7
The Forest as a Myth in Germany

7.1
Why Do Germans Love the Forest So Much? Man Did Originate in the Savannahs

For Germans, the forest has a mythical significance.

The forest is often equated with 'nature'.

Central European forests have a lower degree of importance for species richness than tropical rain forests.

The pathetic evocation of the forest as a pure 'German' landscape began around 1800 in poetry, painting and music and has consequences that can still be recognised today.

The forest became synonymous with the German national identity and embodied the ideals of the nature and heritage conservation movement.

On this ideological breeding ground, it was an easy task for National Socialism to exploit the 'natural phenomenon of forest' by means of extensive propaganda in order to attain its political objectives.

The high esteem in which trees and the forest are held is based on emotions that have nothing to do with the current alarming loss of species richness.

Behind the wilderness model lie aspirations that were interpreted into it later; but these aspirations cannot be achieved.

7.1.1
In the Minds of the Germans, the Forest Occupies a Special Position; They Equate It with 'Nature'

Der Jäger Abschied (The Hunter's Goodbye), a popular poem by Joseph Freiherr von Eichendorff begins with the line, 'Wer hat dich, du schöner Wald, aufgebaut so hoch da droben?' (Who created you, oh beautiful forest, up there so high?) (von Eichendorff, 1964). These words, around 200 years old, sound noble and awe-inspiring. Why is this so?

Since the beginning of the nineteenth century, the 'German forest' was a metaphor and a landscape that aroused a sense of longing; it was found in poetry, fables and legends of romance. Historical treatises declared the forest to be a symbol of German culture and a counterpart to French urbanity (Hillmann, 2005).

Species Conservation in Managed Habitats: The Myth of a Pristine Nature, First Edition. Werner Kunz.

The forest has a special position in the minds of Germans, more than in other nationalities. The forest is simply equated with 'nature'. Nature *is* where the trees flourish and grow. The forest has a prominent role and is regarded as an entity of nature that is more worthy of preservation than other habitats such as meadows, moors and wetlands. The 'forest dieback', a result of 'acid rain' in the 1980s of the last century (Ell and Luhmann, 1996) triggered stronger emotions in the minds of the Germans than did the disappearance of the moorlands and other wetlands, which had occurred only a few decades earlier. If we measure the value of a biotope against the wealth of its species, especially when the latter are vulnerable and need special protection, then the forest in Central Europe comes off poorly. From a species conservation perspective, we have enough forests as habitats in Central Europe. The drainage of meadows and moors after the Second World War was a significantly greater misfortune for species than the so-called *forest dieback*. The drainage programme was about the destruction of an abundance of endangered Central European species, while the forest dieback encouraged, rather than endangered the survival of several species.

The Germans believe the forest is something special, but there is no rational explanation for this. The forest has a mythical significance for the Germans (Schama, 1996). Why do Germans see reforestation as a restoration of nature? The forest is equated with 'nature', although it is just as artificial, that is, human-created as most other habitats in Central Europe.

In hindsight, the drainage of the extended Brandenburg Luch lowlands in the eighteenth century under the Prussian king Frederick the Great and the consequent destruction of species-rich landscapes are not perceived as a destruction of nature, but rather as a great cultural achievement, while on the other hand deforestation drives Germans 'up the wall' (Küster, 2008). The felling of a tree or the creation of whole clearcuts are considered to be a sacrilege: 'They have hurt the forest'. Why is this so?

The relationship of man to forest was not always like this and it differs between peoples, cultures and times. In the Middle Ages in Central Europe, the forests were perceived as a threat, as a place of danger where robbers and wild animals lived and where you could get easily lost (Weyergraf, 1987). Man as an original savannah dweller is a runner who must spot his prey and his enemies in an open landscape. The forest evokes fear because everything is hidden. This picture, however, has changed. The forest was the enemy, but now it has become an object of longing. The status of the forest was raised to that of a myth, particularly in Germany over the last two centuries; it became the embodiment of nature and this idea still prevails today.

Admittedly, criticism of the structure of forests does exist today, and it is rational, not mythic. However, it pertains to the age structure and the species composition of the trees in the recent forests, the call for more deadwood, the modalities of silvicultural interventions and whether the trees are native or introduced. These reform desires are justified, since they aim to increase species richness and the conservation of endangered species. They are concrete demands and well-founded objectives, which, however, have nothing to do with the myth of the forest criticised here.

The German yearning for forests is not directed at specific forest structures that would benefit certain animal and plant species. The myth says that the forest in itself is a good thing. 'Myth is always the result of an unconscious activity and a free product of the imagination' (Cassirer, 1973). Myths accompany the cultures of mankind. They are ideas and narratives with symbolic meaning and they lay claim to truth, not, however, through intellectual justification, but through mere self-understanding. Myth is in direct contrast to Logos, who claims to establish the truth of his allegations by verifiable arguments.

Myths are based on ideas that cannot be substantiated. The myth of forest is dominated by the idea that the forest (because of its alleged naturalness) is a particularly species-rich habitat, in contrast to the 'badlands' devastated by mankind. This concept is quite simply wrong. The idea that the Central European forest is a haven of species richness (which is also the cornerstone of national park ideology) is an unrealistic illusion. Open grasslands (if they are not affected by modern agriculture) are much more species-rich than the forests, at least in insects and vascular plants (Dengler, 2012).

7.1.2
Central European Forests Are Not Natural

Central European forests should not be confused with tropical rain forests, but this mix-up continually happens. Tropical rainforests are the embodiment of species richness and this is especially noteworthy, because the destruction of tropical rainforests is currently progressing rapidly and the species living there are also being destroyed as a result. Nowadays, more than 5 million hectares of rainforest are destroyed every year (Willis, Gillson and Brncic, 2004). The rain forests, however, are not the forests of Central Europe. Why? There are two different reasons here: first, species richness increases with geographical latitude as we move toward the equator and it is for this reason alone that the tropical rainforests are richer in species than the forests of the northern temperate zones (Godfray and Lawton, 2001); and second, the forests of the northern zones (especially the Central European forests) have lost part of their tertiary species richness due to recurring ice ages.

During the Pleistocene, the tertiary tree flora of Central Europe became increasingly species-poorer from warm period to warm period. The horse chestnut, redwood, magnolia, hemlock spruce, chestnut and other original native trees died out during the ice ages in Central Europe and were not able to establish themselves during the interglacials (von Koenigswald, 2004). Other temperate zones of the world, such as North America, south-west Asia (e.g. the forests on the southern coast of the Black Sea) and eastern Asia retained their species richness during the Pleistocene, because they had no predetermined geographical southern boundary. Many plants and animals made extinct by the Ice Ages could therefore be replaced by immigration from the south during the interim warm periods. The species-rich flora from the Tertiary (and probably also the fauna, but no adequate evidence of this exists, since pollen analysis is restricted to plants) had better chances to survive the Ice Age in America and in western and eastern Asia than in Europe.

In addition, the postglacial forest destruction wrought by the increasingly sedentary peoples in the Neolithic and Bronze Ages had a more drastic impact on changes in species declines, extinctions and immigrations in Central Europe than it did in other parts of the world. The Central European forests have been continuously affected by mankind and have not only lost most of their large animals, but probably many other species as well. Part of the former forest fauna of Central Europe was destroyed, while the other part had to adapt to the forest structure that was being reshaped due to human settlement. As a result, the species richness of the Central European forest fauna has been severely thinned out. Central Europe differs from other parts of the world in the species poverty of its forests.

Central European forests are not natural. The myth of the forest is based on the assumption that we have a pristine biotope here in Central Europe, free of human influence and thus natural and native. However, this assumption is not true, because the postglacial reforestation of Central Europe was affected in many places by man, so the forests of Central Europe are neither primary nor virgin. The idea of a natural forestation (unaffected by human beings) in Central Europe in the past millennia is an illusion.

It is true that Europe was covered in forests thousands of years after the last Ice Age. During the Neolithic, Central Europe was neither steppe nor heathland. It was only in the last millennium that we humans deforested Central Europe on a grand scale, making it a steppe landscape. But the forests of 'Germania' were not as Tacitus described them; these were not the primeval forests created by the myth of their unspoiled state (Küster, 2008). As palynological studies have shown, most of the 'Germanic forests' of Tacitus bore the traces of former agriculture. In the large-scale deforestation and river regulation of the seventeenth and eighteenth centuries, evidence of prehistoric settlements was repeatedly discovered. Population numbers in the early millennia were low and many parts of Central Europe were bereft of human beings, but the areas actually used by man for agriculture and cattle breeding were abandoned time and time again after only a short colonisation, after which new areas were cleared and settled. Postglacial man eventually settled in nearly all the areas that were suitable for colonisation even if his sojourn in each area was relatively short. Reforestation occurred on the newly derelict land, but the tree species which had originally grown there were essentially eliminated and that gave other tree species (which had not previously grown there) the chance to gain ground on the original vegetation. Central European forests became secondary forests. The original postglacial tree species such as elm, oak and hazel disappeared in many areas after the humans left. The beech invaded, because it exploited the unrivalled opportunity to propagate on the cleared land – and since the beech is a shade-giving tree, it prevented the re-population of the original tree species.

7.1.3
Central European Forests in the Eyes of the Poets and of the German Nationalists

Why then is the forest so revered, although it is neither a place of exceptional species richness nor naturalness in Central Europe? The reverence for the 'Teutonic oak' as the epitome of freedom, fatherland, heroism and justice goes back a long way (Hürlimann, 1987). In the freedom movement against the Napoleonic occupation of Germany 200 years ago, the terms Fatherland and Freedom were increasingly linked with the 'German oak' in both song and poetry. The veneration of the oak is, however, founded on errors in reasoning. The sweeping form of the gnarled oak tree was a symbol of primordial nature and indomitable originality. In truth, the oaks took their shape (that was so admired) because they could not develop naturally. Their shape was not created in a titanic struggle with nature, but through human use and damage caused by the browsing of animals. The oak does not owe its stretched appearance to struggling with the hardships of Nordic weather, but to continued coppice management (Hürlimann, 1987). What was glorified as venerable originality was in fact a deformity. The growth forms of the 'German oak' do not give us any pointers to primeval forest glory (Küster, 2008).

The pathetic evocation of the forest as a pure 'German' landscape began around 1800 in the poetry, painting and music of German Romanticism. During the wars of liberation against Napoleonic France, a historical reference was made to the Hermann battle (the battle of the Teutoburg Forest) against the Roman occupation. The victorious Teutons emerged from the forest and the German national movement then declared the forest to be a symbol in which Germans should recognise their nationality. The 'German oak' became proverbial and a sign of strength and valour; and the Germans enjoyed identifying themselves with strength and valour. The romantic forest awareness of the Germans has lasted from that period through the generations until today. This is most remarkable and at the same time hardly understandable, given the political and social upheavals of German social life. The Second World War mostly annulled any national yearning for heroism, but the yearning for the forest has remained as strong as ever. At this point, an explanation must be given as to why the forest plays such a prominent role in German nature conservation associations, even today. The overemphasis of the forest as a biotope worth protecting is particularly evident in the fact that the establishment of forest national parks has become a top priority, while the creation of large-scale grasslands and sandy regions is more or less ignored by the German nature conservation associations (in complete contrast to the Netherlands and the United Kingdom).

Remarkably, the Romanticism poets and painters, who made the image of the German forest into an object of sentimental nostalgia 200 years ago actually lived in forests which were different from those dense, dark forests which dominate the landscape of Germany today. On the one hand, the forests of 200 years ago were thinned out much more by cattle grazing and timber extraction than those of today; and on the other hand, monotonous coniferous forests were planted in rows to combat the growing shortage of wood.

The poet Joseph von Eichendorff (1788–1857) repeatedly venerated the 'whispering' forest as a mirror of the soul. His work reflects the fact that man has distanced himself from nature and that he should attempt to regain his lost unity with it. In his political writings, Eichendorff combined the yearning for the unity of the German nation with the desired unity between mankind and nature (Beci, 2007).

The Grimm brothers linked their German folk tales and legends inseparably to the forest. Almost all the figures in Grimms' Fairy Tales (first edition 1812) were forest dwellers. The German nationalists Friedrich Ludwig Jahn (the 'father of gymnastics', 1778–1852) and Ernst Moritz Arndt (1769–1860) linked the forest with the national character of the Germans, while the national characteristics of the British and French were more associated with parks and field landscapes (Elwenspoek, 1925). In his opera 'Der Freischütz' ('The Freeshooter') (1821), Carl Maria von Weber glorifies a fabulously romantic forest as a place of danger and terror, but also as a place of piety and salvation. The German composer Hans Pfitzner wrote in 1926: 'Das Herz des Freischütz ist das unbeschreiblich innige und feinfühlige Naturgefühl. Die Hauptperson des Freischütz ist sozusagen der Wald, der deutsche Wald im Sonnenglanz [...]. Webers Sendung war eine nationale – sie galt der Freiheit und Weltgeltung des Deutschtums, ... '. ('The heart of the Freischütz is the indescribable and totally heartfelt feeling for nature. The main character of the Freischütz is (so to speak) the forest, i.e. the German forest glittering in the sun [...]. Weber's mission was nationalistic – its objective was the freedom and international standing of the German culture, ... ') (Budde, 1998; Busch-Frank, 2003; Bermbach, 2005).

At the end of the nineteenth century, the forest in Germany became the embodiment of the nature and heritage conservation movement. Reformers and hiking nature enthusiasts ('Wandervogelbewegung', 'bündische Jugend') linked walks in the forest with being close to nature and having respect for cultural traditions, especially for Germanic mythology. Walking in the forest made you resilient, gave you contentment and taught you to control your urges, helping to develop standards and values such as 'loyalty', 'camaraderie', 'helpfulness' and 'naturalness' (Helwig, 1998).

After the German defeat in the First World War, it was proposed that a 'national monument' be built to commemorate the dead of the First World War; and in all the political parties of the Weimar Republic, from Social Democrats to the Extreme Right, held the consensus that this monument, as the 'very foundation' and the 'source of strength' of the German people, must only stand in a forest (Ziemann, 2011). During this time the 'Homeland Security' movement was established and it declared the 'German forest' to be the embodiment of the German race. The devotion to all things Germanic was directed against both 'Western decadence', the West's prevailing 'mercantilism' and the 'threat from the East'. Especially active in this regard was the 'Deutscher Wald e.V. – Bund zur Wehr und Weihe des Waldes' ('German Forests Registered Association – for the Defence and Consecration of the Forest') founded in 1923 under the auspices of the former Field Marshal Paul von Hindenburg. This association practised pronounced forest propaganda with the publication of its 'forest books' and 'forest writings'. The 'Association for the Defence and Consecration of the Forest' aimed at providing new impetus to the German people after the defeat in the First World War. It stressed the German national values and was against all things foreign. The forest should also be 'Germanified'; only 'German plants' and 'German animals' should live there.

On this ideological breeding ground, it was an easy task for National Socialism to exploit the 'natural phenomenon of forest' by means of extensive propaganda in order to attain its political objectives. The propagandist Alfred Rosenberg proclaimed that the Germans were a 'forest people' and that they would be as 'eternal' as the forest. The forest was equated with harmony and deforestation was considered to be a destruction of the people. The forest was seen as a Germanic sanctuary and as a 'racial source of power'. It embodied the purity of the German character as a counterpart to progress and urbanisation.

This historical perspective seems to explain why the Germans and their conservation organisations regard the forest as being special even today – without being able to provide a rational justification for this opinion. The relationship of the Germans to the forest has neither anything to do with species richness nor with nature, because the central European forest is almost entirely an anthropogenically created secondary forest; and it is species-poor when compared with other habitats and to the forests of other continents. The popular notion of forest as nature and wilderness is sheer fantasy. The dream of wilderness is interpreted into nature (Schama, 1996). The image many people have of the wilderness is not, however, consistent with the image of a true wilderness. The desired image of the supposed wilderness that many people seek in the forest is more like an English park than a real primeval forest (Küster, 2008). The idea of the forest being nature and wilderness personified is a cultural invention of the last few centuries.

7.1.4
Central European Forests Are Not Particularly Rich in Species

Trees and the forest are held in high esteem; but this opinion is based on emotions that have almost nothing to do with the current and alarming loss of species richness. In contrast to the currently declining species populations in heathlands and sandy regions, most forest-dwelling species are on the increase. Not one single, typical bird species of deciduous woods and forests is threatened with extinction in Germany (Südbeck *et al.*, 2007). The extensive beech forests of Central Europe contain few exceptional species (Fritzlar and Westhus, 2004). So why do we need so many forests?

While grey partridge, skylark and lapwing numbers are declining rapidly and some marsh and meadow birds in Germany are facing extinction, most forest birds such as woodpeckers and owls are not endangered. An investigation by the 'Dachverband deutscher Avifaunisten' (Confederation of German Avifaunists) (www.dda-web.de) found that the numbers of most forest birds in Germany have increased between 1989 and 2003 (Flade and Schwarz, 2004), while the numbers of most birds of the open country are in significant decline – and this trend is ongoing. For some years now, national monitoring programs in 25 European countries have documented the population trends of about 150 breeding bird species. The summarised results are issued yearly in the 'Population Trends of Common European Breeding Birds', published by the 'European Bird Census Council' (EBCC). In the last 10 years, the numbers of 37 selected, European, open landscape bird species have diminished by 52%, while in contrast, the numbers of 33 selected European forest species have risen by 2% (König, 2013).

Those among us who are actually aware of today's species decline would never vouchsafe the propagation of trees and the reforestation of slopes in Germany. Nor can those persons ever understand why the planting of a tree in Germany today is touted as a 'good deed in itself'.

It is true that some forest birds such as the capercaillie and the hazel grouse (Klaus, 2012), the grey-headed woodpecker and the wryneck (Schumacher, 2012) and the numbers of many forest butterflies (Ulrich, 2002) are also currently declining in Central Europe, but this is not happening because our forests are used too much for forestry management; in fact, the opposite is almost true, because our forests are underused and have therefore become too dense. Utilising forests as animal pasture resulted in the species-rich Hude forests; but no farmers drive their animals into the forest nowadays. Open sunlit glades and clearcuts with forest berries are almost only to be found where power lines cut aisles through the forest. Tree pipits, lizards, snakes and several butterfly species can be found there (Plate 8). However, these same power lines are strongly opposed by the nature and environmental protection organisations.

The current disappearance of many forest butterflies, such as the two species of fritillary (*Argynnis adippe* and *Boloria euphrosyne*) and the poplar and white admiral (*Limenitis populi* and *Limenitis camilla*), is rooted in the fact that most woods became too dense. Open and sunny clearings with flowers on the forest floor have become rare. The structures of the forests changed during the last decades. Coppices with sunlit glades were replaced by high-growing dense forests. If we look at the current state of our forests, we might be inclined to reinterpret the beginning of the famous Goethe poem 'Über allen Wipfeln ist Ruh' (Silence reigns above all treetops) in a very different sense – and that is to realise resignedly that 'Unter allen Wipfeln ist Ruh' (Silence reigns below all treetops). The species richness of the former cattle-grazed Hude forests is long gone. The closed canopy of today's dark and dense woods offers no longer a home to many species of butterflies and birds.

The outstanding importance attributed to the forest is not objectively justified in Central Europe today. The eras during which afforestation was economically necessary, not to mention the period during which the afforestation of Germany was prioritised, becoming a 'national-political cultural task' (von Vietinghoff-Riesch, 1936), are no longer with us today. The adherence to the belief that forests are the embodiment of nature is based on the positive attitude that forests prevent soil erosion, maintain the water table, filter the air and bind CO_2. And this makes forests healthy. Forests are also more of a place of peace and quiet than open land habitats. These characteristic values assigned to the forest, however, relate to the environment and the well-being of man. They should not be confused with nature itself.

One objective criterion that would give special priority to the forest in Germany would be the degree of endangerment. Endangered habitats should have a higher level of importance for conservation than common habitats. When we use this as a yardstick, then the forest also performs poorly. The current forests of Central Europe are not at risk, at least not in the state in which they find themselves today. The fairy tale of the nationwide dying forest syndrome is vanquished (Ell and Luhmann, 1996) – in other words, we have enough forests.

Different values are ascribed to different landscapes, with no objective reasons for such actions. Values can be defined, but the defining process always seems to be based on the perspective of the human being. Just as there are no objective criteria to justify why a crane should be more important than a nematode (see Section 3.2), there are no objective criteria to justify why the forest should be more important than a wetland meadow. Both of these examples are based on different, purely human standards of value.

In an interview with the weekly newspaper 'Die Zeit' in 2008, the President of the Federal Agency for Nature Conservation, Beate Jessel, stated the following:

> Behind all the wilderness, cultural and parkland models lie aspirations that were interpreted into those models later; but such aspirations cannot be achieved. There has never been any true harmony between the human being and nature in the cultural landscape ... The free wilderness untouched by humans is an illusion in Central Europe. However, we do need nature as a projection surface for such needs and emotions.
>
> *(Jessel, 2008)*

That statement says it all. The need of the Germans for forest is based on aspirations and emotions – and nothing else. These aspirations have their value, but they should not be reinterpreted in a rationally justified nature and species conservation programme (Tschimpke, 2008).

7.2
Human Encroachment upon the Central European Forests during the Neolithic and Bronze Ages

Prehistoric man encroached on postglacial forests from the outset.

The farmers of the Neolithic and Bronze Ages used enormous amounts of wood.

Livestock herded into the forests to graze changed the structure of the forests and the appearance of the trees.

The sedentary farmers of the Neolithic and Bronze Ages continuously moved their dwellings from one location to another.

The landscape of Central Europe reflects not only soil and climate conditions, it is also a testimony to past historical cycles.

Beech forests are not primeval forests in Central Europe.

In contrast to rain forest protection, nature conservation in Central Europe is faced by the unanswered question regarding which 'nature' it should protect.

The creation of a Central European 'nature' left to its own natural dynamics does not meet the urgent needs of endangered Red List species.

7.2.1
Prehistoric Neolithic and Bronze Age Settlers Wrought Profound Changes in the Forests

Relatively warm temperatures prevailed in Central Europe during the Tertiary era (65 million to 2.6 million years ago). Central Europe was a forest land with species-rich flora and fauna. Almost all recent species of mammals, birds, reptiles, amphibians and invertebrates either already existed at the beginning of the Tertiary, or they evolved during the Tertiary. No new species (or only very few) evolved in Central Europe during the following Quaternary period (2.6 million years ago to the present). To understand the effect of the ice ages on species, we must bear in mind that the drastic climate change between warm periods and ice ages in Central Europe probably destroyed the habitats of many species and that of course led to their extinction; but only a few species were able to evolve in this geologically short period of time.

By the end of the Tertiary period, the climate had gradually cooled and large areas of the northern continents were covered by ice. The Quaternary following the Tertiary is referred to as the *Ice Age* and is divided into the Pleistocene (the change between glaciations and interglacial periods) and the Holocene (12 000 years ago to present). The Holocene is the last interglacial period and the one in which we live today (Table 6.1).

Central Europe has been in a cooling phase since the Tertiary ended, and was characterised by vegetation ranging from savannah to steppe during most of the Quaternary, while the intermediate warm periods (interglacials) with forest cover were significantly shorter. Since most species already existed at the beginning of the Quaternary, Central Europe was predominantly a country of savannah and steppe animals. It is important that this is perceived and taken into account today.

During the cold periods in Europe, there were only forests in the Mediterranean region. In those times, Central and Eastern Europe were steppe landscapes linked to the vast steppes of the Near East and Central Asia. During the interglacial periods, the forests advanced from the Mediterranean region to the north, and the forest species that had adapted to forest life reached Central Europe from the Mediterranean. Forest species endemic to Central Europe probably had no chance of survival and could not re-evolve during the short interglacial periods; as a result we must regard today's European forest species as being non-indigenous immigrants from other countries. The inhabitants of the high mountains were probably an exception in this case; to avoid the increasing cold, they could descend to lower regions, where they found 'empty' habitats; but Central Europe's forest animals had little chance of going south during the cold period, because the habitats in the south were occupied. This assumption could explain why the high-montane butterfly species which presently live in Europe include a considerable number of European endemics, while today you can hardly find endemic species among the inhabitants of the lowlands in Central Europe (Van Swaay *et al.*, 2010) (see Section 5.2.1). In contrast to the constancy of the tropical rain forests, Central Europe was unable to develop a stable biodiversity in the forests.

The biodiversity of the Central European forests was not only depleted by the ice ages, mankind also did its part by drastically changing the forests during the last 10 000 years. The present human being (*Homo sapiens*) migrated from Africa 40 000 years ago and in the period that followed left his mark throughout almost the entire world, with the sole exception of extreme habitats (Willis, Gillson and Brncic, 2004). In the Near East and in the Europe of the Ice Age, immigrant man found ideal sustenance regions comparable to the savannahs of East Africa, which he had to leave due to climate-related forest growth. The mammoth steppes of Europe and western Asia were animal-rich and he thus had plenty of food.

The last ice age ended in Central Europe about 12 000 years ago. This was followed by the Middle Stone Age (Mesolithic) with warmer temperatures (Table 6.1). During the Mesolithic period (12 000–7500 years ago), Central Europe was gradually transformed from steppe into woodland. With the onset of forest cover, the large mammoth steppes of the northern hemisphere disappeared and with the disappearance of the steppes, there was insufficient food for humans. It can be assumed that Mesolithic man – original steppe dwellers – actively resisted the forest. He could keep his living spaces and protect himself from predators by setting forest fires. He could also use fire to drive his prey into tight corners such as cliffs and dead-ends. When the glaciers retreated, prehistoric man already had the ability to impede the gradually increasing forest growth and he used it much to the detriment of the forests (Remmert, 1988). Early man obviously recognised the fact that it was the forest that took away the animals which were his source of food.

In the Mesolithic period (12 000–7500 years ago), Central Europe was covered by birch, pine, hazel and willow. Since it was colder then than it is today and there were many glacial setbacks, it is likely that these tree species did not form thick, closed forests in Central Europe. The landscape probably corresponded to today's northern taiga–tundra transition zone, and there was probably a lot of open country. However, we cannot compare the former Central European landscape with today's taiga–tundra, because the sun's rays were much more intense, the sun's position in Central Europe being much higher than it is further north. The average temperatures rose markedly at the end of the Middle Stone Age and at the beginning of the Neolithic period 7500 years ago. Oak, elm, ash and lime trees displaced the willow, hazel shrubbery, birch and pine and the forests consequently became denser.

At the same time, when the open landscapes of trees and shrubs diminished and the forests began to prevail in Europe, man began to give up his nomadic life and to gradually become settled (Kreuz, 1990; Lüning, 2000). Agriculture, started in the Middle East prevailed, and mankind created village-like structures. This settled lifestyle was accompanied by a human population explosion. Mankind practised agriculture and animal husbandry and cleared the surrounding forests. The newly existent forests were changed, and in some regions the postglacial development of the forests was affected by man from the outset (Pokorny, 2005). The structure of most of today's Central European forests cannot be fully understood unless we take mankind's prehistoric settlement of the Neolithic (7500–4000 years ago) and Bronze Ages (4000–2800 years ago) into account. In every region where human encroachment took place, the forests look different today than what they would have looked like if man had not cleared the forests during the prehistoric period (Küster, 2008). The fact that man changed the forests of the postglacial period is undisputed. Just how far-reaching such encroachments were, however, is difficult to prove, because pollen analysis cannot give us any reliable quantitative information.

Neolithic and Bronze Age farmers had a tremendous demand for wood in Central Europe. In contrast to the Near East, the region where mankind's sedentary way of life and agriculture originated, a sedentary farming life in Central Europe would not have been possible had there been no wood and no forests there. Anatolia and Mesopotamia saw very little rain, but Central Europe's high levels of rainfall meant that there was no mud with which to build houses, since bricks could not harden in frequent rain showers. So forests were not only cleared or thinned to create living spaces and fields, wood was needed as the primary building material for houses. The further agriculture spread in the forests of Central Europe, the more dependent man became on wood. Wood as a raw material was essential for life on the farm. In the vicinity of settlements and fields, the forests were an inexhaustible stock of raw materials, one which was exhaustively exploited.

The greatest quantities of wood were needed for the construction of dwellings and for heating. However, large quantities were also used as a building material for protective palisade fences around the settlements and for roads through swampy terrain (wooden walkways). Wood was used to dry the grain, making it easier to detach it from the husks and to make it last longer. Most of the tools were also made of wood. We should actually refer to the periods of Neolithic and Bronze Age farming in Europe as the 'Wood Ages' (Küster, 2008), but the wooden tools have not lasted out the millennia and are hardly available today as evidence in museums. At the end of the Bronze Age and the beginning of the Iron Age (2800 years ago) the smelting of ore became commonplace and with that the demand for charcoal rose sharply, so the exploitation of the forests became even more intense.

In addition to human encroachment, the forests were also greatly changed by domesticated animals. Pigs, cattle, horses, sheep and goats were driven into the woods that surrounded the settlements. Their browsing and trampling prevented the regrowth of young trees. The grazing of sheep and goats proved to be particularly ominous for the forests, and goats climbing to reach a tasty morsel can also destroy older trees, so the appearance of the forests was gradually but steadily changed. The forests were thinned out, the number of trees decreased, and the girth of the remaining trees' branches steadily increased. In forests where cattle had grazed for decades, there was no more straight tree growth.

The forest also had to feed the animals in winter. There were no meadows where grass could be cut to eventually dry as hay for feeding the animals; but it was possible to make leaf fodder by cutting the branches from the trees. Tree species with particularly nutritious leaves were preferred, especially elm, lime, ash and hazel bushes. The foliage harvested after the leaves appeared on the trees in spring was dried in the summer. The resulting leaf fodder was then brought into the settlement, where it was used for the winter feeding of cattle, sheep and goats if the animals could not find anything else nutritious in the forests.

So the appearance of the forests was gradually but inexorably changed. New branches continually grew from the stumps of harvested branches, with the result that the trees assumed growth forms which had little to do with their natural growth form. Apparently, the elms were the least tolerant of this treatment, with the result that in the Neolithic period in Central European forests, they were forced to retreat more than other tree species (Küster, 2008).

7.2.2
Prehistoric Settlers Changed Their Abodes at Short Intervals, Destroying Forests Throughout Large Areas of Central Europe

Although the population density was very low in the early millennia after the last ice age, this should not be equated with the notion that man only inhabited small areas at this time and that most of the remaining areas of Central Europe were pristine. Despite their sedentary lifestyles, humans had to move to a new location every few decades, because forests and soil no longer yielded enough after a few years of exploitation. Farmland only existed for a few decades. The transition from a nomadic to a sedentary lifestyle was not as momentous as is often assumed, because the prehistoric sedentary lifestyle involved the constant changing of locations. Forest stands were continually being cleared and settled when the humans arrived, then abandoned again when they left after a few decades.

When the humans had abandoned their settlements and agricultural areas, the soil was no longer ploughed and oxygenated and growing wood was not cut back. When man departed from an area, he left behind nutrient-poor soils that had very different characteristics to those that had existed before his arrival. The forest did enclose such areas again, but on those areas that had been burnt down and cleared, the picture of the forest after just a few decades was different from what it had been before the humans came. The original, virgin forest did not return. After a few thousand years the appearance of the forest in Central Europe had completely changed. Man appears to have encroached on the forests of Central Europe for a much longer time and over much larger areas than is generally assumed (Küster, 2008).

The Neolithic and Bronze Age settlements were preferably founded in easily cultivated, agrarian loess (Aeolian sediment) areas. Here, almost all the suitable locations were settled once or even several times during the course of prehistory. Today, if excavations are made in a loess area anywhere in Germany, you can almost be sure that there will be traces of some prehistoric settlement discovered there, dating back to between the fifth millennium BC and Roman times. When large-scale river control, drainage projects and deforestation finally 'conquered' nature in the Central Europe of the seventeenth and eighteenth centuries, those working on the 'land improvement' projects constantly discovered the remains of early human settlements (Blackbourn, 2007).

Man can destroy the true nature around him, but he cannot restore it. The vegetation of an area is only partly characterised by the prevailing local conditions there; it also exists as a consequence of the processes that happened earlier there. The forest landscape is not only a testimony to climatic and soil-related physical parameters, it also reflects the events of the historical past.

When the human settlers left, the forests were able to regenerate to some extent, but they could not revert to their original virgin state. Other competitive relation ships among tree species predominated on the burnt-out forest areas of human settlement after the humans withdrew than those that prevailed at the time of the natural, postglacial primary afforestation. Certain tree species were able to flourish successfully on the cleared areas, because they had no competitors. Under natural conditions, such species would have no chance to propagate due to other competing tree species, but the competition was suddenly gone and they were able to dominate. It was only man who created the conditions for the growth of many tree species in postglacial Central Europe and he did this by eliminating (or at least affecting in some way) the tree species that previously grew there; and as a result, he encouraged the development of new trees that could not have competed in the original forests. In Central Europe, the beech tree was a particularly good example of this; beech forests profited from clear cuttings, and should thus be regarded as secondary and not primeval forests in Central Europe.

In his book 'Geschichte des Waldes' ('History of Forest'), the forest expert and pollen analyst Hansjörg Küster takes the view that one common, basic assumption of nature conservation is wrong; namely that in Central Europe a forest would attain its natural state if it were simply left to its own resources without being encroached upon by human intervention (Küster, 2008). Analyses of tree pollen, traces of which still remain after thousands of years, indicate that natural vegetation no longer occurs in an area that man used for some time. When a cleared forest area is deserted by humans, a forest then develops that differs from the original 'natural' forest that the humans destroyed. The vegetation in an area does not strive to attain a natural final state, but always also reflects its history, which indicates the human encroachments which previously occurred there.

In areas where humans cleared the forests for agricultural purposes, the trees which grew later could never become a truly natural forest. When man encroached upon the forest in a particular area, only what we describe as secondary forest or 'second-hand nature' grew there after his departure. Prior to the founding of settlements, the natural forest in Central Europe during the Neolithic period consisted mainly of elm, oak and hazel bushes; the regrowing secondary forest, however, no longer had many elm trees but many beeches, which grew on the sites where once dwellings had stood and cattle had rested. The belief that 'natural' vegetation will automatically grow again in an area that humans have used is quite simply an illusion. Natural or man-made vegetation that already exists in an area impacts too strongly on the growth of new vegetation. Truly natural vegetation, which grows readily in local soil and climatic conditions, often cannot prevail where other vegetation has already established itself.

Despite their low population density, the humans left traces of their destruction of the forests spread over large areas of Central Europe during the Neolithic and Bronze Ages. Archaeological and palaeontologic studies show that the Neolithic settlers constantly changed their locations and encroached almost everywhere on the natural forests. The stark, black-and-white contrast between wandering nomads and sedentary farmers should not be applied to prehistoric Central Europe. A permanently sedentary lifestyle only came into being with the founding of towns. In the Neolithic and Bronze Ages in Central Europe, virgin forests probably only existed in inaccessible locations in the mountains and swamps; otherwise, the concept of the Germanic primeval forest is simply an illusion.

7.2.3
The 'Conservation of Nature' Cannot be the Reason Behind the Protection of Central European Beech Forests

After humans deserted a prehistoric agricultural area, the trees that settled there first had room to grow and formed a broad crown. The beech really exploited this. Once it had taken control of an area, it achieved a permanent domination over other tree species. It cast very intensive shadows and that prevented the growth of young oak seedlings. Four thousand years ago during the Bronze Age, the beech continued to prevail against the oak in many forests, gradually spreading further and further into Europe. It attained its maximum expansion in Central Europe in the millennium before Christ. Pure beech forests emerged along the Baltic Sea and in many regions of the uplands in Central Europe (Mittelgebirge). In the lowlands of north-western Germany, the oak managed to assert itself along with the beech, and mixed forests of beech, fir and spruce arose in the Alps and in several southern upland areas of Germany (Küster, 2008).

Although these forests grew naturally 2000–4000 years ago (i.e. they were not planted by man), this composition of Central European forests cannot be described as 'natural', because many parts of Central Europe only became copper beech strongholds after the prehistoric methods of creating settlements had impacted upon the new growth of forests in the period before the Roman era. It is probable that the copper beech would never have become a dominant tree in Central Europe without human intervention (Remmert, 1988; Küster, 2008). It could make sense to promote the beech forests in Central Europe purely for aesthetic reasons and from a forestry standpoint; but this cannot be justified by stating that these forests in Central Europe are 'natural'; that is, that they would also have developed if no humans had existed at that time (see Section 3.1.1 for definitions of the terms *nature* and *natural*).

The question that clearly arises here is of fundamental importance for nature conservation in several respects. If 'nature' does not return fully to an area after humans have utilised it, but only partially in the form of 'second-hand' nature, what arguments then exist for its conservation? Without humans paving the way for it, the beech would not have become so widespread in Central Europe. Conservation organisations tend to push the Central European beech forests into the foreground; but these are not virgin forests, they are simply examples of 'second-hand nature'. Although man did not directly intervene in its reforestation (the beeches attained their height without human help), the propagation of the beech in Central Europe was never 'natural'.

The clearing, settlement and management of an area and the subsequent relinquishment of settlement and agriculture are not natural processes. Decades of colonisation and the agricultural management of an area put an end to any natural development of the forests. Once cleared, a natural forest can never return in its original form. The landscape has a 'memory' for what happened in the past – and also for what the culture of mankind brought with it.

As defined elsewhere in this book (Section 3.1.1), what we understand by 'natural' processes are those which take place without the intervention of man. When man interferes in natural processes, he changes or distorts them, giving these processes an artificial character. The only truly natural landscapes on Earth are those upon which man has not yet encroached. When man has encroached upon a landscape, its natural elements change or even vanish. If human intervention is minimal or if it happened a long time ago, the landscape is 'close-to-nature', but it is still not nature. A forest can only be seen as 100% natural if it does not bear any characteristics that pertain to human intervention. Only this would be a real 'virgin' or 'primary' forest. If a forest shows signs of human encroachment, which can be recognised by the dominance of tree species that could not have prevailed without human intervention, that forest is only a secondary forest within the meaning of the term *nature*.

Man has encroached upon the forests of Central Europe for 7000 years. The concept of whether certain tree species are native to their locations or have been introduced there overlooks the fact that the competitive struggle is a main factor in why a particular tree species was able to take hold and dominate in a given area. The Central European beech forests, much vaunted as world heritage sites (http://www.weltnaturerbe-buchenwaelder.de/) are called *primary forests*, that is, forests that arose under natural conditions; but this is quite simply wrong: primary forests do not exist in Central Europe. What are often referred to as virgin forests are in fact secondary forests. These, however, were left in peace and have either been minimally affected by humans or not at all. The process of their natural dynamics therefore continued undisturbed for a long time.

The question remains, what does nature conservation actually want? This question must be asked continually; but we can only answer it objectively if we do not select 'naturalness' as the only reason for protecting a landscape; we must also select its identity, which has been shaped by humans. The concept of whether certain tree species are native or introduced cannot be applied to the protection of Central European forests.

7.2.4 Human Encroachment on Tropical Forests in Prehistoric Times

The concept of the forest as the epitome of unspoilt nature is also based on the fact that Central European forests are mistaken for tropical forests. The crucial difference, however, is that the widespread destruction of tropical forests did not start until the beginning of the colonial period (prior to that human intervention was very sporadic), but the widespread destruction of the central European forests already goes back thousands of years.

Many parts of the tropical forests on our planet are not, however, truly ‘virgin’ (i.e. primary). Archaeological and palaeoecological studies carried out in the three major rainforest areas of Amazonia (all the way down to today’s Xingu region in Brazil), the Congo Basin and the Indo-Malayan region indicate traces of human destruction which are thousands of years old in many locations (Willis, Gillson and Brncic, 2004). Tropical forests are therefore not virgin forests, as is often wrongly assumed (Willis, Gillson and Brncic, 2004). Many parts of the tropical rain forests of all three continents were settled by prehistoric farmers who also caused widespread destruction there, but not as permanent and total as it was in Europe. Prehistoric land utilisation repeatedly led to the rapid collapse of agriculture, forcing the inhabitants to abandon the country after a relatively short period of time. The pioneer populations often disappeared without a trace and may have starved to death.

Even during our Neolithic and Bronze Ages, prehistoric farming settlements with large-scale forest fire clearances existed in the Amazon basin, and these clearances can still be seen today as ‘Terra preta’. In South East Asia, prehistoric finds (dating back 7000 years) of banana crops and of the planting of various agricultural crops point to the existence of large-scale agriculture in, for example, the tropical rain forests of Papua New Guinea and Thailand.

Agriculture damaged the soils in tropical rain forests to a much greater extent than it did in Central Europe, however, because the metabolism of the biomass takes place very rapidly in the tropics. The dead biomass is immediately used by the living biomass. This is why soil nutrient deficiency occurs very much faster in the tropical rain forests than in Europe. In contrast to Central Europe, the clearing of tropical forests left behind soils which remained infertile for a very long period of time. The soils of Central Europe, however, are much more robust. If the vegetation is destroyed, new vegetation springs up quickly. However, a footprint in the vegetation on the surface of a bog in Central Europe’s far north can still be seen after a 100 years.

The prehistoric rainforest settlers also burnt down the forest to a much greater extent than the Central European settlers, in order to fertilise the soil with mineral salts of the ash. In Central Europe, the wood was not destroyed to the same extent, since it was needed for heating (Küster, 2008). The ash fertilisation of tropical forest soils lasted for only a few years, after which the nutrient salts were exhausted and the arable land could no longer be used. After deforestation in the tropics, only barren soil and sand remained. This can also be seen today in many locations such as the Atlantic rain forests of Brazil. Here, mountain forests which are protected in national parks are covered by dense 'jungles', but now and again there are viewpoints from which distant slopes outside the national park can be seen. These slopes can be completely denuded of vegetation. Not one single tree grows there and the sand can be seen over the entire surface of the slope.

Near-catastrophic depopulations often occurred in prehistoric times as a result of these deforestations in tropical regions. The settlers probably died out or emigrated for lengthy periods of time. The colonisation of the Congo basin in the last millennium BC, that is, during the period of the flowering of Greek culture and the beginning of the early Roman Empire ended with a population crash.

The serious crises of prehistoric agriculture in the tropics meant that agriculture could only be carried out in the short term. Peace subsequently reigned in the forests for centuries and even millennia and it was during these long periods of non-encroachment by humans that the rain forest could partly regenerate (though never completely) (Willis, Gillson and Brncic, 2004). In contrast, human impact on the change in nature in Central Europe was much more intensive. There was no long-term disappearance of the human populations. After the last ice age, Central and Southern Europe and the Near East and Central Asia were always colonised in stable fashion, despite the mobility of the farmers and cattle breeders. Here, the difficulties involved in prehistoric agriculture were never so serious that agriculture had to be given up on a large scale. When an area was abandoned, it was soon occupied again by new settlers. This is why postglacial forest destruction throughout the temperate regions of the Old World has been repeated over the millennia at regular intervals. There was almost nowhere else in the world where agriculture could be continually carried out for thousands of years. In this regard, Europe has an exceptional position in the world.

This unique position must be considered when it comes to nature and wildlife conservation in Central Europe. In contrast to rain forest protection, nature conservation in Central Europe is faced by the unanswered question which asks which 'nature' it should protect. In complete contrast to the protection of the rainforest species, endangered species in Central Europe cannot be protected by simply letting nature take its natural course. In order to protect many red-listed species in Central Europe, we must continue to destroy nature at several locations. This will create habitats for particular Red List species which cannot live in pristine nature in Central Europe. We must perpetuate the disruptive impacts on nature which have been occurring in Central Europe for millennia.

7.3
The Forest in Central Europe from Roman Times to the Modern Era

With the advent of Roman colonisation from the first to fourth century A.D., western Central Europe saw the creation of large forest-free zones in which trees could not regrow.

In the Carolingian period in the seventh to ninth century A.D., the colonisation of the forests became vassalage, and steps were taken to ensure that the forest could not again gain a foothold in the former farmlands.

From the twelfth century on, the comprehensive and permanent destruction of forests was launched throughout Central Europe.

Wood was used as raw material for glassworks and for the extraction of ore in the mines.

In the deforested landscape in the eighteenth and nineteenth centuries, the species number in Central Europe reached its peak.

At the beginning of the eighteenth century, the lack of wood in Germany was finally acknowledged and afforestation started.

The idea of afforestation (initially purely pragmatic) was soon exaggerated until it became mythical in proportion, and the German forest became associated with the nationalistic concept and the fatherland.

7.3.1 The Central European Forest in the Roman and Carolingian Periods

Prehistoric settlers created the secondary forest in Central Europe, but they did not establish the permanent forest-free open spaces; the inhabited areas were only briefly occupied and this enabled new forest growth to develop. In the same period, however, completely different circumstances held sway in the more advanced civilisations of the Mediterranean region, because cities were founded there thousands of years ago and the inhabitants were settled in the long term. The populated agricultural areas were no longer abandoned, permanently destroying the forests. To this day, almost none of the forests in the Mediterranean region have been able to regenerate.

In the first century A.D., when the Roman historian Tacitus wrote his famous work 'Germania' about the Teutonic (Germanic) tribes, that is, about the tribes who lived north of the Alps, he compared Germania with his Mediterranean home. On the one hand he glorified the pure culture of the Teutons and held it up as the antithesis of the corrupt and decadent society of his own countrymen. On the other hand, however, the vast forests of Germania filled Tacitus with awe, simply because there were no such forests in his own region. Since these reports were the first historical documents about the country and the culture of the Teutons, this historical source was eagerly adopted in Germany in the eighteenth and nineteenth centuries by nationality-conscious Germans and reinterpreted in their ideology as being pure Teuton. The combination of the purity of the essence of the Teutons with the deep forests in the works of Tacitus found fertile ground, characterising the mythical relationship of the Germans to the forest, a relationship which has continued to this day (see Section 7.1.3).

Tacitus described the forests of Germania from the perspective of a man who was influenced by the Mediterranean landscape. He succumbed to the misconception that the forests of Germania were 'genuine nature' and this view was also taken in Germany in the eighteenth and nineteenth centuries. Nowadays, pollen analysis results tell us that Tacitus was writing about forests that in his time had deviated from their natural evolvement for over 5000 years and which were already at that time 'second-hand nature' (Küster, 2008).

After millennia of relative stability, the impact of humans on the Central European forests changed when the Romans conquered the western region of Central Europe shortly after the birth of Christ and founded cities there such as Mainz, Trier, Cologne and Xanten. The founding of towns encouraged the coexistence of large groups of people and this ended the constant relocation of settlements and agricultural areas practised by the inhabitants of Central Europe for millennia. A long-term sedentary way of life was the result, and from then on, the forests of Central Europe were subject to continuous destruction by humans.

Large forest-free zones were created with the Roman colonisation of western Central Europe and these areas did not recover again for a very long period of time. As recently as 2000 years ago, the forest was emphatically cleared in the Roman-occupied areas of Germania. Large quantities of wood were needed to build towns and it was also required for extensive palisade construction work and the Limes rampart wall. Huge quantities of wood had to be made available for the Romans' domestic heating systems and for the baths with their lavish underfloor heating and hot water tanks. Roman colonisation proved to be a drastic encroachment on the forest ecosystems of Central Europe. The tree species composition in many forest communities was permanently destroyed by selective utilisation and the introduction of new tree species from the Mediterranean region.

Deforestation was again interrupted (albeit briefly) in the fourth and fifth centuries when the Germanic hordes surmounted the Limes wall, driving back the Roman culture (Kemkes, 2006). Pollen analyses from this period show that agriculture came to a standstill in many parts of the region. Permanent settlements were abandoned in favour of the semi-nomadic form of settlement and the forest once again gained the upper hand where Roman castles and manor houses had once stood.

Shortly thereafter, however, the Carolingian period from the seventh to ninth centuries saw the areas annexed by the Romans repopulated and the forests cleared yet again. Clearing work was done mainly on easily accessible areas with nutrient-rich soils. In the western region of Central Europe, the only areas that were spared were the upper ranges of the Central German uplands (Mittelgebirge) because they were inaccessible, and the swampy areas of the alluvial forests, due to the unpredictability of rivers such as the Rhine. In contrast, many forests remained untouched in the eastern region of Central Europe, where the soil was sandy and consequently unsuitable for agriculture.

The landlords of the Carolingian period ensured that the forest would never again regain its foothold on former farmlands. Colonisation was now a vassal's duty. In his 'Capitulare de villis' of 795, Emperor Charlemagne stated that every manorial system had to prevent the forest encroaching on farmland as it had done in earlier times (Küster, 2008). The millennia-long secondary successions of the forest, which were typical of Central Europe, could now be prevented and all agricultural land had to be kept permanently forest-free by order of the kingdom. In stark contrast to the attitude which prevails today, the forest was perceived as a disruptive factor, one which hindered the progress of culture and civilisation.

7.3.2
The Permanent Destruction of Forests in Central Europe from the Early Middle Ages to the Nineteenth Century

The 1000-year-long period in which Central Europe became comprehensively and permanently forest-free began in the early Middle Ages. From the twelfth century onward, human settlements pushed into the formerly deserted and remote valleys of the uplands. Since the settlers needed wood, the forests were first thinned or even completely cleared. The farmers then collected the forest floor litter, spreading it over the floors of their cattle stables. Any acorns and beechnuts that were lying around were fed to their pigs. These rude encroachments on the former forest floors prevented new forest growth from gaining a foothold. On the deforested plots of land, the soil was broken with a hoe or plough and cereal crops were planted. The fields usually yielded little or nothing after just a few years. The land became barren and could only be used as sheep or goat pasture. In the Middle Ages, no plots of land were specifically set aside for cattle grazing; instead, cattle and horses were driven into the woods, where they fed on the new tree shoots, with the result that no new trees could grow. Cattle and sheep found shrubs such as juniper, holly and gorse unappetising in the forest pastures (known as *Hudewälder*), so these grew and spread in place of the trees.

Until 1300 or so, the forests were intensively cleared and used for agriculture, so the entire Central European landscape lost its forest character as a result. By the end of the fourteenth century no new forest growth could attain any height whatsoever. Young wood was collected by farmers and charcoal burners and the forest floor materials were removed and used as litter for cattle stables. The growing human population's increasing demand for wood saw forest clearances penetrate even the most far-flung wooded areas. Wood was increasingly used as raw material for glassworks and for the extraction of ore in the mines. Mining was an industry that consumed vast quantities of wood that was used to shore up the shafts and tunnels. Wood was also needed for the production of charcoal which was used in metal smelting.

The dense forests gradually began to disappear. Central Europe now became almost totally forest-free up to the higher mountain ranges. What remained after the forest clearances was a landscape absolutely destroyed by over-exploitation. Only the human depopulation during and after the 30 Years' War led to a light reforestation of some areas that had formerly been agricultural land. However, in the eighteenth century, the regrowing forests were cleared once again.

The forest floors were looted. The land karstified and soil erosion was rampant. A devastated landscape remained. Barren ridges and seemingly endless heathland dominated the scenery – not a single tree was to be seen. Central Europe became a land of steppes and semi-desert. Due to the lost forests which could no longer regenerate, massive soil erosion took place, especially on the slopes. The country karstified, the wind blew the sand over the open areas, and even the dunes which had been stationary after the glacial epoch for thousands of years, anchored by the roots of pine trees and other plants, were set in motion again. Fields, farms and even villages were buried under the shifting sand of the dunes. The town of Celle was repeatedly threatened by the shifting sand dunes of the Lüneburg Heide (Heath). Large areas of the Thuringian Basin were covered by gravel plains, comparable to today's Crau region in the Rhone delta in southern France. Little wonder that the little bustards bred there.

The heath became the new form of wilderness. In the north-western region of Central Europe and in many mountain ranges, dwarf shrub growth spread relatively quickly, and in the warm regions with calcareous soils it was the steppe-like juniper heath that came to the fore. Areas consisting of steppe to semi-desert, gravel plains and sand dunes were prevalent in northern Germany, on the high plateau of the Central German uplands (Mittelgebirge), in Rheinhessen and elsewhere. Central Europe was dominated by a landscape that was more reminiscent of the Mediterranean area, the steppes of the eastern lands or the tree line of northern Europe than the heart of Central Europe today. As a result, the species of the north, the east and the Mediterranean regions colonised Central Europe almost entirely; and these are the species that are on the Red List in Central Europe today.

Paintings and watercolours from the nineteenth century (and even the twentieth) show a northern Germany consisting of bleak plains reaching to the far horizons with only a few lonely bushes and isolated trees to break the monotony. On the old pictures depicting the landscape of the 'Mittelgebirge' in Central Europe, we can see well-known castle ruins that stand on the hills, almost unchanged to this day. However, the surrounding environment is unrecognisable today. The eighteenth century saw the castles on the Rhine, in Hesse and Baden-Wuerttemberg perched high above bare rock slopes; today, however, mountains and hills covered with bushes and forests dominate these same areas. In the eighteenth century, the view from these castles was still open and stretched to the horizon, but today it is obstructed by forests, and the vastness of the landscape remains hidden. On Germany's mountain hiking trails, there are still many viewpoints and benches to be found; and it is obvious that these were created and placed in these locations to allow hikers to enjoy panoramic views of the then vast plains. Now tall forests block these vistas.

7.3.3
The Richness of Biodiversity in the Ruined Landscape during the Nineteenth Century

During the atmosphere of change in the nature conservation objectives that prevailed in the early 1980s of the last century, conservation ideology failed to recognise the fact that the species richness of the past centuries was not generally based on intact ecosystems, but (quite the contrary) on their destruction (Häpke, 1990; Engels, 2006). The exploited and ruined landscapes of Central Europe hosted an absolute maximum number of species. This means we may safely draw the conclusion that today's quest for an intact and undamaged nature will *not* promote the species richness of Central Europe.

It is difficult today for most people to understand that the deforested, exploited landscape of Central Europe became an ideal habitat for a variety of plants and animals, including many species that are on the Red List of threatened species today. Today, however, these species cannot find a habitat, because we simply have too much vegetation. Our modern-day landscape in Central Europe is as overgrown as an unkempt garden.

The reason why it is so difficult to associate the former biodiversity with the destroyed landscape can be found in the fact that no distinction is made between nature conservation and species protection. It is regarded as a paradox that a destroyed nature was the prerequisite for the occurrence of many species. If, however, we can understand that nature conservation and species protection are two different objectives, the abundance of species in the earlier ruined landscape of Central Europe will no longer be perceived as a contradiction. The landscapes of Central Europe have been anthropogenously overformed for millennia and the nature that has been destroyed there cannot be equated with any corresponding destruction of biodiversity.

Biodiversity probably reached its peak in Germany in the eighteenth and nineteenth centuries (Plachter, 1997; Schulze-Hagen, 2005). A species diversity and richness existed that in retrospect seems hardly credible to us today. Hen harriers and short-toed eagles were common breeding birds in Germany in these times because they found enough prey on the sparsely vegetated surfaces, and the Thuringian Basin was interspersed with gravel plains where little bustards nested. The poor grasslands and open valleys of the Central German Uplands were rich in butterflies; they sheltered great numbers of blue and fritillary species, most of which have since disappeared. Although they are mostly heat-loving species, the peak of butterfly population occurrence still existed even into the final phase of the 'Little Ice Age' and this indicates that the quality of the habitats of that time played a much more significant role than the fluctuations in the climate (Warren *et al.*, 2001; Segerer, 2012).

In the eighteenth and nineteenth centuries, the black grouse of north-west Germany and of the reedy areas of the alpine foothills lived on open heaths and moors with an all-round, unobstructed view. These birds are extinct today because the landscape has become overgrown. The black grouse will also soon die out in the Netherlands (http://www.natuurmonumenten.nl/over-natuurmonumenten/). It is probably scrub encroachment and rampant vegetation that are to blame for this and not the bicycle tourists and joggers (Reichholf, 2000; Anonymous, 2012).

7.3.4 Afforestation as a National Ideology in Germany Since the Eighteenth Century

The conquest of nature and the devastation of the landscape continued in Central Europe until the twentieth century (Blackbourn, 2007). It was not because people realised how ruthless the exploitation of the landscape was that brought them to their senses, it was the lack of raw materials. Wood was man's most important energy source. In the Central Europe of the nineteenth century, the demand for wood did not slacken, even when coal gradually replaced wood as the common fuel. Enormous quantities of wood were necessary to shore up the shafts and tunnels of the underground coal mines.

The beginning of the eighteenth century saw the first movements and regulations against the unrestrained use of wood. Reference was made to the economic danger that would threaten if no more wood were available one day for heating, house building and particularly for the smelting of ore. Sod cutting, the cutting of timber from the forests and the peeling of bark from the oak trees were prohibited and extensive afforestation measures gradually got underway.

Captain Carl von Carlowitz (1645 – 1714, a mining captain in the Erz Mountains of Germany) was one of the first to realise that there was a real dearth of wood. He subsequently promoted the reforestation of clearcuts to ensure that wood would be available for mining in the future (Küster, 2008). In his main work of 1713, 'Naturmäßige Anweisung zur wilden Baum-Zucht' (Instructions for Natural Wild Tree Breeding), von Carlowitz summarised the knowledge of forest science at that time, expanded it through his own experiences and formulated the principle of sustainability for the first time. In many locations, afforestation was subsequently decreed as law by Frederick the Great (*inter alia*).

After the mid-nineteenth century, the demand for wood gradually dwindled. By then, the over-exploitation of the forests had diminished and the number of forested areas had increased. Nevertheless, afforestation continued unchecked. The strong demand for wood had been the motive behind afforestation for almost 200 years, but the original economic demand was now transformed into a kind of national ideology. Afforestation had been an economic necessity for a long period of time, but now it became a 'good deed in itself', exactly like the draining of the swamps a century earlier in Prussia: not an absolute necessity, but an ethical duty.

From the mid-nineteenth to the mid-twentieth century, the forests doubled in size in Germany, because coal reduced the demand for wood and sheep farming declined due to cotton imports. The erstwhile vast sheep pastures in Germany, such as the Eifel, the Sauerland and the Lüneburger Heide regions were no longer utilised. They were not, however, left to themselves; intensive reforestation was carried out. Wood from the boreal coniferous forests of Canada, Finland and Russia could be delivered much cheaper than the wood from Central Europe. Instead of wood and charcoal, fossil coal was now available for the smelting of ores, the operation of steam engines and heating in the houses. When the coal was eventually transported by rail in many regions, the economic value of the forests began to decline even further.

However, the rate of afforestation in Germany remained high and the landscapes changed rapidly. The forests also became denser, because the introduction of mineral fertilisers made the use of forest floor litter and sod cutting no longer financially viable. Paradoxically, the growth of the forests in Germany increased exactly at the same time as the economic need for them was gradually diminishing.

Unlike in other countries, where land that was no longer used was left to become wasteland, afforestation took place on all the available land in Germany. Many more forested regions developed in Germany than in the countries that bordered Germany to the west. The increase in the number of forests meant that the inhabitants of one village could no longer see the neighbouring village. It was quite different in France and England: no planting took place on the land around the towns, with the result that the views from the towns today overlook wide landscapes. There is a huge amount of widespread reforestation in Germany and this characterises the identity of Germany's countryside to a great extent.

Even today, if we compare the Black Forest on the German side of the Rhine river to the Vosges on the French side, there is a clear difference in the intensity of reforestation. The Black Forest High Road (Schwarzwald-Hochstraße) was originally planned as a scenic route and to this end was provided with parking spaces from which tourists could look down upon the Upper Rhine plain. Today, however, the road leads through dense forest that obstructs the view of the landscape. Many of us are familiar with benches and seats for hikers and walkers in our local environments; these were originally placed there because of panoramic vistas, but nowadays tourists can see for about 50 yards at most until the forest obstructs their view.

The question arises about the reasoning behind the unrestrained afforestation in Germany. There are ideological and mythological reasons for this. A myth-based exaggeration of the purely pragmatic motivation that created the need for reforestation soon developed in Germany. The sustainability principle, originally validated by the demand for wood, was transferred to nature conservation and seen more in ecological terms than in an economic sense. Since ecological motivations are much more prone to mythical exaggerations than economical and commercial motives, the principle of sustainability became a myth; this explains why the forest has been raised so high in the esteem of the German people for more than 200 years.

A quest for freedom became popular thanks to the influence of the French Revolution of 1789. In Germany, this quest was associated with the forests (unlike in France). In searching for national unity, the Germans found their identity in the Germanic tribes, a free forest-dwelling race in Central Europe and one which had successfully resisted the Romans. The German forest now became inexorably linked with the nationalistic concept and the 'Fatherland', and poets and songwriters soon dedicated as much of their time and skills to the forest as did the foresters themselves (Küster, 2008).

Afforestation was given the traits of a moral postulate. The concept of tree planting was continually raised on high until it was seen as an ethically sound and praiseworthy deed. This attitude continues to prevail in Germany even today. Many Germans still believe that afforestation will bring nature to their doorsteps and that the forest is synonymous with nature; but they do not understand that this is Central Europe, not the Sahel or the rainforest. It is fundamentally wrong to believe that the German forest is 'nature pure' – and it is a belief to which the German people have habitually succumbed, even up to the present day.

References

Anonymous (2012) Sanfter Tourismus gefährdet Birkhuhn – NABU erneuert Beschwerde bei der EU-Umweltdirektion. *Naturschutz heute - NRW*, **3**, 3.

Beci, V. (2007) *Joseph Von Eichendorff – Biographie*, Patmos Verlag, Düsseldorf.

Bermbach, U. (2005) *Opernsplitter*, Königshausen & Neumann, Würzburg.

Blackbourn, D. (2007) *Die Eroberung der Natur*, Deutsche Verlags-Anstalt (DVA), München.

Budde, E. (1998) in *Musikwissenschaft zwischen Kunst, Ästhetik und Experiment: Festschrift für Helga De La Motte-Haber Zum 60. Geburtstag* (ed R. Kopiez), Königshausen & Neumann, Würzburg, p. 47.

Busch-Frank, S. (2003) *Den alten Heroismus treu bewahren – Anmerkungen zu Pfitzners politischer und ideologischer Weltsicht in den Jahren 1933–1945*, GRIN Verlag, München.

Cassirer, E. (1973) *Philosophie der Symbolischen Formen. Zweiter Teil. Das mythische Denken*, Wissenschaftliche Buchgesellschaft, Darmstadt.

Dengler, J. (2012) Europäische Trockenrasen schlagen tropische Regenwälder. *Biologie in unserer Zeit*, **42**, 148–149.

von Eichendorff, J. (1964) *Gedichte, Eine Auswahl*, Philip Reclam jun, Stuttgart, pp. 63–64.

Ell, R. and Luhmann, H.-J. (1996) Von den Schwierigkeiten der Entdeckung des Waldsterbens in Deutschland. Man sieht nur, was man versteht oder: Schäden ohne Ursache und Ursachen ohne Schäden. *Forstarchiv*, **67**, 103–107.

Elwenspoek, C. (1925) *Das Ernst-Moritz-Arndt-Buch – Eine Auswahl Der Werke*, Hädecke, Stuttgart.

Engels, J.I. (2006) *Naturpolitik in der Bundesrepublik: Ideenwelt und politische Verhaltensstile in Naturschutz und Umweltbewegung 1950–1980*, Ferdinand Schöningh Verlag, Paderborn.

Flade, M. and Schwarz, J. (2004) Ergebnisse des DDA-Monitoringprogramms, Teil II: Bestandsentwicklung von Waldvögeln in Deutschland 1989–2003. *Vogelwelt*, **125**, 213.

Fritzlar, F. and Westhus, W. (2004) *Naturschutz und Biologische Vielfalt*, Bundesamt für Naturschutz, Bonn, pp. 255–265.

Godfray, H.C.J. and Lawton, J.H. (2001) Scale and species numbers. *Trends in Ecology & Evolution*, **16**, 400–404.

Häpke, U. (1990) Die Unwirtlichkeit des Naturschutzes – Böse Thesen. *Kommune*, **2**, 48–53.

Helwig, W. (1998) *Die blaue Blume des Wandervogels – Überarbeitete Neuausgabe*, Deutscher Spurbuchverlag, Baunach.

Hillmann, H. (2005) in *Europäische Lyrik seit der Antike* (eds H. Hillmann and P. Hühn), Hamburg University Press, Hamburg, pp. 197–233.

Hürlimann, A. (1987) in *Waldungen. Die Deutschen und ihr Wald* (ed B. Weyergraf), Nicolaische Verlagsbuchhandlung, Berlin, pp. 62–68.

Jessel, B. (2008) Mehr Räume fürs Leben. *Die Zeit*, **20**, 39.

Kemkes, M. (2006) *Der Limes – Grenze Roms zu den Barbaren*, 2. Neuauflage edn., Thorbecke, Ostfildern.

Klaus, S. (2012) Lückenbewohner im Natur- und Wirtschaftswald: Auerhuhn und Haselhuhn. *Der Falke*, **59**, 26–29.

von Koenigswald, W. (2004) Das Quartär: Klima und Tierwelt im Eiszeitalter Mitteleuropas. *Biologie in unserer Zeit*, **34**, 151–158.

König, C. (2013) Alarmierend: Europaweite Bestandstrends häufiger Brutvogelarten 2012. *Der Falke*, **60**, 20–21.

Kreuz, A.M. (1990) Die ersten Bauern Mitteleuropas. Eine archäobotanische Untersuchung zu Umwelt und Landwirtschaft der ältesten Bandkeramik. *Analecta Praehistorica Leidensia*, **23**, 1–156.

Küster, H. (2008) *Geschichte Des Waldes. Von der Urzeit bis zur Gegenwart*, 2nd edn, C.H. Beck, München.

Lüning, J. (2000) *Steinzeitliche Bauern in Deutschland – Die Landwirtschaft im Neolithikum*, Habelt, Bonn.

Plachter, H. (1997) Naturschutz im Abseits? *Biologie in unserer Zeit*, **27**, 306–316.

Pokorny, P. (2005) Role of man in the development of Holocene vegetation in Central Bohemia. *Preslia*, **77**, 113–128.

Reichholf, J. (2000) Ende der Aussperrung. *Der Spiegel*, **50**, 256–259.

Remmert, H. (1988) *Naturschutz. Ein Lesebuch nicht nur für Planer, Politiker und Polizisten, Publizisten und Juristen*, 1st edn, Springer-Verlag, Berlin/Heidelberg.

Schama, S. (1996) *Der Traum von der Wildnis. Natur als Imagination*, Kindler, Berlin.

Schulze-Hagen, K. (2005) Allmenden und ihr Vogelreichtum – Wandel von Landschaft, Landwirtschaft und Avifauna in den letzten 250 Jahren. *Charadrius*, **40**, 97–121.

Schumacher, H. (2012) Lieber licht als dicht: Ziegenmelker, Wendehals und Grauspecht. *Der Falke*, **59**, 21–25.

Segerer, A.H. (2012) Die physikalisch-geochemischen Grundlagen des planetaren Klimas und die Auswirkungen auf die öffentliche Diskussion – potenzielle Fallstricke für Ökofaunisten. *Nachrichtenblatt Bayerischer Entomologen*, **61**, 32–45.

Südbeck, P., Bauer, H.-G., Boschert, M., Boye, P., and Knief, W. (2007) *The Red List of Breeding Birds of Germany*, 4th edn, vol. **44**, Berichte zum Vogelschutz, pp. 23–81.

Tschimpke, O. (2008) Naturnahe Wälder mit heimischen Baumarten. *Naturschutz heute*, **4**, 3.

Ulrich, R. (2002) Vom Naturschutz vergessen: Die Lichtwaldarten. *Naturschutz im Saarland*, **3**, 22–25.

Van Swaay, C., Cuttelod, A., Collins, S., Maes, D., Lopez Munguira, M., Sasic, M., Settele, J., Verovnik, R., Verstrael, T., Warren, M., Wiemers, M., and Wynhof, I. (2010) *European Red List of Butterflies*, Publications Office of the European Union, Luxembourg.

von Vietinghoff-Riesch, A.F. (1936) *Naturschutz*, Eine nationalpolitische Kulturaufgabe, Neudamm.

Warren, M.S., Hill, J.K., Thomas, J.A., Asher, J., Fox, R., Huntley, B., Roy, D.B., Telfer, M.G., Jeffcoate, S., Harding, P., Jefcoate, G., Willis, S.G., Greatorex-Davies, J.N., Moss, D., and Thomas, C.D. (2001) Rapid responses of British butterflies to opposing forces of climate and habitat change. *Nature*, **414**, 65–69.

Weyergraf, B. (1987) in *Waldungen. Die Deutschen und ihr Wald* (ed B. Weyergraf), Akademie der Künste, Berlin, pp. 6–12.

Willis, K.J., Gillson, L., and Brncic, T.M. (2004) How 'virgin' is virgin rainforest? *Science*, **304**, 402–403.

Ziemann, B. (2011) Wald-Gewalt. Wald und Krieg, in *Unter Bäumen – Die Deutschen und der Wald* (eds U. Breymayer and B. Ulrich), Sandstein, Dresden.

8 The Apocalypse of Global Extinction of Species

8.1 How Many Species Live in the World?

No one really knows how many species there are in the world.

When we count species, we must decide what exactly we are counting: reproduction communities, descent communities or organisms that have similar group characteristics.

Integrative taxonomy tries to unite reproduction communities, descent communities and more groups (combined through other criteria) into one relevant species concept.

In the case of bacteria, however, the different groups cannot be combined into common workable units.

Can we even draw conclusions about how many species there are in the world from the known number of species on limited areas?

8.1.1 The Uncertainty About What a Species Actually Is

Contrary to all other allegations, conjecture and the frequent presentation of statistics, it should be clearly stated at this point that no one knows how many species there really are in the world. The estimates range from 3 to 100 million (Mora *et al.*, 2011). Around 1.2 million species on Earth have currently been described and catalogued, and there could be more, by several powers of 10, but no one knows. Coral reefs, the mud of the seabed and tropical soils with an unknown number of species are the richest habitats yet to be explored.

The number of species of organisms without cell nucleus (prokaryotes = bacteria and viruses) is a totally unknown quantity. However, a completely different factor plays a role here and not just unexplored regions of the world. Species are defined according to different grouping principles (Reydon, 2005; Kunz, 2012). A species is not a species, THE species does not exist. Instead, there are reproduction communities, descent communities or groups of organisms that share characteristics.

These groups are not completely congruent with one another, since two organisms can belong to the same descent community, but not to the same reproduction

Species Conservation in Managed Habitats: The Myth of a Pristine Nature, First Edition. Werner Kunz.

community. So when we count species, it comes down to what we mean. Do we mean reproduction, descent or characteristics communities? The term *species* is homonymous; this means that it is applied simultaneously to different things that exist in nature, which, however, are partially independent of one other and are not completely superimposed on one another (Reydon, 2004, 2005, 2008).

Nucleated organisms (eukaryotes) almost succeed in superimposing the different species concepts on one another and integrating them into one single synthetic structure. Although this integrative concept of species (Dayrat, 2005; Padial *et al.*, 2010; Schlick-Steiner *et al.*, 2010; Kunz, 2015) no longer describes a unit that really exists in nature, it is a practical instrument with which we can work successfully. The given number of about 950 000 animal species which have currently been described and catalogued worldwide (Mora *et al.*, 2011) is not the number of animal species that actually exists in nature; it is a man-made classification grid, one which integrates various group cohesions that occur in nature into units, but the groups that actually exist in nature are reproduction communities, descent communities or groups of organisms that share characteristics; and on Earth, each of these groups consists of a different number of animal species than 950 000. We must be aware that the integrative concept of species is indeed a reproducible and verifiable expression of the nature of biodiversity, but it is a unit of biodiversity (in this case of animals) that has been selected from pragmatic considerations; it does not mean that 950 000 animal species really exist in nature. Species numbers are determined in accordance with conventions; a pertinent international consensus does not always exist. Nevertheless, the integrative concept of species is now a common concept which is increasingly used in taxonomy. The integrative concept of species is beginning to increasingly replace the 'Barcode' unit of the concept of species, because the 'Barcode' type only acknowledges one single principle of biodiversity, namely the phylogenetic distance of two groups from one another. However, biological diversity in nature is also realised through other manifestations and not only through phylogenetic distances.

The illusions of those who firmly believed that only ONE species existed in nature (despite all the conflicts between the various concepts of species) have been shattered in the past few years. Ironically, it was in the case of the so well-known and much-studied birds that a drastic change in the weighting of values occurred, a change that was acknowledged in order to grant certain groups the status of species. Since many groups that had been previously considered as subordinate had now become species, the number of globally recognised bird species increased from approximately 10 000 to a total of 11 000, without any new species having been discovered anywhere in the world (del Hoyo and Collar, 2014).

The reason behind the increase in the number of species was a re-assessment of species barriers. Reproductive barriers and diagnosable differences in the morphology, ecology and behaviour of each species were weighed against one other according to new quantifiable benchmarks, and that led to a different kind of species classification (Tobias *et al.*, 2010). The classic divisions according to reproduction communities (Mayr, 1942) or descent communities (Hennig, 1966) revealed the problem that the species had to be segregated differently in some instances. Reproduction communities were not the same as descent communities.

If we grouped the members of the common redstart (*Phoenicurus phoenicurus*) according to the criteria of the species as a descent community, for example, there would be several species of common redstart in Central Europe. The barcode taxonomy would force a split of the 'common redstart' species into several separated species, because they are groups that are genetically sufficiently different from one another and are only slightly related to one another. This classification of species, however, cannot be accepted by any taxonomist who takes reproductive barriers and diagnosable phenotypic traits into consideration when segregating species. However, the phylogenetically diverse Central European common redstart groups are not distinguished by phenotypic characteristics from one another, and they are also linked reproductively.

One quite opposite example is the species pair the common swift (*Apus apus*) and the pallid swift (*Apus pallidus*). As a descent community, these two species are one and the same, with the result that the barcode taxonomist would have to unite them into one species. However, they are clearly distinguishable from one another, and they do not hybridise with one other, so these two swifts must be classified as two separate species by the followers of other species classification concepts. Authors who believe that species are groups of organisms that share characteristics and those authors who believe that species are reproductive communities come to the clear conclusion that the swifts are indeed two separate species.

These examples show that different species concepts do not lead to congruent units and that one single species concept (as in the case of barcode taxonomy) does not meet the many real existing group segregations. If we limit ourselves to only one species concept, other boundaries are disregarded, such as reproductive connections and diagnosable units.

This conflict is being discussed in a current reform of bird nomenclature. Together with 'BirdLife International', the successful publisher Lynx Edicions of Barcelona – the company which published the 'Handbook of the Birds of the World' – worked out a completely new system of bird nomenclature in its new work, 'Illustrated Checklist of the Birds of the World' (del Hoyo and Collar, 2014). This new classification is an integrative taxonomy. The principles of classification are based on the 'Tobias criteria' (Tobias *et al.*, 2010). This classification takes reproductive species barriers and diagnosable differences in morphology, ecology and behaviour into account. This new bird nomenclature is a total departure from the barcode taxonomy approach. The results significantly deviate from previously valid taxonomy and from barcode taxonomy. According to barcode taxonomy, we would have nearly 20 000 species of birds, and according to the previously valid taxonomy, there are around 10 000 species of birds worldwide, but according to the authors J. del Hoyo (Lynx Edicions) and N.J. Collar (BirdLife International), there are now approximately 1000 species more, meaning a total of 11 000 species (del Hoyo and Collar, 2014).

As shown here by the example of bird nomenclature, it is after all possible (albeit imperfectly and controversially) to unite the different groupings that occur in nature, namely reproduction, descent or groups that consist of organisms with similar characteristics. The units thus created match the maximum of group boundaries that exist in nature. However, these units are man-made artificial creations, which are based on subjective measures of value because man decides the degree of importance that is allocated to reproductive boundaries compared with diagnosable characteristics differences or phylogenetic differences. However, the practical benefit of integrative taxonomy is of great importance, because this taxonomy approach addresses the greatest possible number of group boundaries that exist in nature (Dayrat, 2005; Padial *et al.*, 2010; Schlick-Steiner *et al.*, 2010).

In the case of bacteria, however, the question is whether or not we can actually use an integrative taxonomy. Here, it is difficult to pool the various existing groups in nature into common units. The different units in bacteria that can be recognised as groups of mutual gene exchange, groups of common descent or groups of organisms that share characteristics cannot be superimposed on one another, as can be done in the case of eukaryotes. As a result of this, some authors arrive at the most frustrating conclusion, stating that: 'Species don't really mean anything in the bacterial world' (Hollrichter, 2007). Put that way, it is simply not true, because there are real, existing species in the world of bacteria. As in the case of the eukaryotes, these are groups of mutual gene exchange, groups of common descent or communities of organisms that share the same characteristics. However, it is not possible to combine these groups into practical common units; this is probably a difference between the prokaryotes and the eukaryotes.

8.1.2
Different Methods of Assessing the Species That Actually Live in the World

Roughly 950 000 animal species, 215 000 species of plants and about 64 000 other eukaryote species have been catalogued to date (Mora *et al.*, 2011). This amounts to a total of about 1.2 million species on Earth. Other authors arrive at slightly different numbers, namely 1.6 – 1.9 million species (May and Harvey, 2009; Wirth and Glaw, 2009). All authors, however, are confident that the majority of species has still to be discovered.

In his eighteenth-century work *Systema Naturae*, Linnaeus describes a total of only 549 species; almost 3300 times more species have been described today. The almost 10 000 bird species which have been globally determined by conventional classification methods number more species than the total number of known animals and plants that Linnaeus knew. The most species-rich order is that of the beetles with 400 000 classified species to date. However, this number should be taken with a large pinch of salt, because many of these species descriptions date back to one or just a few sightings. About 40% of the roughly 400 000 known species of beetle are described with only one site of discovery (Gibbs, 2002). This means that it is not possible to make any decisions about a large number of species, whether it is a new species or merely a variant of a previously known one.

Ongoing discoveries and the use of modern genetic test methods are currently resulting in a steadily increasing number of species – and the recent new descriptions are not just about microscopic life forms. Numerous species of all vertebrate classes are being discovered and described every year, and recent years have seen greater numbers of reptilian and amphibian species being discovered and described since the introduction of binomial nomenclature by Linnaeus 250 years ago. Within only two decades, the number of amphibian and reptilian species has increased by almost 40% (Wirth and Glaw, 2009). In just 15 years from late 1992 to late 2007, the number of amphibian species discovered has increased from 4533 to 6296. The reptiles have also achieved a new record, with at least 149 new species in 2007, and a flattening of the new description curves for these two animal classes is not in sight.

The mammals are also not as completely known as is often assumed. The number of mammalian species is constantly increasing. Research projects in locations that are difficult to access and the use of new genetic testing methods are responsible for the many new discoveries being made. Between mid-1992 and mid-2006, the number of mammalian species increased by 341 new species and came up to a total of 5421 species: but estimates forecast that more than 7000 species of mammals live on Earth. Even in the case of the (supposedly) well-researched primates, no less than 36 new species were described between 1992 and 2006.

At present, two opposing processes are converging, both of which are running at high speed. On the one hand, the number of species is decreasing due to the deforestation of the rainforests in tropical areas and the seamless, blanket-coverage of land used by agriculture in the economically prosperous countries of the Western world; but on the other hand, the number of known species is propagated by an increasing number of new discoveries. This leads to the paradoxical situation that the number of registered species is increasing, though the number of species in nature is on the decline.

Different authors have often tried various methods to calculate the actual number of species from the number of species known today. Since the degree of probability of the existence of still undiscovered species in different taxonomic groups varies, and varying degrees of discord occur concerning the splitting up of new species into different animal and plant species, we cannot use the speed of new discoveries as a basis for any accurate extrapolation about how many species there really are on Earth. Predictions about the total number of species have no firm foundation; they imply a spectrum ranging from 10 up to 100 million currently existing species (May and Harvey, 2009; Mora *et al.*, 2011).

Mora *et al.*, used a special procedure to estimate the global number of species in all the animal and plant groups and in the protists (Mora *et al.*, 2011). They started with the concept that the number of existing higher taxa (genera, families, orders and classes) is more fully known than the number of species belonging to these higher taxa, especially among the lesser-known plant and animal groups. In the well-researched places on Earth, therefore, if we know how many species the genera, families and so on, of a particular group of animals contains, then we can use this same group of animals to conclude the number of species in other places, if we know the number of genera, families and so on, there, but not (yet) the number of species.

Using this extrapolation method, the researchers calculated a global figure of 8.7 million species that existed on Earth. This number entails seven times more species than those currently described. The plants have the smallest gap between known and unknown species. Here, 216 000 (of the almost 300 000 calculated species) are already known and catalogued. The gap is much larger in the case of the animals, where the calculated actual number of species is around 7.8 million – but only around a million of these has been catalogued. In the case of fungi, the percentage of known species is even lower: only about 43 000 of an estimated 600 000 species are known.

If these figures are correct, this throws an alarming shadow of doubt over the Red Lists. The roughly 60 000 species described in the IUCN list ('International Union for Conservation of Nature and Natural Resources', see Chapter 4) do not even account for 1% of the total life on Earth – consequently, many unknown species are probably threatened with extinction. Many species may vanish before we even learn of their existence.

T.L. Erwin and J.C. Scott carried out a completely different method for determining the actual number of species in the tropical forest of Panama (Erwin and Scott, 1980; Erwin, 1982). They selected a medium-large tropical tree species for this, *Luehea seemannii*. The number of treetop beetle species on 19 of these trees (which were covered with some epiphytes and lianas) were counted for three seasons. Erwin and Scott found approximately 1200 beetle species on the 19 *L. seemannii* trees.

Of the beetle species examined, several were monophagous; that is, they could only feed and breed successfully on this tree species. Other species are polyphagous; that is, they could feed on other tree species and breed there. It is not known how many of the 1200 beetle species were really monophagous. Erwin and Scott assumed that the number was around 14%. This would mean that 14% of all the beetle species found on *L. seemannii* depended totally on this host plant. Erwin and Scott therefore classified 163 of the 1200 beetle species found on the 19 *L. seemannii* trees as host-specific. The remaining 86% (1037 species) were then species that were coincidentally present and which also lived on other trees.

An average of 70 different species of trees grows in 2.5 acres of tropical Panamanian forests; and each individual tree species harbours its own beetle species which is dependent on it. The subsequent conclusion of Erwin and Scott was that approximately 11 400 host-specific beetle species (163 × 70) therefore lived on 2.5 acres of tropical forest. Since 1037 polyphagous (non-host-specific) species were found on the original 19 *L. seemannii* trees, Erwin and Scott assumed that this number would not increase on the remaining tree species. Consequently, the treetops on 2.5 acres of Panamanian tropical forest harboured 12 447 beetle species (1037 + 11 400).

It is estimated that the beetles constitute about 40% of all arthropod species – so there are 31 120 arthropod species in the treetops of 2.5 acres of tropical rainforest in Panama. In addition to the species that live in the treetops, there are the species that occur on the forest floor, and these are different species. This is why Erwin and

Scott added 33% more species, arriving at the conclusion that 41 390 arthropod species lived in 2.5 acres of tropical rainforest in Panama.

There are 50 000 species of trees in the tropical regions of the Earth. If we take into account that most arthropod species are not particularly mobile and live in small distribution areas, then there should be 30 million tropical arthropod species, far more than the usual estimates. Today, the number of arthropod species is estimated at between 1.5 and 10 million worldwide.

Both of these estimates are hotly disputed. In the case of extrapolation of the number of species where the number of genera and families is known (Mora *et al.*, 2011), it is questionable whether or not the basic assumption is true that certain genera and families evince a similarly high number of species in all comparable areas. In the case of extrapolation of the number of species of monophagous beetles that live on a particular tree species at a particular location, on other tree species and throughout the entire tropical zones of the world (Erwin and Scott, 1980), it is questionable whether or not the assumption is true that various tropical tree species always have a similarly high number of species and that this number can then be applied to all tropical regions of the world.

8.2
Does a Global Extinction of Species Exist Today?

Germans tend to cover their land with an abundance of precautions and regulations that stifle some species conservation initiatives.

The population seems to have a need for horrific visions – and nature conservation propaganda has exploited this.

Although forest dieback in Central Europe did actually happen, it was never as catastrophically intensive as the conservationists have propagated.

In Germany, politicians and the various associations primarily see nature conservation as technical environmental protection; species protection often falls by the wayside.

Over the past 500 million years there have been five major mass extinctions on Earth, and three-quarters of all species fell victim to each of these.

The breathtaking speed at which the present plant and animal species are dying out is nourishing the fear that the sixth great species apocalypse is indeed starting on our Earth.

The cause of the sixth mass extinction is quite simply man, whose populations are expanding too rapidly.

When we see that the chaffinch (the most common bird in Central Europe) with a total of approximately 8 million breeding pairs in Germany barely attains the human population of one single metropolis, this alone should be a frightening wake-up call for us.

The current species extinction caused by humans did not start in the twentieth and twenty-first centuries.

It is estimated that man has destroyed around a fifth of all bird species since the last ice age.

The tropical regions of the Earth are those which are most affected by today's species extinction.

Basing its findings on the eradication of mammalian, avian and amphibian species in the past 100 years, the IUCN has calculated that the current loss of species is probably proceeding 100 times faster than it did during the five major mass extinctions of Earth's past.

Europe is the region least affected by the global extinction wave of species richness in recent and current decades.

Almost all the predictions about the upcoming sixth mass extinction are based on vulnerable extrapolations.

The need to preserve the biodiversity of tropical rainforests cannot be justified by stating that they are the reservoir for the medicines of the future.

The true level of extinction does not simply consist of the moa, dodo, great auk and the Californian condor. True species extinction is neither due to hunting nor to overfishing – habitat destruction is the real killer.

8.2.1
Germany – A Country of Caution and Precaution against Deplorable Circumstances That Occur Either Never or Only Rarely

Nature conservation and species protection in Germany are difficult issues. Money is not the problem, since Germany is one of the richest countries in the world, but the root cause is often social–psychological.

Grievances are often a matter of definition. In Germany, problems arise that are not seen as such in other countries. It is not only individuals that can develop neurotic traits, entire cultures can be affected. The Americans' obsession with hygiene, for example, is a trait which is lightly ridiculed in Europe, while the Germans tend to cover the land with an abundance of small and even tiny precautionary measures that smother many initiatives (Jessen, 2005).

The Germans also tend to feel continually guilty. A panic-driven fear of possibly being held responsible for anything that happens anywhere at all rules the roost. This leads to a fanaticism for precautions in German society, which in turn manifests itself in a myriad of regulations and laws; and these laws are constantly being passed to remedy abuses that have either occurred very rarely or not at all. Authorities and lawmakers develop a high degree of fantasy regarding what could happen to the citizens or what they themselves could perhaps do wrong. An exaggerated idea of health and order lies at the root of this. The French do not suffer from the American hygiene craze, nor are they familiar with the German fear of a lack of regulations.

Many Germans are convinced that cleanliness and hygiene in the fields would also benefit the animals. They do not believe that a little more 'sloppiness' could suit the animals much better. Many Germans want ruins, vacant woodsheds, rubbish heaps, piles of stones and wooden boards left lying around after construction to be removed as quickly as possible, because they see them as being eyesores in the environment. Many Germans simply cannot understand that these eyesores are actually habitats for many rare species of animals and that rubbish dumps and discarded heaps of slaughterhouse waste are valuable feeding areas for many animals in nature (Meyer, Eilers and Schnapper, 2003). The German population cannot accept this. Many Germans want their landscape as clean as their homes. Steep escarpments are valuable habitats for many insects and birds, but the risk that someone might fall from the escarpment making someone else liable for damages has resulted in legal regulations that escarpments should be levelled when their economic use has ended; and this process continually destroys irreplaceable habitats.

Doomsday scenarios get special attention. It is predicted that current polar bear populations will be reduced by two-thirds in just a few decades and could become extinct as a consequence of global warming. This fear is motivated by the feeding behaviour of polar bears in winter. They must travel over solid ice covers to find small ice-free holes where they can wait for seals to come up for air. If there were no solid ice, the polar bears could not approach individual holes and would have no chance of killing seals. Those who predict this scenario, however, disregard the fact that current global warming is nothing new.

Polar bears are a very young species. They only split off about 200 000 years ago to become a new species of Siberian brown bear (Talbot and Shields, 1996; Breiter, 2008). The Eemian warm period occurred during this time around 120 000 years ago, during which large areas of polar ice melted and the ice retreated much more then than it will in the next few decades. The polar bears survived these periods, and this tells us that they can change their hunting methods and eating habits. Those who forecast that the polar bear will die out in the foreseeable future ignore the adaptability of many species; their apparent morbid pleasure in horror scenarios comes to light instead (Schuh, 2008).

Even the notorious forest dieback, which provided many people with horrific visions in the 1980s and 1990s of last century, is an example of the needs of the population for visions of disaster; and the propaganda machine of conservation organisations has exploited such instances to the full (Ell and Luhmann, 1996). Forest dieback was never cured; it was only displaced by other imaginary or actual ailments. Although forest dieback in Central Europe did actually happen, it was not such a catastrophic level of destruction as the conservationists propagated. German environmental protection has exploited forest dieback for itself, and that was a big political mistake. Statements of impending doom such as, 'In 2000, we will no longer have trees along our streets' (Bölsche, 1982, p. 185) captured the attention of the public during the first phase of nature conservation. When the predicted catastrophe never even looked like happening, however, it put a large dent in the credibility of such statements and made many turn away from the idea of nature conservation. This is something that nature conservation has not been able to offset to this day. Swimming on the wave of disastrous predictions and events generates many short-term donations, but after a while it reveals itself as a kind of fashion trend; the effect then dulls, putting people off for the long term. The consequences of such exaggerations are potentially serious, because the general public becomes less willing to openly confront new issues, an attitude that is now taking shape with the current climate change, for instance.

At first glance, we get the impression that nature conservation is solidly established in Germany. We have Environment Ministers, nature conservation agencies and many more such bodies, but Environment Ministers primarily care about protecting the environment. They are more like 'Green Economy Ministers' – and species protection often falls by the wayside. In Germany, technical environment protection has a much higher priority level than the protection of species. This is also amply shown by the German political party 'The Greens'. Its members seem to care more about a clean environment than about the technical engineering of the landscape for the preservation of endangered species. Even many nature conservation organisations attach a greater value to technical environmental protection than to actual nature conservation (Zöckler, 2007). Windmills are easier to sell to party members and voters than dry grasslands and wetlands.

8.2.2
The Lifespan of Species, and the Five Major Mass Extinctions of Species in the Past 500 Million Years

The extinction of species is a 'normal occurrence' and a natural companion of evolution; but it is not just about the dinosaurs, the dodo or the Neanderthals. The great majority of all the species that have ever evolved on Earth have become extinct again (May, Lawton and Stork, 1995; May and Harvey, 2009). It is probable that not one of the species that existed 500 million years lives on Earth today. Most existing species are different from those which existed 10 million years ago. It is a fundamental principle of evolution that species continually disappear and new ones evolve.

We cannot estimate how long a species will live. It normally takes a considerable amount of time before new species evolve; in fact, we can assume a minimum of 2 million years in the case of many species (Avise, Walker and Johns, 1998). The rate at which speciation occurs is accelerated by greatly changing environmental conditions. The changes between ice ages and warm periods during the Pleistocene probably favoured speciation; however, most species living in the present Holocene also lived in the early days of the Pleistocene (over a million years ago). At constant external conditions, such as in the deep ocean, species can live for hundreds of millions of years without being split into new branched species or being replaced by new species.

The lifetime of a species also depends on its geographical distribution; species that are distributed across entire continents survive longer. Ubiquitists can live for more than 10 million years (May, Lawton and Stork, 1995). Endemic species are much more specialised and consequently have a lower life expectancy. Very young species such as the East African cichlids (family Cichlidae) are rare. In the East African lakes, hundreds of cichlid species evolved less than 200 000 years ago (Verheyen *et al.*, 2003). It is even thought that the cichlids living in Lake Victoria have a species age of only 15 000 years, since Lake Victoria was dry 15 000 years ago. This presumption, however, is very uncertain, since some animals may have survived in the lake's residual water bodies during the dry period (Verheyen *et al.*, 2003).

There are also components in some genomes (e.g. transposable elements), which can influence speciation, so it is hardly possible to specify the evolutionary rate of speciation in a generally applicable biological law.

The extinction of species is a natural process of evolution. Extinction is normally offset by the emergence of new species and the rate of species extinction is usually offset by speciation. The balance is equally weighted (roughly speaking); in terms of magnitude, the species that die out are roughly the same in number as those that newly evolve. However, this is a very approximate assumption, because we do not know how many species exist today, and we certainly do not know how many species existed a million years ago. If the balance is negative, we call it a period of species extinction, meaning that species die out much faster than speciation takes place. Palaeontologists speak about the mass extinction of species when the Earth loses more than three-quarters of its species in a geologically short interval. When a mass extinction like this has occurred, it takes a very long time until species richness has recovered.

Since the Cambrian period, there have been five such mass extinctions of species in the past 500 million years on Earth. More than three-quarters of all species have become extinct in five relatively short time intervals of only a few million years; and today we have to ask ourselves if we are now at the beginning of the sixth mass extinction. Authors differ in their opinions as to how this question should be answered (Gibbs, 2002; Barnosky *et al.*, 2011). It is difficult to find a clear answer to this. First of all, we would have to wait for a few hundred thousand years before we could make a final judgement, and second, this question generates a tendency to descend into the 'doom and gloom' scenario (see earlier discussion).

Barnosky *et al.* speak about a current sixth major mass extinction, because the species extinction of the last few centuries is proceeding at a faster pace than the extinction of species during the five major mass extinctions in the past history of the earth (measured by means of the rate of disappearance of fossils) (Barnosky *et al.*, 2011). However, we must remember that current species extinction cannot be measured quantitatively, because we do not know most of the species (discussed later). All these doom-laden forecasts are based on estimates. However, the fear is justified that today's rate of species extinction is so rapid that it can no longer be offset by the rate of speciation. The world today seems to be experiencing a net loss of species, one that is probably higher than we have hitherto experienced in many tens of millions of years (Baillie, Hilton-Taylor and Stuart, 2004).

The first major mass extinction occurred about 450 million years ago at the end of the Ordovician period, between the Cambrian and the Silurian. Over the course of about 10 million years, nearly 60% of all genera and 90% of all species died out on Earth. The suspected causes were alternating ice ages and interglacials, combined with dramatic variations in sea levels.

The second major mass extinction also took place in the Paleozoic, during the late Devonian some 350 million years ago. Thirty five per cent of genera and 75% of species died out and the Earth cooled down significantly.

The third mass extinction ended 250 million years ago during the Permian period at the end of the Paleozoic; it lasted about 3 million years, and destroyed more than 50% of all genera and more than 90% of all species. During this time, the entire class of arthropod trilobites died out. These were crab-like organisms that populated all the oceans with tens of thousands of species. They attained a size of up to 70 cm. During this period, the entire Earth warmed up, and opinions differ whether volcanic activity or a large meteorite impact were the reasons behind the climate change.

The fourth major mass extinction occurred in the Triassic period during the Mesozoic, some 200 million years ago. During this time, violent volcanic eruptions caused the Earth to once again become warmer, with elevated CO_2 content and the partial drying out of the oceans. Half of all genera and more than three-quarters of all species were destroyed.

The best-known mass extinction, the fifth, occurred 65 million years ago at the end of the Cretaceous period. Similarly high extinction rates took place, and the dinosaurs also fell victim to this mass extinction. It is believed that either the impact of a large meteorite on the Yucatan or strong volcanic eruptions resulted in the rapid and intensive cooling of the Earth.

In the past, the majority of species have become extinct five times in succession, but a minority of species survived each time and brought forth new species again, albeit over a very long period of time. New species were able to evolve, because previous mass extinctions created competition-free areas, making room for them. No one can answer the question as to whether or not man, who rules the earth today, would ever have arisen if he had not been given the chance to evolve by the past mass extinctions. The mammals could not prevail against the saurians, even at the end of the Cretaceous period.

8.2.3
The Extinction of Many Species by Prehistoric and Recent Humans

One question that interests almost everyone is whether or not our Earth today is seeing the beginning of the sixth great species apocalypse. Is Earth threatened today with a new mass extinction, as occurred at the end of the dinosaurs' era? The suspected breathtaking speed at which the present plant and animal species seem to be disappearing nourishes this fear (Barnosky *et al.*, 2011). It is feared that the evolutionary emergence of new species can no longer match the pace at which species are becoming extinct today. However, there are contrary opinions that warn of the consequences of our doom and gloom reaction (Gibbs, 2002).

Those who advocate that the sixth major mass extinction has already begun rely on the Red Lists, published by the International Union for Conservation of Nature (IUCN) (Baillie, Hilton-Taylor and Stuart, 2004). The authors compare the speed of the current species extinction with the palaeontological data of earlier mass extinctions, and consequently believe that the speed at which species are currently vanishing is significantly higher than the speed at which species disappeared during the previous disasters, such as can be ascertained from the fossil finds (Barnosky *et al.*, 2011). Many of the species still living today will be gone from the Earth after a few more millennia. These sound like very long periods of time; but in comparison to the known major mass extinctions of the past, they are geologically – and frighteningly – short. The extinction of species caused by man took place within a few hundred to a thousand years, and the extinction process is continuing. Basing its calculations on the eradication of mammalian, avian and amphibian species in the past 100 years, the IUCN has estimated that the current loss of biodiversity is occurring with unusual speed (Baillie, Hilton-Taylor and Stuart, 2004), because the last large global extinction in Earth's history did not happen abruptly, it took place over a period of up to millions of years (Weber, 2011). The causes were varied and not identical for all extinction events. If we assume that our present knowledge of some well-known groups (such as mammals and birds) is representative, the resulting balance shows us that the extinction rate today is more than 100 times more than the average extinction rate of Earth's past history, going by the beginning of fossil finds that date back to more than half a billion years ago (Hoffmann *et al.*, 2010). The rate at which species are currently vanishing appears to exceed the average by a great margin.

While earlier species extinctions were caused by cosmic events and volcanic eruptions, it is now man who is destroying species. The question is whether man could also halt the disaster. Sceptics doubt this, because all previous international conferences on the conservation of species richness have shown that man (entirely in the Darwinian sense) thinks only of his own survival and that he is not ready to scale back his own prosperity in favour of the conservation of biodiversity. In 'man', evolution has produced a species that will destroy the majority of the other species.

It is a sign of human short-sightedness that we destroy a diversity of life about which we know too little. In contrast to the current environmental damage, the extinction of animals and plants cannot be undone. The main problem is that man occurs too frequently in relation to the animal species. When we see that the chaffinch (the most common bird in Central Europe) with a total of approximately 8 million breeding pairs in Germany barely matches the human population of one single metropolis on Earth (Südbeck *et al.*, 2007), this statistic alone should act as a frightening wake-up call for us. The correct balance no longer exists.

Elsewhere in this book, a simple calculation was used to present the almost astronomical extent at which organisms proliferate when no factors deter their reproduction (see Section 3.7.4). During its lifetime, a single lapwing pair would produce nearly 30 million descendants in the course of 15 generations (i.e. children, grandchildren, great-grandchildren, etc.), if the great majority of its descendants did not die through diseases or enemies relatively quickly after birth. In one of his famous phrases, Darwin pointed out that organisms naturally multiply so intensively that the earth would soon be covered by the progeny of a single pair, if the vast majority of these were not destroyed before they themselves could propagate (Darwin, 1859).

Man breaks this fundamental biological law by not allowing his descendants to die before they can reproduce (for the most part). Perhaps there is no way to appreciably prevent the current species extinction. When confronted with the population explosion of his own species, man is presumably almost as powerless as he is against volcanic eruptions or meteorite impacts, although the present increase in the human population is only partially attributable to a force of nature to which we must be passively exposed without any possibilities of intervention. Here, a role is also played by archaic Old Testament religious commandments, the fatal destructive consequence of which remains unrecognised, with the result that they are not reformed, but instead are practised in many religions today.

The man-made extinction of species is not a phenomenon that started in the twentieth and twenty-first centuries; it harks back to prehistoric times. Evidence that man is to blame for the extinction of most species exists – and it dates back to the Bronze Age, 3500 years ago (Baillie, Hilton-Taylor and Stuart, 2004). Whenever man entered new countries and continents in prehistoric times, he first destroyed all the species that could not escape in time, did not hide themselves well enough or did not put up enough of a fight. This applies to the first settlements in America, Australia and Madagascar, but it is particularly true of the invasion of the first people into the Oceanic islands. Even in prehistoric times, man settled the tropical Pacific islands and eradicated more than 2000 species of birds there, many of which were flightless rails. It is estimated that man has destroyed around a fifth of all bird species since the last ice age (Baillie, Hilton-Taylor and Stuart, 2004).

The IUCN list includes 784 historically extinct species (dating back to 1500) the names of which are known (Baillie, Hilton-Taylor and Stuart, 2004). Most of them died on islands in the eighteenth and nineteenth centuries when the Europeans invaded their domains, bringing with them cats, dogs and rats on their ships. Since 97% of all species known to science today are not included in the IUCN Red List, and the extinction of most species took place in more or less unnoticed manner, the number of 784 documented extinction cases lies far below the actual number. We have no idea how many species have actually become extinct in recent centuries and millennia, but the number is probably in the region of several hundred thousand.

A more concrete idea of the extent of the current extinction has only existed since 1988, when the IUCN recorded all species of birds at risk. Since then, there has been a continuous decrease in species. The same goes for amphibians, but in their case the rate of decline is significantly larger. A comparison of the latest IUCN lists of 2000 and 2004 shows that 18 registered species have become extinct during these 4 years.

Most vertebrate species killed off by humans in past centuries lived in small, isolated habitats – in other words, they were endemic species. These include the dodo (a relative of the pigeon) on the islands of Mauritius and Reunion in the Indian Ocean (species extinction 1690), the great auk in the North Atlantic (species extinction 1844) and the steller sea cow on the coasts of Behring Island (species extinction 1768). One of the last examples of the disappearance of an endemic bird is the Atitlan grebe (*Podilymbus gigas*). This bird only inhabited one single location in the world, namely the Lago de Atitlan in south-western Guatemala (Baillie, Hilton-Taylor and Stuart, 2004). It became extinct in 1989.

In current decades, the exterminations have shifted away from the islands to the mainland, because the islands have already been cleared of most species that were inflexible and unable to escape from humans. Overall, the extinction of birds and mammals seems to be on the decrease. However, a comparison of all species between then and now cannot be made, simply because we have hardly any data (from past centuries) on amphibians and fish and almost none on invertebrates. In the age of the current destruction of tropical rainforests, we have no concrete ideas about how many of the small and very small species are being exterminated by humans on a daily basis. The extinction of a large variety of tiny insects in the tropical rainforests is hardly noticed, because humans perceive these species as being insignificant and they are largely unknown. In addition to the tropical rainforests, the main core centres for this unknown diversity are the oceans, especially the coral reefs and the mud of the seabed (Mora *et al.*, 2011).

Several amphibian species have become extinct in recent decades. It is a fact that the amphibians (at least among the vertebrates) are affected most by the current species extinction. The future existence of 32% of all amphibian species worldwide is threatened. Mammals and birds are not as sensitive as amphibians, the survival of which depends on their ability to breathe through their skins and the quality of the water. Another reason for the high level of risk to amphibians seems to lie in the (often) very small distribution areas of these less mobile animals. Endemic species are found among amphibians to a greater extent than among mammals and birds. The majority of amphibians have no major geographical distribution areas and they have specifically adapted to certain habitats. Amphibian species are often only found in small numbers in very small areas.

Chytridiomycosis is a deadly illness caused by fungi of the genus *Batrachochytrium* – and it is killing amphibians worldwide (Baillie, Hilton-Taylor and Stuart, 2004). A whole family of amphibians, the stomach-brooding frogs (Rheobatrachidae) of Australia, has already become extinct because of this mycosis. The northern gastric brooding frog (*Rheobatrachus vitellinus*) was still common in its distribution area in Queensland, Australia in January 1985, but it had died out by May (species extinction within just 5 months) (Baillie, Hilton-Taylor and Stuart, 2004). A similar fate befell the jambato toad (*Atelopus ignescens*) in the mountain rainforests of Ecuador. This toad was very common, even in the 1960s and 1970s, but its numbers declined between 1984 and 1988, when the last individual died as the result of the chytrid fungus infection.

8.2.4
Is the Earth Suffering Its Sixth Mass Extinction of Species Today?

Faced by so many human-induced species exterminations triggered by habitat destruction and the global spread of infectious diseases, the question arises: Is our Earth sliding into a sixth era of mass extinction? Researchers have on several occasions expressed the fear that this process has already begun. Many scientists believe that the sixth major mass extinction is based on facts and they rely on the Red Lists, published by the IUCN, for their evidence (Baillie, Hilton-Taylor and Stuart, 2004).

Overfishing pushes many marine animals to the brink of extinction. Species are losing their homes everywhere in the world where primeval forests are being cut down – the Orang-utans in Borneo are just one prominent example of this. Every fourth mammalian species is classified as threatened, while in the case of amphibians this number is as high as 40%. The IUCN Red List of 2004 includes assessments for 38 047 species, of which 41% (15 589) are threatened (listed as critically endangered, endangered or vulnerable) (Baillie, Hilton-Taylor and Stuart, 2004).

The most transparent data are available for vertebrates. Here, the extinction rate was estimated for the first time globally for 25 780 species (Hoffmann *et al.*, 2010). The result was that 20% of vertebrate species are currently on the IUCN Red List. They had to be classified as 'vulnerable', 'endangered' or even 'critically endangered'. These are

- 25% of mammals
- 13% of birds
- 22% of reptiles
- 41% of amphibians and
- 33% of cartilaginous fish and 15% of bony fish.

However, in considering these numbers, we must bear in mind that there are also species that have not only recently slipped into a risk category due to habitat changes. These are species which were always rare and endangered from the outset, even if man had not created the threat to them by his intervention.

Europe is least affected by the global extinction wave of species richness in our recent decades (Baillie, Hilton-Taylor and Stuart, 2004), probably because hardly any unspoiled nature still exists in the region and species cannot therefore be eradicated by the destruction of primary habitats. The ice ages and postglacial man already eradicated many species in Europe in earlier times. The extinction of species is currently affecting other parts of the world where human populations are increasing in number and man is destroying nature at a rapid pace; but these processes have already taken place in Europe hundreds or even thousands of years ago.

Earth's tropical areas and the subtropical regions with humid forests, for example, Latin America, West Africa and South East Asia are most affected by today's species extinction. The majority of the world's terrestrial and freshwater species live in these regions, and this is where most species are endangered worldwide; these are also the areas with the highest percentages of endangered species compared to the total numbers of the species. The various countries of the world are consequently not affected equally by the extinction of species. There are certain countries where the number of critically endangered species is particularly great. The major drawback here is that it is these countries that are the poorest in the world. The rich countries of Europe, which can raise the money for species protection, have the lowest proportion of globally threatened species. So far, there is no evidence that the destruction of habitats in tropical countries is slowing down, so the extinction of species will continue. It is hard for Western affluent societies to argue that poor countries should now stop destroying nature, when it is blatantly obvious that these Western countries owe part of their prosperity to such past interventions in nature.

If humanity does not want to be the species that is responsible for a great mass extinction, funds for biodiversity protection must be provided and the legislation amended accordingly. So far, however, the balance is negative: there are some successes, but the target set by the European Community (EU) to stop the loss of biodiversity by 2010 has clearly been missed. When mass extinction actually occurs between the next couple of 100 years to a 1000 years, it will take an extremely long time for biodiversity to recover its strength. The period of recovery takes place over millions of years and probably presupposes that one certain species becomes extinct: man, *Homo sapiens.*

8.2.5
Conflicting Figures and Scepticism Regarding Today's Mass Extinction of Species

The major flaw inherent in all forecasts regarding the alleged, current sixth major extinction of species is the extremely uncertain statistics. The IUCN documents the extinction of 784 species in historical times (after 1500); but no one knows how many species there really are and how many die each year. Taxonomists have so far identified nearly 2 million species, but biologists know hardly anything about most of these. Neither do we know whether or not the comparison with the extinction rates of Earth's past is legitimised, since it is a comparison based on fossil discoveries. No one knows, however, how many of the currently existing mammals and birds will show up later in fossilised form.

The widest gaps in our knowledge are actually those that apply to the creatures which possess the greatest species numbers in the animal kingdom, namely insects, nematodes and crustaceans – but no particulars can be furnished for the insects, because only 1000th of all registered insect species are listed in the IUCN Red List of Threatened Species (May and Harvey, 2009); so we do not know how many species of insects there are, nor do we know how many of the known species are at risk. All we do know is that the number is almost unimaginably great. There are approximately 400 000 known species of beetles, but around 40% of these are found on only one exploration site; but if we have no idea about the distribution area of a species, that species' extinction rate can hardly be determined with any degree of accuracy. When the passenger pigeon became extinct in North America in 1914, it took two species of parasitic lice with it to pigeon paradise; but these species have still to appear in the lists of the IUCN (Gibbs, 2002). If we look for precise evidence to show that dozens, hundreds or thousands of species disappear each year, we will find none. This may of course be due to the insufficient database, but we cannot ignore the possibility that the situation may be not as devastating as many fear.

Almost all predictions about the upcoming sixth mass extinction are based on extrapolations which are vulnerable. Barnosky *et al.* make the following predictions on the basis of possible forthcoming scenarios (Barnosky *et al.*, 2011). They assume (and this is the vulnerable part) that other species (which are not yet categorised in risk levels, because their populations are still stable) are gradually disappearing, although not all at the same speed. Here are two of the predicted scenarios:

1) If all of the species classified by the IUCN as 'critically endangered' were to die out in the next 100 years, the concentrated mass extinctions would impact on other species (which are yet categorised in lower risk levels) in the next 900–2300 years.
2) If all of the species classified by the IUCN in the weaker categories 'EN' ('endangered') or 'VU' ('vulnerable') were also to die out in the next 100 years, the concentrated mass extinctions would impact fully on other species in the next 240–540 years.

These extrapolations about impending mass extinction are founded on conditions which cannot be verified (May, Lawton and Stork, 1995; May and Harvey, 2009). They are primarily based on the assumption that the species which are at little or no risk today will be at risk in the future.

Furthermore, drawing conclusions from our knowledge on vertebrates and applying these conclusions to lower invertebrates can lead to incorrect results, because the life spans of the different species are not the same (Gibbs, 2002). We possess the most knowledge about the extinction of the mammals (apart from the birds and amphibians), and it is the mammals which appear to have relatively short life spans as species when compared with the lower animals.

Neither is it easy to predict the extinction of a species. Some species survive longer than expected. A species may not become extinct as quickly as expected, because it often finds remote refuges where it can survive even after a long time of living on the brink. A long interval of time often elapses between the decline of a species and its definitive disappearance, and recovery periods even exist for species which had previously almost been given up as extinct. There is a lot of scope for interpretation here. We often forget that there are many species which are 'naturally' rare from the outset, and not every one of these will become extinct in the foreseeable future.

A certain amount of care is lacking when we deal with such numbers. The well-intentioned bad tidings lose their intended effect if their credibility is built on a house of cards. There are a number of populist writers who make apocalyptic prophecies, which are eagerly ingested by the media. In his 1979 book 'The Sinking Ark', Norman Myers estimated that 1 million species would become extinct by the year 2000 (Myers, 1979; Myers, 1997). Predictions that no wild animals will be left in the world in just a few decades are most certainly wrong. We also tend to forget that there is a whole range of plants and animals that benefits from the destruction of the environment – the Central Europe of past centuries is a shining example of this.

Probably the best known and high-profile author who has also published a number of internationally acclaimed books about the mass extinction of species is E.O. Wilson. In the 1990s, he stated that the world was losing 74 species every day (Wilson, 1992). This alarming figure has been widely quoted, but it is based on quantitative extrapolations that have no foundations whatsoever. If it were correct, we would have lost nearly a million species in only the past 20 years. If we distribute this 1 million throughout different classes of animals, a few dozen bird and mammalian species would have died out in the last 20 years. According to the IUCN, however, only one bird and one mammalian species have become extinct in the wild (but both have thankfully survived in zoos) (Baillie, Hilton-Taylor and Stuart, 2004). If we simply extrapolate Wilson's extinction numbers to birds and mammals, there is a significant discrepancy between the generally assumed millions of mass deaths and the cases already proven. When scientists evangelise a little too much, it can get really dangerous. Partisanship easily undermines the impression of neutrality and objectivity.

The justifications of many nature conservationists are also greatly exaggerated, when they state that the species richness of the Earth contains an enormous genetic reservoir that man will need in the future. They assume that the tropical plants and animals are a potential source of new medicines because they contain the raw material for many biotechnological products. However, it is more likely that the future of new medicines does not lie in newly discovered plants or microorganisms, but in the bioengineering laboratory, where targeted 'protein engineering' is used to combine reactive side groups together with structural protein molecules, achieving a predicted reaction against pathogens. Several companies (such as Merck) abandoned their plans to win new pharmaceutical substances from the plants of the rainforest towards the end of the last century already. Today most of the known cultures of the most important crops are stored in seed banks. Given these circumstances, national parks are hardly the cheapest insurance against a possible disaster in agriculture.

The fears of many nature conservationists that the structures of entire ecosystems would collapse if just a few 'pillar' species were to become extinct are more or less irrelevant. Recent studies have shown that biodiversity in most ecosystems has a certain degree of redundancy and that the functionality of the system is not dependent on the existence of a few specific species.

It is also very important to realise that there is a very different reason why the extinctions of many species of the past cannot be transferred to the present day. There are other reasons behind the current species extinctions than the destruction of species caused by the conquest of the South Pacific islands centuries ago, where species were destroyed because they were defenceless and as such they were killed out of hand, shot or eaten. The dominant, current, main reason for biodiversity loss is, however, quite different – it is the loss of habitats. We cannot simply take the exterminations perpetrated centuries ago and the current loss of species, add them together and derive future forecasts from the result. The trapping and killing of animals is a fundamentally different danger, one that has other dimensions than habitat destruction (Baillie, Hilton-Taylor and Stuart, 2004).

On top of that, the killing of animals can be prevented in principle, but the loss of habitats as a result of population growth cannot be prevented for the most part – and it cannot be undone. In this respect, the clock cannot be turned back.

The true level of extinction does not simply consist of the moa, dodo, great auk and the Californian condor. True species extinction is neither due to hunting nor to overfishing; the real killer is habitat destruction. In earlier times, the species were killed by man, but today we take their habitats away. Furthermore, there is a great difference between recent species decline in the rainforests and species decline in Europe: in South East Asia, Africa and Latin America, species die because their (largely) still-primary habitats are being destroyed by human overpopulation and the intrusion of humans into areas that previously saw little or no utilisation. The situation is quite different in Europe: in Central Europe, there is no destruction of pristine habitats; here species are disappearing because agricultural land is being exploited down to the last square yard by means of our 'perfect crop yield technology', which in turn is generated by the demand for cheap products.

Unfortunately, these important differences are not being sufficiently emphasised in the current national species protection policies in Central Europe (see Section 3.4). Political measures can be used to prevent direct killing through hunting or trapping; but it is much harder to use these measures to prevent habitat destruction. As a result, politicians (and unfortunately the nature conservation associations) adorn themselves to an unmerited degree with their alleged successes in nature conservation. However, these successes do not really evince any positive, upward trend because they are only based on firearms and hunting prohibitions. We experienced this in Germany in the 1980s of the last century, when picking wild flowers, capturing butterflies and entering nature reserves were prohibited. This was hailed as a great success of nature conservation; but while the accolades were showering down on the nature conservation associations, habitat destruction and species decline were continuing on an almost unhindered scale. Central Europe in the last few decades has seen positive population trends for raptors, herons, ducks, geese, crows and beavers, as well as the return of the moose and wolves. These trends, however, are due to controlled shooting bans – they mostly cannot be interpreted as being signs of habitat improvement. The current recovery of these species' populations has almost nothing to do with the improvement of their living conditions – and it is simply wrong to proclaim this as an indication of successful nature conservation.

References

Avise, J.C., Walker, D., and Johns, G.C. (1998) Speciation durations and Pleistocene effects on vertebrate phylogeography. *Proceedings of the Royal Society B: Biological Sciences*, **265**, 1707–1712.

Baillie, J.E.M., Hilton-Taylor, C., and Stuart, S.N. (2004) *2004 IUCN Red List of Threatened Species – A Global Species Assessment*, International Union for Conservation of Nature and Natural Resources (IUCN), Gland and Cambridge.

Barnosky, A.D., Swartz, B., Quental, T.B., Marshall, C., McGuire, J.L., Lindsey, E.L., Maguire, K.C., Mersey, B., and Ferrer, E.A.

(2011) Has the Earth's sixth mass extinction already arrived? *Nature*, **471**, 51–57.

Bölsche, J. (1982) *Natur Ohne Schutz – Neue Ökostrategien Gegen Die Umweltzerstörung*, Spiegel-Verlag, Hamburg.

Breiter, M. (2008) Überlebensspezialist der Arktis – Der Eisbär: Postertier mit ungewisser Zukunft. *Biologie in unserer Zeit*, **38**, 86–93.

Darwin, C. (1859) *On the Origin of Species by Means of Natural Selection or the Preservation of Favoured Races in the Struggle for Life*, Murray, London.

Dayrat, B. (2005) Towards integrative taxonomy. *Biological Journal of the Linnean Society*, **85**, 407–415.

Ell, R. and Luhmann, H.-J. (1996) Von den Schwierigkeiten der Entdeckung des Waldsterbens in Deutschland. Man sieht nur, was man versteht oder: Schäden ohne Ursache und Ursachen ohne Schäden. *Forstarchiv*, **67**, 103–107.

Erwin, T.L. 1982. Tropical forests: their richness in Coleoptera and other Arthropod species. *Coleopterists Bulletin*, **36**, 74–75.

Erwin, T.L. and Scott, J.C. (1980) Seasonal and size patterns, trophic structure, and richness of Coleopteran the tropical arboreal ecosystem: the fauna of the tree *Luehea seemannii* Triana in the Canal Zone of Panama. *Coleopterists Bulletin*, **34**, 305–322.

Gibbs, W.W. (2002) Gibt es ein unsichtbares Artensterben? *Spektrum der Wissenschaft*, **1**, 62–71.

Hennig, W. (1966) *Phylogenetic Systematics*, University of Illinois Press, Urbana, IL.

Hoffmann, M. *et al.* (2010) The impact of conservation on the status of the world's vertebrates. *Science*, **330**, 1503–1509.

Hollrichter, K. (2007) Species don't really mean anything in the bacterial world. *Lab Times*, **2**, 22–25.

del Hoyo, J. and Collar, N.J. (2014) *Illustrated Checklist of the Birds of the World – Non-Passerines*, Lynx Edicions, Barcelona.

Jessen, J. (2005) Im Land der Vorsorger – Hypochondrie ist auch eine Frage der kulturellen Befindlichkeit. Jede Nation leidet auf ihre Weise, die deutsche besonders intensiv. *Die Zeit*, **20**, 39.

Kunz, W. (2012) *Do Species Exist? – Principles of Taxonomic Classification*, Wiley-VCH, Weinheim.

Kunz, W. (2015) Integrative Art oder Barcode-Art – welche ist die richtige Art? – Ergänzende Bemerkung zum Artikel "Integrative Taxonomie", ein neuer Ansatz, Fragen der Biosystematik zu beantworten von H. Rajaei [in German] Integrative species or barcode-species – What is the correct species? – Additional note to the article "Integrative taxonomy", a new approach to answer the questions in the biosystematics by H. Rajaei). *Entomologie heute*, **27**, 177–181.

May, R.M. and Harvey, P.H. (2009) Species uncertainties. *Science*, **323**, 687.

May, R.M., Lawton, R.H., and Stork, N.E. (1995) in *Extinction Rates* (eds J.H. Lawton and R.M. May), Oxford University Press, Oxford, pp. 1–24.

Mayr, E. (1942) *Systematics and the Origin of Species*, Columbia University Press, New York.

Meyer, W., Eilers, G., and Schnapper, A. (2003) *Müll Als Nahrungsquelle Für Vögel Und Säugetiere*, Westarp Wissenschaften, Hohenwarsleben.

Mora, C., Tittensor, D.P., Adl, S., Simpson, A.G.B., and Worm, A.G.B. (2011) How many species are there on earth and in the ocean? *PLoS Biology*, **9** (8), e1001127.

Myers, N. (1979) *The Sinking Ark: A New Look at the Problem of Disappearing Species*, Pergamon Press, London.

Myers, N. (1997) Mass extinction and evolution. *Science*, **278**, 598.

Padial, J.M., Miralles, A., De la Riva, I., and Vences, M. (2010) The integrative future of taxonomy. *Frontiers in Zoology*, **7**, 16.

Reydon, T.A.C. (2004) Why does the species problem still persist? *BioEssays*, **26**, 305.

Reydon, T.A.C. (2005) On the nature of the species problem and the four meanings of 'species'. *Studies in History and Philosophy of Biological and Biomedical Sciences*, **36**, 135–158.

Reydon, T.A.C. (2008) Species in three and four dimensions. *Synthese*, **164**, 161–184.

Schlick-Steiner, B.C., Steiner, F.M., Seifert, B., Stauffer, C., Christian, E. & Crozier, R.H. 2010. Integrative taxonomy: a multisource approach to exploring biodiversity. *Annual Review of Entomology*, **55**, 421–438.

Schuh, H. (2008) Klimawandel – Seeräuber in Eisnot. *Die Zeit*, **17**, 35–36.

Südbeck, P., Bauer, H.-G., Boschert, M., Boye, P., and Knief, W. (2007) *The Red List of Breeding Birds of Germany*, 4th edn, vol. **44**, Berichte zum Vogelschutz, pp. 23–81.

Talbot, S.L. and Shields, G.F. (1996) Phylogeography of brown bears (*Ursus arctos*) of Alaska and paraphyly within the Ursidae. *Molecular Phylogenetics and Evolution*, **5**, 477–494.

Tobias, J.A., Seddon, N., Spottiswoode, C.N., Pilgrim, J.D., Fishpool, L.D.C., and Collar, N.J. (2010) Quantitative criteria for species delimitation. *Ibis*, **152**, 724–746.

Verheyen, E., Salzburger, W., Snoeks, J., and Meyer, A. (2003) Origin of the superflock of cichlid fishes from Lake Victoria, East Africa. *Science*, **300**, 325–329.

Weber, N. (2011) Massensterben im Blitztempo – Menschheit beschleunigt neue Arten-Apokalypse. Spiegel-ONLINE, 3 March 2011.

Wilson, E.O. (1992) *The Diversity of Life*, Harvard University Press, Boston, MA.

Wirth, R. and Glaw, F. (2009) Globales Unwissen contra Biodiversität: Gedanken zum Schutz der biologischen Vielfalt. *Biologie in unserer Zeit*, **39**, 42–48.

Zöckler, C. (2007) Waldsterben und Wiesen-Limikolen: Naturschutzbewegung in Deutschland zwischen Katastrophen und technischem Umweltschutz. *Der Falke*, **54**, 257–263.

Animal Index

Species Conservation in Managed Habitats: The Myth of a Pristine Nature, First Edition. Werner Kunz.

Subject Index

Colour Plates

Plate 1 The corn bunting (right) and the meadow pipit (left) have disappeared from many parts of western Central Europe, where they were very common breeding birds a few decades ago. They find their food on the ground, which must have only sparse growth and open areas with warm soil (a). Today's fertilised meadows are fully covered with high, dense grass (b), and these two species, formerly typical birds of the agricultural landscape, simply cannot find food on such ground. Most insects can no longer live in the cold environment at the bottom of a dense and wet grass cover, and the few insects that do still exist there cannot be seen and caught by the birds under the thick grass cover. Today's agricultural land in Central Europe is lost to species such as the corn bunting and the meadow pipit. It is unrealistic to try and deploy 'organic farming' to return meadows and pastures to their former meagre crop yields of earlier centuries. The former agricultural and meadow areas which existed as important bird areas for centuries have meanwhile lost their relevance as breeding grounds for these species. Open-cast mining excavations, airfields and military and industrial sites have replaced them. The top photo shows a meagre path edge at a location where earthwork has been freshly heaped up after brown coal excavation (a: Königshovener Höhe, Grevenbroich, Germany).

(a)

(b)

Plate 2 The swallowtail is a typical Central European butterfly, which has also become rare. For its morning warm-up (b, left) and its egg-laying (b, right), the swallowtail needs bare surfaces on open grass terrain. It can only deposit its eggs on young carrot plants on warm ground (b, right). Today's arable land, meadows and pastures in over-fertilised Central Europe are in many cases no longer suitable for the propagation of this butterfly. The swallowtail has sought new habitats which have taken the place of the formerly species-rich Central European agricultural landscape, such as brown coal mining excavations (a), military training areas, airfields, industrial wastelands, highway embankments and disused railway lines. As far as species protection is concerned, these artificial habitats have even been outstripping several nature reserves (Königshovener Höhe, Grevenbroich, Germany).

(a)

(b)

Plate 3 Many butterflies in Central Europe are dependent on the warming of the earth. These include the common copper (a), the silver-studded blue (left) and the small heath (right). Only open ground with sand and stones heats up to a sufficient extent in sunlight (a). Today's meadows and other grasslands in Central Europe do not have enough sandy and stony open spaces. Instead, dense, high grass growth almost completely covers the entire ground (b), so no sunlight can penetrate and the ground stays moist and cool. These three butterflies, which were common in many locations 60 years ago, have now abandoned several of their former areas of distribution in western Central Europe.

(a)

(b)

Plate 4 Our human longing for original, untouched nature often blinds us to the kind of Central European habitats that are actually home to many rare species. For example, here on a small brownfield site in the indus trial area, the six-spot burnet (left) and the cinnabar moth (imago and caterpillar right) live together with many other species not often seen in Germany nowadays. This raises the issue of how intensively conservation organisations need to rethink their ideology, turning their minds to technically engineered habitats when it comes to saving our rare species.

Plate 5 Bare ground areas on steep escarpments have become rare today because flood damage has been contained and sunken defiles have disappeared. In Central Europe today, valuable habitats like this are almost exclusively created through the economic exploitation of certain areas in the countryside. As a rule, the steep escarpments are quickly machine-levelled, sown with crops and even afforested when their exploitation (such as mining) stops. Escarpments are referred to in public as 'ugly landscapes' that have to be 'beautified' and walkers are assiduously warned that there is a danger of falling there. This is why escarpments are in most cases quickly destroyed for aesthetic and legal reasons as soon as they are no longer economically useful, and this process is euphoniously called *re-naturalisation*, although it has nothing to do with nature. Species conservationists know that the term *re-naturalisation* is in itself a paradox. As a result of this 're-naturalisation', highly valuable habitats for bees, butterflies and other insects and birds are being destroyed. Photo: male and female wall brown butterflies (Königshovener Höhe, Grevenbroich, Germany).

Plate 6 The wheatear (male left, female right) is on the Red List in several German states and is classified as a critically endangered species. The wheatear does not need any nature conservation or nature protection measures to prevent its disappearance, it simply needs stones. In past centuries, stones were available almost everywhere for the bird, because the land was relatively bare in contrast to today's overgrown landscapes. The slopes of the hills and the areas between the fields and meadows used to be only sparsely covered with grasses, bushes and trees, because wood was collected and the topsoil was removed and used as litter for the cattle shed in many locations. Numerous areas were also intensively grazed by sheep and goats, and the vegetation was also not so lush, because nitrogen did not rain down from above the way it does today. Nowadays, all these areas and the piles of stones are overgrown. Nature conservation will not save the wheatear; only man with machines can achieve this. Hardly any species can be so easily encouraged and saved from extinction by technical means than the wheatear. We only have to deposit large cairns in the countryside and keep them free of vegetation at regular intervals. It is not the lack of financing that prevents such species protection measures from being carried out, because the creation of a pile of stones (as can be seen in the centre of the photo) is considerably cheaper than the creation of a national park. The problem is that a species protection measure like this has nothing to do with nature conservation, so it finds very little support. The photo shows a cairn near the ash landfill in the brown coal open-cast mining excavation area west of Cologne (Königshovener Höhe, Grevenbroich, Germany). This site looks like a rubbish dump and in this state, it would never be envisaged as a candidate for a nature conservation area – but it is nevertheless the breeding ground of several wheatear pairs. The photo of the wheatear male on the left was kindly contributed for publication by Christine Jensen/naturgucker.de.

Plate 7 In earlier centuries and millennia, high waters and severe floods cut steep escarpments into the land. For several bird species and especially for insects, these proved to be unique habitats that would not have existed in Central Europe without the occurrence of natural disasters. Nowadays, the consequences of natural disasters are under control and valuable habitats have disappeared as a result. Today, escarpments are almost exclusively created through massive industrial encroachments upon the landscape (a: Königshovener Höhe, Grevenbroich, Germany). Without these encroachments, many bird and insect species would no longer find the habitats they need to propagate; and even more regrettable is the common practice of destroying these habitats immediately after they have been economically exploited. The steep walls are levelled and sown with crops, a process which is called *re-naturalisation* (b: Fortuna-Garsdorf, Grevenbroich, Germany). This term is misleading, because the result resembles nothing at all that deserves to be called *nature*. It is in truth only a 'gentle' artificial landscape that looks pleasing to a majority of the population (panel (b) in comparison to panel (a)); but 're-naturalisation' is anything but desirable for the preservation of rare species. The 're-naturalised' landscape is home to considerably fewer rare species than the earlier escarpment. However, the process of 're-naturalisation' is still continuing in many cases and the flattened escarpment is then afforested (c: Vollrather Höhe, Grevenbroich, Germany). Hardly any rare animal and plant species can be found on these forested slopes. Many people find the forest beautiful, but they are not aware of how many Red List species are decimated by the afforestation of their habitats.

(a)

(b)

(c)

Plate 8 Half a century ago, the tree pipit was still a common breeding bird throughout all of north-western and West Germany, inhabiting open forest landscapes with wide paths. It found its food on the insect-rich path edges which were only sparsely covered with vegetation (a: Königshovener Mulde, Grevenbroich, Germany). Today, forest growth has increased tremendously and this has continually compressed the free forest aisles between the trees, so forest paths are now narrow and shaded. Nowadays, it is the overland power lines that have frequently preserved (or created) many tree-free, sunlit and consequently species-rich areas in the dense forests. Even where the edges of the paths between the forested areas are wide enough, they are overgrown by dense vegetation, permitting no sunlight to warm the ground where insects can still live (b). The edges of forest paths and lanes in Central Europe are today overgrown with lush, moist and green vegetation. A habitat like this is particularly insect-hostile, so it no longer offers the tree pipit any insect nourishment.

(a)

(b)

Plate 9 For centuries and even millennia, nitrogen deficiency characterised Central European agriculture. The fields of the Neolithic and Bronze Ages often had to be abandoned after only 10 years of cultivation, because they had ceased to provide sufficient yield for the early farmers. Nitrogen deficiency plagued agriculture up until roughly 100 years ago. Throughout the millennia, the fields of Central Europe often had only sparse growth; agriculture was accompanied by poor yields and famine. However, it was in areas like these where animal and plant species flourished – species that are on the Red List today. The richness of the species was nourished by the poverty of the population. In the Central European forest landscape of the Holocene, agriculture had created habitats that simulated the animal-rich steppes of last Ice Age. Today's fields and meadows are vastly different. They are too densely overgrown and consequently do not provide a habitat for most of the animals and plants. The ortolan bunting, for example, needs arable land with sparse stem growth, as shown by the photo of a cornfield in north-eastern Poland in 2010, before the agricultural policy of the European Union gained a foothold there (a). The ortolan bunting (b) is now almost extinct in western Central Europe. Today's Central European efficient agriculture has no room for the ortolan bunting. The survival of this bird depends on fields with only sparse vegetation, but a habitat like this cannot be created by organic agricultural reform. Healthy organic agriculture, which is the objective of consumer protection politics, has very little to do with the rescue of endangered species – and it will not bring back the ortolan bunting. Today's environmental and consumer protection bodies cannot ask any farmer to revert to a field like the one shown in the top photo, although this is the only way to bring back the ortolan bunting again. The future of this bird in Central Europe does not lie in the organic movement and agricultural reform. Agricultural areas are probably lost forever as far as species richness is concerned, because agriculture cannot be scaled back to the level it was at a hundred years ago. To save the biodiversity of the former Central European agriculture, artificial habitats must be created and sustained by technical engineering. Such habitats should simulate the state of the fields and meadows of the past with their sparse growth. The photo of the ortolan bunting was kindly contributed for publication by Angela Najak/naturgucker.de.

(a)

(b)